SOCIETY FOR EXPERIMENTAL BIOLOGY
SEMINAR SERIES: 34

ACID TOXICITY AND AQUATIC ANIMALS

SOCIETY FOR EXPERIMENTAL BIOLOGY SEMINAR SERIES

A series of multi-author volumes developed from seminars held by the Society for Experimental Biology. Each volume serves not only as an introductory review of a specific topic, but also introduces the reader to experimental evidence to support the theories and principles discussed, and points the way to new research.

1. Effects of air pollution on plants. *Edited by T.A. Mansfield*
2. Effects of pollutants on aquatic organisms. *Edited by A.P.M. Lockwood*
3. Analytical and quantitative methods. *Edited by J.A. Meek and H.Y. Elder*
4. Isolation of plant growth substances. *Edited by J.R. Hillman*
5. Aspects of animal movement. *Edited by H.Y. Elder and E.R. Truman*
6. Neurones without impulses: their significance for vertebrate and invertebrate systems. *Edited by A. Roberts and B.M.H. Bush*
7. Development and specialisation of skeletal muscle. *Edited by D.F. Goldspink*
8. Stomatal physiology. *Edited by P.G. Jarvis and T.A. Mansfield*
9. Brain mechanisms of behaviour in lower vertebrates. *Edited by P.R. Laming*
10. The cell cycle. *Edited by P.C.L. John*
11. Effects of disease on the physiology of the growing plant. *Edited by P.G. Ayres*
12. Biology of the chemotactic response. *Edited by J.M. Lackie and P.C. Williamson*
13. Animal migration. *Edited by D.J. Aidley*
14. Biological timekeeping. *Edited by J. Brady*
15. The nucleolus. *Edited by E.G. Jordan and C.A. Cullis*
16. Gills. *Edited by D.F. Houlihan, J.C. Rankin and T.J. Shuttleworth*
17. Cellular acclimatisation to environmental change. *Edited by A.R.Cossins and P. Sheterline*
18. Plant biotechnology. *Edited by S.H. Mantell and H. Smith*
19. Storage carbohydrates in vascular plants. Edited by *D.H. Lewis*
20. The physiology and biochemistry of plant respiration. *Edited by J.M. Palmer*
21. Chloroplast biogenesis. *Edited by R.J. Ellis*
22. Instrumentation for environmental physiology. *Edited by B. Marshall and F.I. Woodward*
23. The biosynthesis and metabolism of plant hormones. *Edited by A. Crozier and J.R.Hillman*
24. Coordination of motor behaviour. *Edited by B.M.H. Bush and F. Clarac*
25. Cell ageing and cell death. *Edited by I. Davies and D.C. Sigee*
26. The cell division cycle in plants. Edited by *J.A. Bryant and D. Francis*
27. Control of leaf growth. *Edited by N.R. Baker, W.J. Davies and C. Ong*
28. Biochemistry of plant cell walls. *Edited by C.T. Brett and J.R. Hillman*
29. Immunology in plant science. *Edited by T.L. Wang*
30. Root development and function. *Edited by P.J. Gregory, J.V. Lake and D.A. Rose*
31. Plant canopies: their growth, form and function. *Edited by G. Russell, B. Marshall and P.G. Jarvis*
32. Developmental mutants in higher plants. *Edited by H. Thomas and D. Grierson*
33. Neurohormones in invertebrates. *Edited by M.C. Thorndyke and G.J. Goldsworthy*
34. Acid toxicity and aquatic animals. *Edited by R. Morris, E.W. Taylor, D.J.A. Brown and J.A. Brown*

ACID TOXICITY AND AQUATIC ANIMALS

Edited by

R. Morris

Zoology Department, University of Nottingham

E.W. Taylor

Department of Zoology and Comparative Physiology, University of Birmingham

D.J.A. Brown

Central Electricity Research Laboratories, Leatherhead, Surrey

J.A. Brown

Zoology Department, University of Hull

CAMBRIDGE UNIVERSITY PRESS

Cambridge

New York New Rochelle

Melbourne Sydney

CAMBRIDGE UNIVERSITY PRESS
Cambridge, New York, Melbourne, Madrid, Cape Town, Singapore, São Paulo

Cambridge University Press
The Edinburgh Building, Cambridge CB2 8RU, UK

Published in the United States of America by Cambridge University Press, New York

www.cambridge.org
Information on this title: www.cambridge.org/9780521334358

First published 1989
This digitally printed version 2008

A catalogue record for this publication is available from the British Library

Library of Congress Cataloguing in Publication data
Acid toxicity and aquatic animals.
(Seminar series/Society for Experimental Biology; 34)
Includes index.
1. Acid precipitation (Meteorology) – Environmental
aspects. 2. Aquatic animals – Effect of water pollution
on. 3. Acid pollution of rivers, lakes, etc. –
Environmental aspects. 4. Acid precipitation
(Meteorology) – Toxicology. I. Morris, R. (Ronald)
II. Series: Seminar series (Society for Experimental
Biology (Great Britain); 34.
QH545.A17A234 1987 591.5'2632 87–24237

ISBN 978-0-521-33435-8 hardback
ISBN 978-0-521-05762-2 paperback

CONTENTS

CONTRIBUTORS

P.C.M. Balm
University of Nijmegen, Toernooiveld 25, 625 ED Nijmegen, The Netherlands.

D.J.A. Brown
Central Electricity Research Laboratories, Kelvin Avenue, Leatherhead, Surrey
KT22 7SE, UK.

J.A. Brown
Department of Zoology, University of Hull, Hull HU6 7RX, UK.

J.N. Cameron
Department of Zoology/Marine studies, University of Texas at Austin, Port Aransas
Marine Laboratory, Texas 78373, USA.

T.R.K. Dalziel
Department of Zoology, University of Nottingham, University Park, Nottingham
NG7 2RD, UK.

C.H. Dempsey
Central Electricity Research Laboratories, Marine Biology Unit, Fawley,
Southampton SO4 1TW, UK.

N. Heisler
Abteilung Physiologie, Max-Planck-Institut für Experimentelle Medizin, D-3400
Göttingen, West Germany.

A.G. Hildrew
School of Biological Sciences, Queen Mary College, Mile End Road, London
E1 4NF, UK.

G.D. Howells
Department of Applied Biology, University of Cambridge, Pembroke Street,
Cambridge CB1 3DX, UK.

John Mason
Centre for Environmental Technology, Imperial College of Science and Technology,
48 Princes Gardens, London SW7 1LU, UK.

D.G. McDonald
Department of Biology, McMaster University, 1280 Main Street West, Hamilton,
Ontario L8S 4K1, Canada.

B.R. McMahon
Department of Biology, University of Calgary, 2500 University Drive N.W., Calgary, Alberta T2N 1N4, Canada.

P.G. McWilliams
Zoological Laboratory, University of Bergen, Allegt.41. N-5014, Bergen, Norway.

R. Morris
Department of Zoology, University of Nottingham, University Park, Nottingham NG7 2RD, UK.

W.T.W. Potts
Department of Biological Sciences, University of Lancaster, Lancaster LA1 4YQ, UK.

J.P. Reader
Department of Zoology, University of Nottingham, University Park, Nottingham NG7 2RD, UK.

K. Sadler
Central Electricity Research Laboratories, Freshwater Biology Unit, Ratcliffe-on-Soar Power Station, Nottingham NG11 0EE, UK.

S. Stuart
Department of Biology, University of Calgary, 2500 University Drive N.W., Calgary, Alberta T2N 1N4, Canada.

D.W. Sutcliffe
The Freshwater Biological Association, Ferry House, Ambleside, Cumbria LA22 0LP, UK.

E.W. Taylor
Department of Zoology and Comparative Physiology, University of Birmingham, Birmingham B15 2TT, UK.

R.C. Thomas
Department of Physiology, The Medical School, University of Bristol, Bristol BS8 1TD, UK.

A.W.H. Turnpenny
Central Electricity Research Laboratories, Marine Biology Unit, Fawley, Southampton SO4 1TW, UK.

O.L.J. Vanderborght
Belgian Nuclear Centre, Department of Radiobiology, Laboratory for Mineral Metabolism, 24000 MOL, Belgium.

L.D.H. Vangenechten
Belgian Nuclear Centre, Department of Radiobiology, Laboratory for Mineral Metabolism, 24000 MOL, Belgium.

S.E. Wendelaar Bonga
University of Nijmegen, Toernooiveld 25, 625 ED Nijmegen, The Netherlands.

H. Witters
Belgian Nuclear Centre, Department of Radiobiology, Laboratory for Mineral Metabolism, 24000 MOL, Belgium.

C.M. Wood
Department of Biology, McMaster University, 1280 Main Street West, Hamilton, Ontario L85 4K1, Canada.

UNITS, SYMBOLS AND FORMULAE

Standard chemical formulae and certain SI units, prefixes and symbols have been used in this volume. The following, widely accepted, units have also been used which are not strictly SI units

Quantity	Unit	Symbol	SI equivalent
Concentration of a substance	molarity equivalent	M E or Eq. eq or equiv.	mol, where 1 M = 1 mol dm^{-3}
Mass (concn)	mg 100 ml^{-1}	mg%	
Volume	litre	l or L	m^3, where 1 l = 1 dm^3 = 10^{-3} m^3
Pressure	torr	Torr mm Hg	where 1 Torr = 133.322 Pa where 1 mm Hg = 133.322 Pa
Enthalpy (H)	calorie	cal	where 1 cal = 4.184 J
Time	day hour minute second	d h min sec	 second (s)

PREFACE

Whilst the subject of 'acid rain' has received a great deal of attention from the mass media, focusing on dramatic toxic effects and the ensuing political arguments, the scientific studies have been slowly forging ahead. This book concentrates on one aspect of the problem, that of the effect of acid toxicity on the group of animals most severely affected - the aquatic animals living in fresh water.

The book results from a Symposium held at the University of Nottingham in March 1986 by the Society of Experimental Biology, when a number of scientists with expertise in this particular subject met to discuss the results of recent research into the problems. They also provided a series of articles on which this book is based. The conference was greatly helped by a financial contribution from SWAP (Surface Waters Acidification Project) – a research organisation jointly administered by the Royal Society of London, The Norwegian Academy of Science and Letters and the Royal Swedish Academy of Sciences.

The aim of the book is to cover a wide range of investigations taking place in both field and laboratory and to provide the necessary scientific background information for present and possible future approaches to the problems facing animals living in acidified fresh water. Starting initially with the environment in order to assess why problems have arisen in particular areas, the volume then deals with field and survival studies on invertebrates and vertebrates in order to examine the extent of the biological problem and the attempts which have been made to relate water quality and the susceptibility of animals. Major advances in this area have included the realisation that declining populations are often the result of acid waters and their interactions with traces of other ions, and that these situations can produce the most severe physiological problems. The natural progression of environmental and field studies, toxicity and survival tests provide the background information to the physiological studies which follow. These form the major component of the book and whilst some chapters provide the basic information for present and future studies, the majority seek to analyse the toxic effects of acid waters and trace metals on acid-base balance, respiration and ionic balance together with cardiovascular and endocrinological effects. Whilst it is inevitable that most contributions concentrate on fish as the dominant aquatic animals of economic importance, invertebrate studies are also included.

There are many gaps in our knowledge and it is hoped that this book will stimulate others to help to fill them and join in the solution of this vital environmental problem.

Ron Morris

SIR JOHN MASON

Introduction
The Causes and Consequences of Surface Water Acidification

Introduction

Acid rain is a short-hand term that covers a set of highly complex and controversial environmental problems. It is a subject in which emotive and political judgements tend to obscure the underlying scientific issues which are fairly easily stated but poorly understood. In this article I shall deal solely with the scientific problems involved in the acidification of surface waters, attempt to establish the facts, describe the present state of knowledge and understanding and discuss what research is needed to provide a firm basis for remedial action.

Although the term *acid rain* is commonly used to describe all acid deposition from the atmosphere that may cause damage to trees, vegetation, fisheries, buildings, etc., in fact rain (and snow) brings down only about one third of the total acids over the UK, two thirds being deposited in the dry state as gases and small particles. But wet or dry, there is little doubt that acid deposition from the atmosphere poses an ecological threat, especially to aquatic life in streams and lakes on hard rocks and thin soils in southern Scandinavia and in some parts of Scotland and North America. A great deal of research is being undertaken in these three areas but this account is based largely on work in Scandinavia and the United Kingdom where more than 30 research groups from a wide variety of disciplines and institutions are working in a closely integrated and coordinated programme under the author's direction. A complete coverage of the relevant scientific problems includes studies of the transport and chemical transformations of emitted pollutants in the atmosphere; the wet and dry deposition of the resulting acids; the acidity and chemical composition of the rain and snow; modification of the chemistry of the rainwater as it percolates through and interacts with the soil and rocks; the toxic effects of the modified water chemistry on aquatic biota in streams and lakes.

The atmospheric chemistry of acid depositions

Understanding the problem of acid deposition requires knowledge of the distribution in space and time of the major acidifying pollutants, SO_2, NO_x and HCl,

Table 1. *Annual emissions of S or N (Mt/yr)*

		1900	1950	1960	1970	1980	1984
UK	SO_2	1.4	2.3	2.8	3.0	2.33	1.77
	NO_x	0.21	0.30	0.41	0.50	0.54	0.56*
Europe SO_2 (exc. USSR)			10.0		18.4	20.0	

*recent measurements of NO_x emissions from car exhausts suggested this figure is too low by at least 10%.

their chemical transformation in the atmosphere and their removal by deposition on the Earth's surface, either directly in gaseous or particulate form (dry deposition), or after incorporation into cloud and raindrops (wet deposition). The chemical reactions involved in both the gaseous and liquid phases are complex and incompletely understood but are the subject of much active research involving the measurement of the concentrations and conversion rates of chemical species in the atmosphere, laboratory measurements of key reaction rates, and the use of complex models to simulate the many simultaneous, interactive, chemical reactions.

Emissions

The total annual emission of SO_2 and NO_x in the UK (expressed in millions of tonnes of S or N), together with the figure for Europe are shown in Table 1.

Thus the UK contributes < 2% to the total input of sulphur into the global atmosphere and < 10% of the man-made sulphur produced in western Europe. The UK emissions of SO_2, 60% of which come from power stations and about 30% from industrial plants (e.g. refineries), have fallen by 40% since 1974 and by 24% since 1980; but the emission of NO_x, about 45% of which comes from power stations and 30% from motor vehicles, continues to rise. The total deposition of sulphur on the UK in 1980 was 0.7 Mt, about 30% of the emissions, two thirds being dry deposition and one third in precipitation. About 80% of this total deposition was estimated to come from UK sources. The rain bearing westerly winds ensure that the UK emissions make a significant but not a predominant contribution to total acid deposition in Sweden and Norway, the contributions being about 8% and 16% respectively in 1980 but the latest estimates indicate lower values of about 5% and 10% respectively.

Chemical transformations

Once emitted into the atmosphere, the pollutants are carried and dispersed by atmospheric motions, the plume from a point source such as a power station spreading out into an expanding cone which meanders with fluctuations in the wind. The plume is largely confined to within the atmospheric boundary layer, in the lowest 1–2 km, unless it is carried up into cloud systems. A good deal of the acid deposition reaches the ground in dry (gaseous or particulate) form close to the source but the rest may travel for hundreds of kilometres during which time the gases SO_2 and NO_x are oxidized and converted into lowly volatile products such as sulphuric and nitric acid either in gas phase reactions or, more effectively, by becoming captured by cloud and raindrops (where the chemical transformations proceed much more rapidly in the liquid phase) and are eventually brought to the ground in rain or snow.

Although acid production proceeds much more rapidly in the liquid phase, clouds and rain are present only a small fraction of the time, so gaseous transformations and deposition are important and account for about two thirds of the total acid deposition in the UK and about one third in Norway.

The rates of conversions of SO_2 and NO_x to H_2SO_4 and HNO_3 are determined by measurements from the Hercules flying laboratory of the Meteorological Office which can locate and follow a chemically marked plume from a particular power station, sample the air inside and outside the plume as it crosses the North Sea, analyse it for all the relevant chemical species, e.g. SO_2, NO, NO_2, oxidizing agents such as O_3, H_2O_2, hydrocarbons, and aerosols, collect cloud and rain water and analyse these for pH, all main ionic species, H_2O_2, etc. In order to explain the observed conversion rates of SO_2 and NO into acids it is necessary to invoke photochemical reactions involving highly reactive oxidizing agents such as O_3 and O* leading to the formation of the important radical OH which is unreactive to oxygen and therefore relatively stable. Some of the more important chemical reactions may be summarized as follows:

Gaseous reactions in a dry atmosphere

(a) Sulphuric acid

$SO_2 + HO + M \rightarrow HSO_3 + M$ (M is a third molecule, usually N_2)

$HSO_3 + O_2 \rightarrow HSO_5$

$HSO_5 + H_20 \rightarrow H_2SO_4 + HO_2$

with HO resulting from

$O_3 + hv \rightarrow O^* + O_2$

$O^* + H_2O \rightarrow 2HO$

The aircraft measurements indicate a conversion rate of about 16% d^{-1} in summer, when solar ultraviolet radiation permits ready photolysis of ozone, but this is reduced to about 3% d^{-1} in winter.

(b) Nitric acid

$NO + O_3 \rightarrow NO_2 + O_2$ (fast)

$NO_2 + OH + M \rightarrow HNO_3 + M$

The conversion rate is about 20% per *hour* in summer and about 3% h^{-1} in winter so that conversion would be complete in a 24 h traverse of the plume across the North Sea in summer.

Liquid phase reactions in clouds and rain

$2SO_2 + 2H_2O \rightarrow SO_3^{2-} + HSO_3^- + 3H^+$

$HSO_3^- + H_2O_2 \rightarrow HSO_4^- + H_2O$

with H_2O_2 resulting from

$HO_2 + HO_2 \rightarrow H_2O_2 + O_2$

and HO_2 from gaseous reactions such as:

$OH + CO \rightarrow CO_2 + H$

$H + O_2 + M \rightarrow HO_2 + M$

The conversion rates are very fast, almost 100% per hour in summer and 20% per hour in winter in the presence of sufficient concentrations of the oxidants O_3, OH, HO_2, H_2O_2 and hydrocarbons which may be the limiting factor.

Nitric acid

$NO_2 + NO_3 + M \rightarrow N_2O_5 + M$

$N_2O_5 + H_2O \text{ (liq.)} \rightarrow 2HNO_3$

This reaction is believed to be rapid but conversion rates have not been established.

Both the aircraft measurements and the photochemical models indicate that the rates of production of acids from the precursor gases are often limited by the availability of oxidizing agents, hydrocarbons and solar ultraviolet radiation.

Acidity of precipitation

In order to assess the effects of changes in the emissions of SO_2, NO_x and hydrocarbons on the acidity of precipitation it is necessary to have long term, reliable, accurate and representative measurements of pH, alkalinity, and concentrations of the main ionic species SO_4^{2-}, NO_3^-, HCO_3^-, etc., so that one may study their variations in space and with time. Unfortunately rather few series of measurements satisfy these criteria. Accurate measurement of the pH of poorly buffered waters in the field is particularly difficult. When the same sample of rainwater was divided among 18 different reputable European laboratories, differences of one whole pH unit were obtained and the standard deviation of the 18 measurements was 0.75 unit. In addition to analysis errors, changes may occur during collection, storage and transport, resulting in contamination, evaporation and biological activity of the sample. Urgent attention is being given to the improvement of analytical techniques and to their standardization, intercomparison and intercalibration.

The pH of uncontaminated rainwater in equilibrium with atmospheric carbon dioxide is 5.6. Rain and snow almost everywhere, even in places as remote as Hawaii, the southern Indian Ocean and the polar regions, are more acidic than this with average pH values of 5.0 or lower. This testifies to the ubiquity of acidic pollutants, the oceans being a major source of sulphur compounds resulting from biological activity and sea spray.

In 1978/80 the annual average pH of rain falling over the UK was almost everywhere between 4.5 and 4.2, the rain being more acidic on the eastern side of the country downwind of the main industrial conurbations where the pH values are very similar to those encountered in southern Scandinavia.

Unfortunately, there are few reliable, long term records with which to assess recent trends in the acidity of precipitation. Perhaps ther best record, maintained by the Freshwater Biological Association in Cumbria, shows that the annual mean pH remained sensibly constant at 4.4 between 1955 and 1975 during which period the total sulphate deposition also remained roughly constant. This is consistent with the fact that the UK emissions of SO_2 and NO_x increased only slightly over this period and that 80% of the total sulphur deposition in the UK comes from local sources. There is now some evidence, notably from measurements made by the Freshwater Fisheries Laboratory at Pitlochry, that the acidity of rainfall has decreased (pH increased by about 0.2 unit) since 1979 in conformity with the 34% reduction in UK SO_2 emissions since then.

In Europe there is also a dearth of reliable long term measurements and some of the reported rather sharp increases in acidity of precipitation appear to have coincided with changes in measuring techniques. On balance the evidence indicates that the acidity increased gradually from 1955 to 1970 during which time European emissions of SO_2 doubled but there are indications of a slight reduction in the acidity and sulphate content of rainwater since 1980 concurrent with reduced emissions.

There is some evidence of a linear correlation between the sulphate content of rainwater and emissions of SO_2 based on average annual values but the implications of this are complicated by the fact that, in many parts of the UK and southern Scandinavia, a large fraction of the total annual acidic deposition occurs on only a few days of heavy rainfall. In order to detect these high deposition episodes, which have a major ecological impact, it is necessary to have frequent or continuous monitoring of the rainfall and its chemistry.

Moreover, the annual averages hide large seasonal variations in the acid and sulphate content of UK rainfall, with the highest values occurring in summer when the emissions are least. This is probably a consequence of the fact that the conversion of SO_2 and NO_x is limited by the availability of oxidants such as O_3, OH, H_2O_2, that are produced by photolysis more effectively in the summer time (see Figure 1).

The acidification of streams and lakes

Measurements in many lakes in southern Scandinavia suggest that the pH has decreased by between 0.5 and 1.0 unit over recent decades. However there is some doubt about the reliability of the measurements and the comparability of past and recent data. A reduction of one whole pH unit is unlikely to be due solely to increases in European emissions of SO_2 and NO_x which only doubled between 1950 and 1970 and have increased only slightly since then. Part of the decrease may have resulted from additional acidification processes at work in the catchment or from the acid neutraliziang mechanisms in the soil not being able to keep pace with the acidic deposition (see next section). By contrast, biologically inactive lakes in Cumbria, subject to acidic rainfall, with hard bedrocks and thin soils very similar to those prevailing in southern Scandinavia, have shown no significant change in acidification over the last 50 years, during which period UK emissions of SO_2 doubled between 1930 and 1970 and have fallen by 40% since then. In biologically active lakes major changes of pH are caused by respiration, by photosynthesis and by decomposition of vegetation and these often show strong diurnal and seasonal variations.

A chemical survey of some hundreds of lakes in southern Norway in the early 1970s revealed that 40% of them had pH values of <5.5 and 16% with pH <5.0. It will be interesting to determine whether there have been significant changes in response to the marked reductions in emissions of SO_2 from the UK and Sweden in the meantime.

Evidence for the gradual acidification of lakes over longer periods comes from the analysis of acid-sensitive species of diatoms from radioactively dated lake sediments.

Figure 1. Seasonal variation in H^+ and non-marine SO_4^{2-} concentration in Northern Britain: average three-month running mean concentration expressed as a ratio to the annual mean. (Courtesy of Warren Spring Laboratory.)

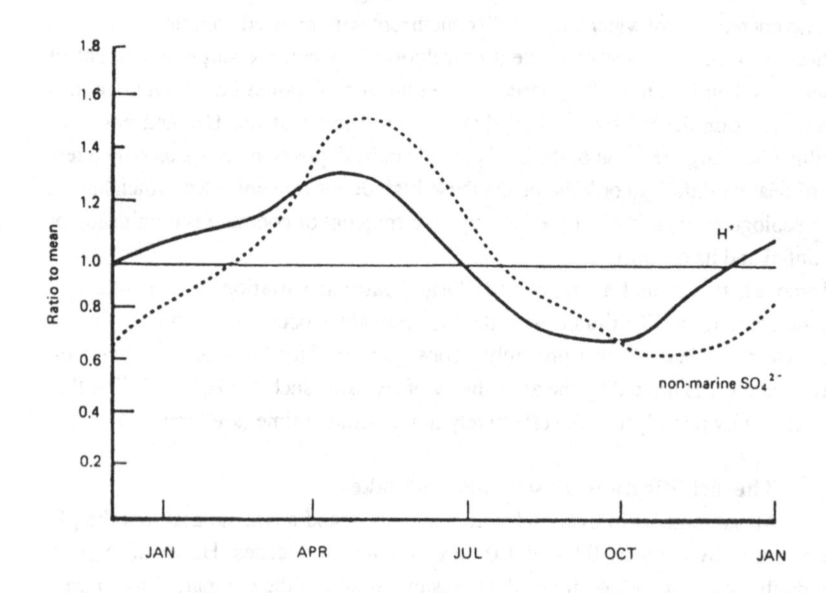

The layers of sediment, laid down over several hundred years, are dated using radioactive lead isotopes and the diatoms found in the layers are correlated with similar populations found in the uppermost layers of lakes of known pH. It is thus possible to reconstruct a pH-age profile of the sediments. By this method it has been found that some lakes in Galloway which are situated on hard granitic bedrock and thin soil have tended to become acid in recent decades. Figure 2 shows that the pH of Round Loch remained sensibly constant at 5.5 between 1600 and 1850 but thereafter steadily declined, with fluctuations, to 4.8 in 1973, the decline being particularly rapid since 1900, but this cannot be attributed to afforestation or changes of land use. A slight recovery to pH = 4.9 is discernible between 1973 and 1980, associated perhaps with reduced emissions. Loch Grannoch showed little change in pH at 5.5 between 1840 and 1930, but in 1930/40 it apparently fell to 5.0. From 1940 to 1960 there were even larger changes in the diatom population suggesting a further fall in pH to 4.5. Again these changes could not be ascribed to afforestation or changes in land use.

A similar study on lakes near the southwestern coast of Sweden, all afforested with

Figure 2. The history of acidification of Round Loch of Glenhead, Galloway as deduced from diatom records and radioactive lead dating. (Courtesy of Dr R. Battarbee.)

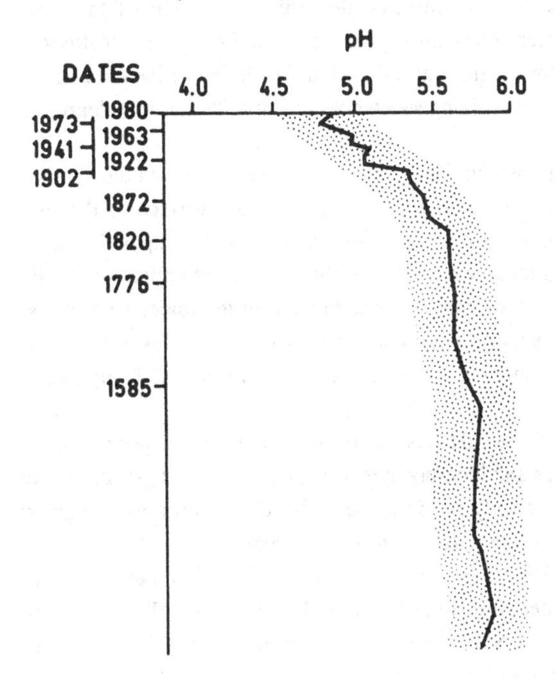

hard bedrock and thin soil, also indicated a fall in pH from 6.5 to 4.5 since 1950. Diatom records from lakes in southern Norway also indicate a decrease in pH over recent decades but a lake on the west coast with pH ≈ 5 showed no significant change over 200 years.

The decreases in pH since the industrial revolution are likely to be due mainly to increased acidic deposition from the atmosphere but more extensive measurements involving a wider variety of acid-sensitive species are required to establish the historical trends and help resolve the apparent differences between the results for different locations.

Apart from the effects of man-made depositions, there appears to be a natural tendency for lakes to become more acidic with time due to the natural acid-producing mechanisms in the soil (see next section) and a gradual decline of their neutralizing capacity. These natural acidification processes are normally slow but may be accelerated by changes in land use.

The importance of hydrogeology and soil chemistry

There is undoubtedly a correlation between rainfall acidity and the occurrence of acid streams and lakes in susceptible areas. However, little of the water feeding streams and lakes arrives directly from the atmosphere; most does so only after flowing over the land surface or through the soil. The details of the flow pathways and of the soil–water interactions are not well known. Hydrological conditions and water flow pathways through the soil and rock determine the contact time of the acidified water with neutralizing substances and thus the ultimate degree of water and soil acidification.

When the acidified rainwater reaches the ground some seeps through the soil but some, especially during continuous heavy rain, runs over the surface or through macro-pores and may enter the nearest stream with its chemical composition little changed. If, however, the rainwater penetrates the soil its chemistry may be profoundly modified as the result of complex reactions with the underlying rocks, soil and vegetation involving many processes such as dissolution and weathering of minerals, cation exchange, accumulation and release of contaminants including humic and other organic acids.

If the rain or melt water of low pH passes through sandy soils poor in base minerals (e.g. Ca/Mg carbonates and bicarbonates) or passes over hard granitic rocks then the rain/melt water will be only partially neutralized by cation exchange or weathering. The H^+ in the percolate will tend to be associated with SO_4^{2-}. In most catchments any NO_3^- ions which are present in the rain are largely taken up by the roots of growing plants and trees, thus immobilising the associated H^+. (This may change if the increasing trend in NO_x emissions continues). The acidified water therefore passes through the soil with only a moderate pH change and the soil itself undergoes little acidification.

In base-rich soils, H^+ in the acid water are either neutralized by bicarbonate ions during the dissolution of limestone or removed by exchange with cations such as Mg^{2+}, K^+, Na^+, Al^{3+} on the surfaces of mineral particles or with cations on organic macromolecules in humus. H^+ may also be consumed in the soil by the reduction of nitrates and sulphates.

Chemical weathering of primary and secondary minerals is the ultimate means by which inputs of acids are counteracted. Chemical weathering transfers basic cations from primary minerals to the pool of exchangeable cations. A fraction of these are retained on the surfaces of soil particles and the remainder are lost in the runoff or percolate. Hydrogen ions from the acidic inputs are exchanged for the basic cations on the surfaces of the soil so that the acidity of the percolate decreases and that of the soil tends to increase. If the release of base cations by weathering keeps pace with the rate of increase of acid input, no soil acidification occurs. However, if the rate of chemical weathering is not increased sufficiently, soil acidification and subsequently water acidification results. Soil acidification partly caused by acid deposition seems to have occurred both in Central Europe and Scandinavia.

However, besides the deposition of acid substances from the atmosphere, several natural processes produce H^+ in the soil and reduce the pH of the percolating water – see Figure 3. These include the production of carbonic acid by CO_2, the hydrolysis of minerals, the decomposition and nitrification of ammonium produced by bacterial decomposition of vegetation and by fertilisers, the oxidation of sulphur in dry soil, the action of organic acids from decaying humus, and the release of H^+ from the roots of plants and trees to compensate for the take up of Mg^{2+} and Ca^{2+}.

The acidity/alkalinity of the percolate is the net result of all these processes which act at different rates depending, in many cases, on the pH. In general, base-rich, weatherable soils tend to reduce the acidity of the percolate whilst growing plants and trees, aided by fertilisers, tend to acidify both the soil and soil water. The deposition of acidifying components is also in general larger in a coniferous forest than in an unforested area. The rate of water acidification and the time lags involved probably depend very much on the amount of sulphate accumulated in the soil and the rate at which the SO_4^{2-} are released to accompany the H^+ in the percolate. After a long dry spell followed by heavy rain, or during snow melt, there is often a heavy transient release of mobile anions including NO_3^- which cannot all be taken up by plants and so is available for mobilizing the H^+, thereby increasing the acidity of the soil water.

Studies are underway in Scandinavia and Scotland to investigate the above mentioned processes: chemical, biological, and hydrogeological, that determine the quality of the surface waters and therefore the populations of fish and the organisms that provide their food. Each experimental catchment will require detailed surveys of the geology, soils and vegetation and the mapping of the hydrogeological pathways of the water over and through the ground on its way to the streams and lakes. Of key importance is the modification of the chemistry of the water as it percolates through the soil by a variety of acidifying and neutralizing processes. This will be studied by extraction and detailed chemical analysis of the percolate from different levels in the soil profile and relating these changes to the physical and chemical properties of the soils, measured weathering rates of the minerals, the rates of water flow (residence times) and the measured input/output budgets of the whole system. By making parallel studies in highly acidified, 'clean' and intermediate catchments, it should be possible to deduce how the contributions and balance of the various processes and agencies are affected by changes in acid deposition.

Considerable effort is also being devoted to the development of hydrochemical models to help identify the key processes, design the field experiments and interpret the results.

Figure 3. Examples of ion transfer processes involved in the acidification/alkalization of soil and soil water.

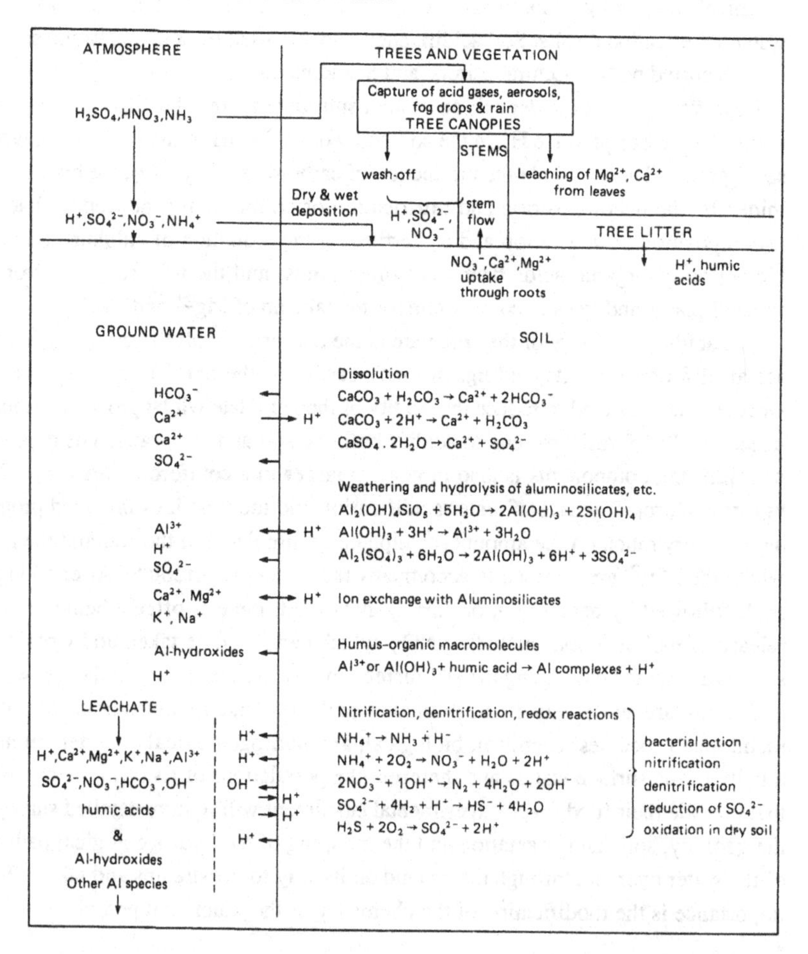

Effects of acidity and aluminium toxicity on fishes

A decline in fisheries, or loss of certain species since the 1930s has been reported for many lakes in S. Norway, S. Sweden, parts of UK, Ontario and NE USA. In view of the (not entirely convincing) evidence that the pH of some lakes has declined by as much as 1 whole unit and that low pH (<4.5) is known to be lethal to some fish, especially in early life stages, it is reasonable to attribute decline in fish populations to an increase in acidity of surface waters. Unfortunately, well documented, long term records of fishery status and of water quality exist for only a relatively few lakes in Scandinavia and NE America. For a sample of lakes in S. Norway over half of those with pH <5 were found to be fishless compared with about one in seven of those with pH >5. This is illustrated in Figure 4 which also shows the importance of calcium in the water, higher concentrations of which tend to compensate for lower pH (higher acidity). An extensive survey of fishery status and water quality for some 700 lakes in S.Norway failed to establish a direct relationship between fishery status and sulphate concentration because the calcium concentration and pH of the lakes, which are correlated with their altitude and distance from the coast, are dominant factors. More recently the toxic effects of aluminium leached from the soil and lake sediments have become firmly established. The toxicity of aluminium is a function of pH and is greatly moderated by the presence of calcium in solution (see Brown & Sadler, this volume).

Figure 4. The fishery status of a large number of Norwegian lakes in relation to their acidity (H^+ concentration) and the calcium content.

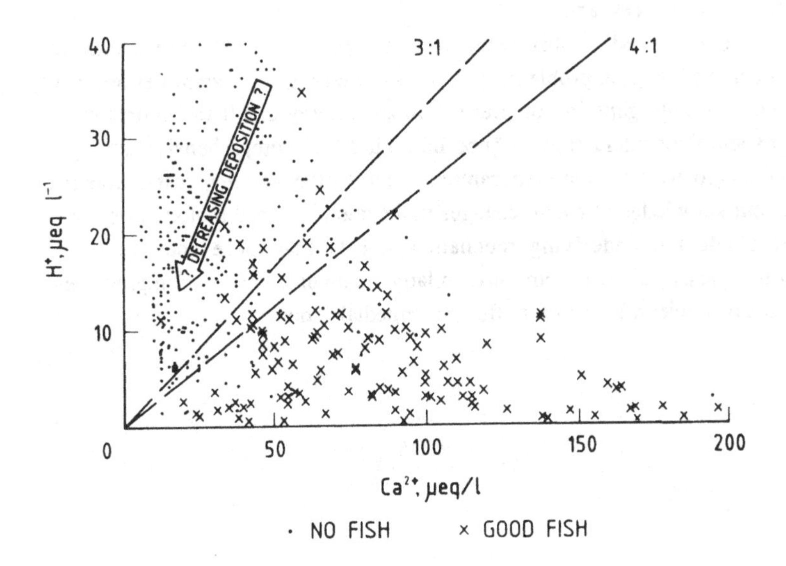

In the pH range below 4.5, acidity *per se* may be a direct cause of mortality in fish depending on the species, (salmon and trout are especially vulnerable), the age, size and genetic origin of the fish and the degree of acclimatization. Acidity may also affect growth, fertility, egg mortality and recruitment of fry to the population.

Attempts to establish clear cut relationships between water acidity and fishery status have been hampered by conflicting evidence between field observations, which are usually concerned with long term and fluctuating acid exposures, and laboratory studies which almost always involve short term, constant exposures. It is important to investigate the response to both types of exposure, long term and episodic, since it is not clear which is the more important in the depletion of fish stocks. However there is increasing evidence that large kills are produced by short episodes of high acidity, lasting only a few hours and associated with snow melt or rain after a long dry spell. It appears that these occasional or seasonal pulses of acidity, often accompanied by high concentrations of aluminium, are more lethal than sustained exposure to which the fish may adjust more easily. (See Reader & Dempsey, this volume).

The toxic effects of aluminium on fish are complex. Certain inorganic aluminium ions appears to be particularly potent whereas organic complexes of aluminium formed by interaction with organic acids derived from humus in the soil appear relatively innocuous. The detailed mechanisms controlling the salt balance of fish and involving ion exchange across the gill membranes will have to be further elucidated before the relative toxicity of the various aluminium species and their moderation by calcium and organic complexation can be assessed. These matters are discussed in much greater detail in later chapters.

Concluding remarks

I have attempted, in this paper, to demonstrate that the acidification of surface waters is a complex problem spanning a wide range of disciplines and that there are considerable gaps in our present understanding of all the contributory mechanisms and their interactions. I hope that our rather comprehensive, strongly coordinated and focused research programme will, over the next few years, do much to increase our knowledge of recent changes in the chemistry and biology of surface waters, elucidate the underlying mechanisms, establish more accurately the contribution made by man-made emissions relative to natural acidification processes, and provide a firm scientific basis for effective remedial action.

D.W. SUTCLIFFE AND A.G. HILDREW

Invertebrate communities in acid streams

Introduction

Stony streams normally contain a characteristic fauna of benthic macroinvertebrates, prominent among which are insects, especially Ephemeroptera, Plecoptera, Trichoptera and Diptera. Also present are flatworms, oligochaetes, a few malacostracan crustaceans and molluscs. In streams with moderate alkalinity and a pH of about 6 or more, the fauna is rich in total numbers of species (usually there are some 70–90 taxa) and many of these may be abundant. In acid streams where the mean pH is below 5.7–5.4, however, there is a distinct change in faunal composition. Some taxa disappear or become scarce, particularly mayflies, some caddisflies, crustaceans and molluscs. The fauna is thus impoverished in species and may contain only half the number of taxa found in less acid (soft water) streams.

For instance, in the upper River Tywi in west Wales a diverse fauna of 60–78 taxa occurs in streams containing mean concentrations of at least 85 μequiv l^{-1} calcium and magnesium, with mean pH 5.6 or above. At mean pH 5.3 or below, and slightly less calcium, only 46 taxa were found and these were generally sparse (Stoner, Gee & Wade, 1984). A similar reduction in numbers of taxa occurred at reduced pH in lowland streams of the Ashdown Forest in southern England (Townsend, Hildrew & Francis, 1983; Hildrew, Townsend & Francis, 1984). This pattern seems to be repeated wherever acid streams have been studied in Europe and North America (Sutcliffe & Carrick, 1973a,b,c; Harriman & Morrison, 1982; Otto & Svensson, 1983; Burns *et al.*, 1984; Engblom & Lingdell, 1984; Aston *et al.*, 1985; Burton, Stanford & Allan, 1985; Simpson, Bode & Colquhoun, 1985).

Three main hypotheses can account for these observations; they are not necessarily exclusive and factors within all three probably interact to determine the composition of invertebrate communities. Firstly, the chemical conditions in acidified waters are intolerable to some taxa or have sublethal physiological effects; some animals may actively avoid such waters. Secondly, the chemical conditions affect invertebrates indirectly via their food supply. Thirdly, the absence of fish from many acidified waters removes predation pressure and produces ramifying community effects. In this chapter, we first describe the particular chemical characteristics of soft waters and then deal with each of these three hypotheses in turn.

The chemical composition of acid and soft waters

The alkalinity of water is defined as the net concentration of bicarbonate, carbonate and hydroxyl ions minus hydrogen ions:

$$[HCO_3^- + 2CO_3^{2-} + OH^-] - [H^+]$$

Soft waters generally have pH around 6 to 7 and alkalinity less than 500 μequiv l^{-1} (Figure 1). pH normally falls rapidly from about 5.6–5.7 to about 5.1–5.2 as alkalinity passes through zero, equivalent to a mean pH of 5.4 ± 0.16 SD in the English Lake District. Above 0 μequiv l^{-1} the alkalinity concentration determines the mean pH in soft waters. Small tarns and streams containing 100–200 μequiv l^{-1} alkalinity are less prone to acid episodes (pH < 5.5) than waters containing less than 100 μequiv l^{-1}. Acid episodes of about one to three days' duration are tolerated by most benthic insects. Notice that alkalinity has a very wide range of values relative to pH (see Figure 1 and the equation in the legend). For instance a hard water with 2300 μequiv l^{-1} alkalinity would have a calculated mean pH of about 8.0. For a 2000 μequiv l^{-1} fall in alkalinity the pH falls only to about 7.2 (at 300 μequiv. l^{-1}) and then to about 6.7 at 100 μequiv l^{-1} in soft waters. Alkalinity is therefore very useful

Figure 1. Curve (solid line) relating observed mean pH to alkalinity in Lake District streams and lakes, based on >1000 samples (mean pH = 4.916 + 0.9164 $\log_{10}$ alkalinity in the range 1–5475 μequiv l^{-1}). Alkalinity was determined by Gran titration (Makereth *et al.*, 1978; Sutcliffe *et al.*, 1982). The broken portion of the pH curve is a plot of pH against H^+ concentration (pH 6.0 = 1.0, pH 5.0 = 10.0, pH 4.0 = 100 μequiv l^{-1}). The dotted line represents the mean calcium concentration relative to alkalinity, calculated from a regression of calcium versus alkalinity.

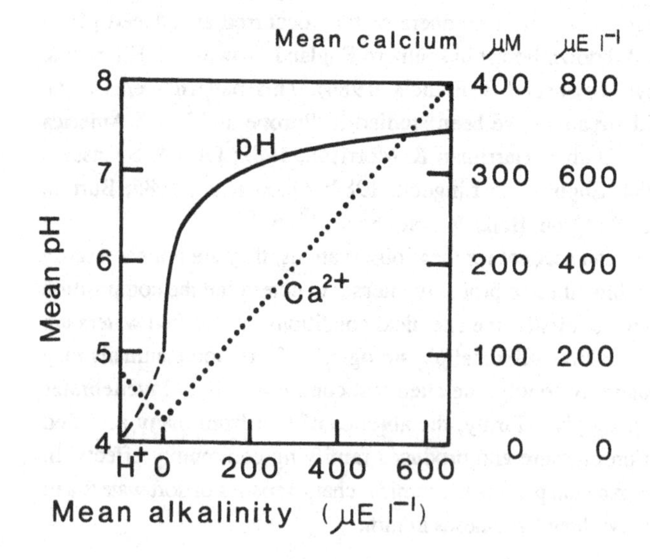

Table 1. *Calculated mean concentrations of ions (μequiv. l^{-1}) at pH 5.0 and 6.5 for 49 streams in Wales and 23 in the Peak District*[(a)]*, and means for the Estaragne stream in the Pyrenees*[(b)] *at pH 8.25.*

	Na	Ca	Mg	K	Cl	SO_4	NO_3	Alkalinity
pH = 5.0 (H^+ = 10 μequiv l^{-1})								
Wales	144	47	47	9	74	190	8	1
Peak	223	154	209	21	386	253	28	5
pH = 6.5 (H^+ = 0.3 μequiv l^{-1})								
Wales	190	107	97	14	107	220	13	53
Peak	296	374	339	35	449	302	36	120
pH = 8.25 (H^+ < 0.01 μequiv l^{-1})								
Estaragne	23	1140	17	13	21	42	15	936

[(a)]Calculated from regressions of $\log_{10} C$ versus $\log_{10} H^+$, where C is the mean concentration of each ion (μequiv l^{-1}) in each stream (Sadler & Lynam, 1984).
[(b)]From Lavandier (1979).

for determining water quality. Further details of the complex relationship between pH, bicarbonate, dissolved CO_2 and other weak acids are discussed by Henriksen & Seip (1980), Norton & Henriksen (1983), Lucas & Berry (1985), and Skeffington (1985).

Calcium concentrations are frequently lowest at zero alkalinity, rising as pH falls below 5.0 (when calcium is balanced by additional sulphate) and rising (with magnesium) at pH above 5.4, when normally balanced by bicarbonate (Figure 1). Thus concentrations of calcium, magnesium and bicarbonate alkalinity are all strongly correlated with pH in soft waters: potassium, sulphate, sodium, chloride and nitrate are also weakly correlated with alkalinity and pH. Most importantly, concentrations of these major ions are suboptimal and sometimes limiting for invertebrate animals in soft and acid waters at pH below 7.0 and especially below 6.0 (they reach minimum concentration at about pH 5.4). It is therefore very difficult to identify the individual physiological effects of any ion from natural distributions or from tolerance in the field.

The physiological hypothesis

Chemistry and invertebrate communities

As we have seen, acid waters (pH <5.5) normally have impoverished faunas. However, there is some circumstantial evidence that chemical factors other than pH may modify aquatic communities in acid waters. For instance, calculated concentrations of major ions in English Peak District streams running off Millstone Grit and Carboniferous Limestone are relatively high, even at pH 5.0, in comparison with a group of Welsh streams (Table 1). The faunas of these two groups are generally similar at a given pH (Figures 2,3). However, notice (Figure 2) that two Peak District streams contained mayflies (*Baetis rhodani* and *Baetis* spp.) at mean pH less than about 5.2–5.3. In the Peak District the adverse effects of high acidity (low pH) seem to be reduced by high concentrations of calcium, sodium, potassium and chloride. Magnesium is probably not important and the relatively moderate concentrations of sulphate and nitrate (Table 1) are not harmful to invertebrates.

In other cases the fauna may be impoverished even in hard water and at high pH.

Figure 2. The numbers of mayfly (Ephemeroptera) species per stream (a) and the numbers of mayfly larvae per 0.1 m^2 of stream bed (b), related to mean pH and calcium concentrations of 49 streams in Wales (closed symbols) and 23 in the Peak District (open symbols). (Based on data tabulated by Sadler & Lynam, 1984, and Aston *et al.*, 1985.) Curves are drawn from regressions of $\log_{10}$ calcium versus $\log_{10}$ hydrogen for mean concentrations in each stream in Wales (——) and the Peak District (----); ...... is drawn from Figure 1 where calcium is indirectly related to pH via alkalinity.

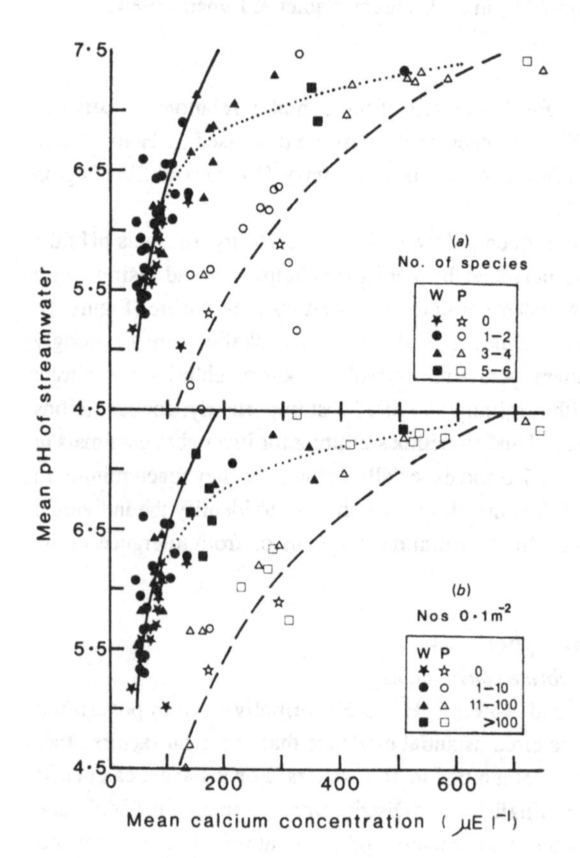

In Britain, except where they are close to the sea, upland streams usually have mean sodium and chloride concentrations below 200 μequiv l^{-1} but rarely much below 50 μequiv l^{-1}. However, in the hard water streams of the Estaragne in the Pyrenees, mean concentrations of sodium and chloride are far lower than those in Britain (Table 1). These streams have a diverse fauna, including mayflies, typical of hard waters, but molluscs and crustaceans are conspicuously and remarkably absent. It is difficult to see how such apparently anomalous distributions could be explained other than by the direct effects of water chemistry.

Physiological limitations of invertebrates in acid and soft waters

Ion uptake from water is required to maintain internal acid–base balance and ionic equilibrium between blood and tissues for those ions that are continuously lost by diffusion across permeable parts of the external body surface (Burton, 1973). A major feature is the excretion of ammonia or ammonium ions via the gills, during which other positively charged cations move inwards to maintain electroneutrality (Houlihan, Rankin & Shuttleworth, 1982). Because electrical neutrality has to exist between cations and anions, both externally and internally, the concentrations of ions in water are best expressed as equivalents (μequiv l^{-1}). However, the relative

Figure 3. Wet weight of benthic invertebrates (per 0.1 m^2 stream bed) related to mean pH of 49 streams in mid and north Wales (O) and 23 in the Peak District (Δ). The mean regression and 95% confidence limits are shown for total mass versus pH as a solid line and curves ($r = 0.38, p < 0.01$); a regression mean for log $_{10}$ total mass versus $\log_{10}$ H^+ is drawn as a broken curve ($r = 0.37$, $p < 0.01$). (Based on data tabulated by Sadler & Lynam, 1984 and Aston *et al.*, 1985.)

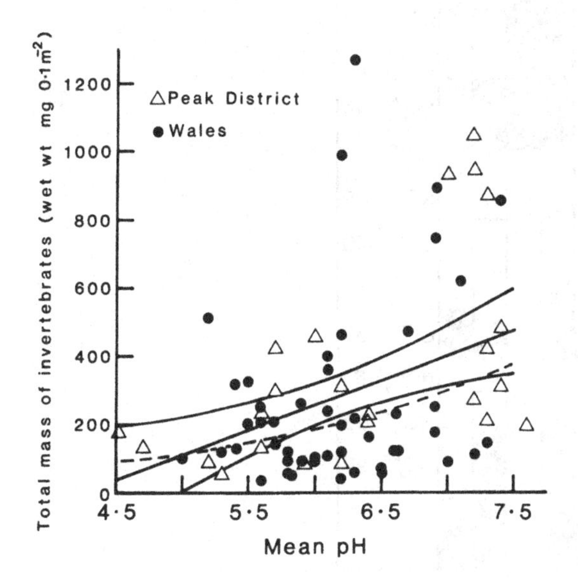

mobility of ions and their attraction to ion transporting sites depends on their mass, those with low atomic weights being more readily transportable. Therefore molar concentrations are also useful (μmol l^{-1}). Both these measures of concentration are given for Lake District waters at different pH in Figure 4.

Continuous uptake of sodium, chloride, potassium and calcium is necessary for survival. Active uptake of ions is dependent on the external concentrations; transport systems display Michaelis–Menten kinetics. Ion uptake is well below saturation level at concentrations normally found in British freshwaters and is severely limited (< 50%) below 100–200 μequiv l^{-1} for sodium and chloride, and below 200 μequiv l^{-1} for calcium. Less is known about potassium requirements but concentrations below 10 μequiv l^{-1} may be suboptimal. Comparing these concentrations with those in Figure 4 shows that acid waters may be inimical to the survival of many invertebrates.

Figure 4. Typical values for concentrations (μmol. l^{-1}) of major ions in Lake District waters on volcanic rocks. Examples are given for very soft waters at pH 6.0, slightly acid (alkalinity absent) at pH 5.0, and strongly acid at pH 4.0. The equivalent concentrations (μequiv. l^{-1}) of divalent ions are shown (light stipple) as well as the molar concentrations (dark stipple). Atomic weights for the common cations and anions are listed on the right.

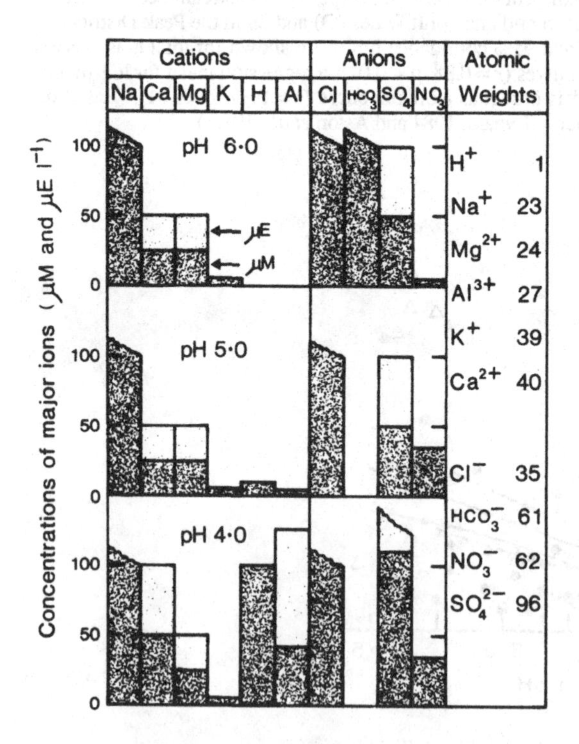

At pH below about 5.5 bicarbonate virtually disappears, sulphate usually increases and so may nitrate. The effect of bicarbonate removal on invertebrates has not been studied. Change in the relative concentrations of cations are thought to be more important. Calcium has the biggest change, from pH 7.0 (not shown in Figure 4) to pH 6.0. In Britain, calcium is approximately equal to sodium in waters with a mean pH of 7.0 but considerably less in waters with mean pH 6.0 and 5.0. Potassium uptake deserves more attention from physiologists (Shaw, 1959; Stobbart, 1967, 1974; Sutcliffe, 1971; Willoughby & Sutcliffe, 1976); potassium is notably scarce during summer months in headwater streams where even the winter concentrations are relatively low (<10 μequiv l^{-1}). Otherwise the biggest changes occur at pH 5.0 and below, when hydrogen and dissolved aluminium assume prominence and may exceed the concentrations of all other cations. Where sodium is well below 100 μequiv l^{-1}, as in some continental mountain areas, hydrogen and aluminium are completely dominant. As they are relatively small and mobile they may be transported inwards across the gills instead of sodium, potassium or calcium. This would upset the normal internal equilibrium and could lead to a fatal net loss of vital ions from blood and tissue.

Compared with insects, benthic macrocrustaceans (crayfish, *Gammarus, Asellus*) and molluscs, especially gastropods, are less tolerant of very low sodium and chloride concentrations, probably because these animals are more permeable. All arthropods are most susceptible to low pH and low ionic concentrations when they moult, when permeability to water and ions is greatly increased. At this time mortality is high, especially in juveniles which moult frequently.

Natural populations, or at least some individuals, invariably seem to tolerate concentrations of ions and pH that are lower than the minimum levels obtained in laboratory experiments. The discrepancies may imply the existence of physiological races and 'super-tolerant' individuals but another explanation needs careful consideration in future experimental work. Physiological studies are usually done on starved animals. Active ion uptake is faster in well fed mosquito larvae compared with starved animals; a regular intake of food is required to provide energy for ATP-coupled ion uptake (Stobbart, 1967, 1974). In addition to this requirement, natural food in some instances may provide some or all of the daily uptake of calcium, sodium and potassium, by absorption through the gut, whereas ion uptake directly from the surrounding water may predominate at the moult.

More attention should be given to metabolic aspects of ion uptake and 'osmotic stress' in acid and soft waters (Sutcliffe & Carrick, 1975; Otto & Svensson, 1983; Sutcliffe, 1984). An increased energetic demand for ionic regulation at low concentrations might reduce growth rates where the food supply is barely adequate for other requirements. In arthropods that are sensitive to acid water the natural rhythm of regular moults, which are sometimes quite frequent in small (young) specimens, might also become irregular and result in premature death.

Some benthic stream invertebrates, including arthropods, tolerate acid water in the range pH 4–5. More investigations on these taxa are needed to explain how they deal with high concentrations of H^+ and Al, especially when both are supposed to be toxic. Taxa that do not tolerate low pH values might then be examined in a new light. One aspect needing study is absorption of ions through the very permeable gut wall, as distinct from the gills. Freshwater animals usually drink small amounts of water but when they are exposed to saline (brackish) water, osmotic and ionic stress results in uncontrolled drinking and expulsion by vomiting through the mouth; this is rapidly fatal (Sutcliffe, 1962, 1967). Controlled uptake of water is particularly important for moulting arthropods. If large amounts of water are imbibed and ejected when animals are exposed to acid water, vital body ions can be lost and toxic ions have a ready means of ingress.

A final aspect, almost totally ignored so far, is the tolerance of invertebrate eggs to acid water and the egg-laying habits of adult insects. Sutcliffe & Carrick (1973a) could not find eggs of *Baetis* mayflies in acid tributaries of the River Duddon, although they were common elsewhere in the catchment. Thus the impoverishment of acid streams could be due proximately to simple behavioural avoidance by egg-laying adults. The underlying selective advantage of such behaviour remains to be demonstrated, however.

Field experiments

There have been a few attempts to test the effects of acidification experimentally in streams. In one case Hall *et al.* (1980) reduced the pH of a New Hampshire stream from about 6.0 to about 4.0 by adding sulphuric acid over a period of four months. The density of invertebrates drifting in the water column increased immediately although there was subsequently little difference between the experimental and control sites (Figure 5(*a*)). Overall, invertebrate species diversity was decreased in the experimental section. These changes occurred rapidly and Hall *et al.* (1980) concluded that they were due to direct behavioural and physiological effects. From the same experiment Fiance (1978) found that the growth of the mayfly *Ephemeralla funeralis* was slower in the acidified section of stream. In August, after three months of acidification, the size of larvae differed from the control stretch and in September there were far fewer larvae belonging to the next generation (Figure 5(*b*)). Fiance (1978) ascribed reduced growth to the extra metabolic cost of ion exchange at low pH.

More recently, Hall *et al.* (1985) carried out a short term addition of aluminium to a stream. As well as increasing the aluminium concentration, the pH and concentration of dissolved organic carbon were reduced and the surface tension of the water was lowered, producing the copious foam characteristic of many acidified waters. Again the drift of stream invertebrates increased and drift nets also captured larger numbers of adult terrestrial insects (prominently those forms often observed in dancing swarms close over the surface of streams) and aquatic larvae living at the surface film, such as meniscus midges of the family Dixidae. Hall *et al.* (1985) ascribed these latter changes to the reduction in surface tension.

The food hypothesis

In order for populations to persist, animals must not only be able to tolerate the physical and chemical environment but also find food. Evidence that the reduced fauna and low productivity of acidified waters is at least partially due to the nature of food available is, at present, largely circumstantial. For instance, Sutcliffe & Carrick (1973a) found that it was particularly the herbivorous taxa which were absent in streams of low pH and suggested this might be due to lack of suitable food. Townsend *et al.* (1983) categorised the invertebrates of some English streams into a

Figure 5. Experimental acidification of a New Hampshire stream resulted in (*a*) a short term increase in invertebrate drift and (*b*) reduced growth and recruitment of the mayfly ***Ephemerella funeralis*** (solid histograms show the newly recruited generation, open histograms the cohort recruited in the previous year); (*a*) after Hall *et al.* (1980), (*b*) after Fiance (1978).

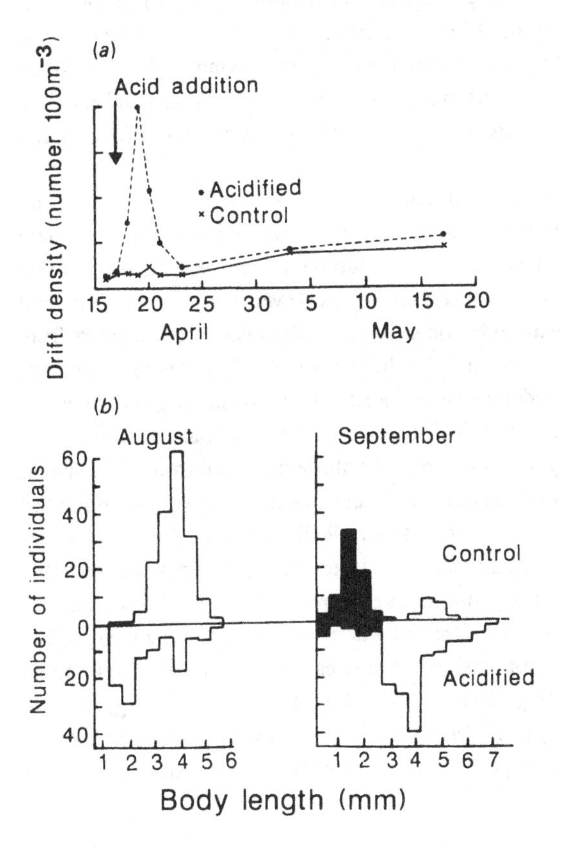

number of feeding guilds (Cummins & Merritt, 1984). These were: 'shredders', which feed on large particles of dead organic matter (often tree leaves); 'grazers/scrapers', which rasp attached material from stone surfaces; 'collectors', which feed on deposited fine particles of detritus; 'filter feeders', which feed on fine particles in suspension; 'predators', which eat other animals. Townsend *et al.* (1983) found that streams with a mean annual pH less than about 5.7 had a community dominated by shredders, predators and collectors only. The more diverse community of circumneutral streams contained, in addition, grazer/scrapers and filter feeders. In four of the guilds animals were also more numerous in circumneutral streams but shredders actually declined in abundance with increasing pH. Whilst such patterns could evidently arise if there were systematic differences in physiological tolerance betwen the guilds, it seems likely that there are also differences in the food resources available in streams of contrasting pH.

The food of the grazer/scraper guild is the material building up on stone surfaces in streams, usually called the 'epilithon' (Lock *et al.*, 1984). This consists of attached algae, bacteria and fungi in a slimy matrix. Comparison of the epilithon in the southern English streams previusly surveyed by Townsend *et al.* (1983) showed that at sites with higher conductivity and pH, layers were dominated by algae and other living components (Winterbourn, Hildrew & Box, 1985). At the most acid site, however, the layer was a rather structureless film with few living organisms. Much more extensive research on these layers in acidified streams is badly needed because it is by no means clear that algae are always reduced in acid waters (Wiederholm, 1984).

Winterbourn *et al.* (1985) offered epilithic layers from various acid and circumneutral streams to a number of common stream invertebrates, not all of them usually considered 'grazer/scrapers'. They all produced copious faeces indicating that they were able to ingest such material (Figure 6). However, the weight of material removed from stone surfaces was higher on the microbially impoverished layers from the most acid site and was substantially lower where algae were abundant. Winterbourn *et al.* (1985) considered this supported the view that organic layers on stones from acid streams were a poor food source because the organic fraction of the faeces produced was extremely low compared with those produced from other layers.

The growth of invertebrates fed on tree-leaf litter is usually fastest when microbes are already growing on the leaves. For instance, Willoughby & Sutcliffe (1976) showed that *Gammarus pulex*, the common freshwater amphipod, grew best on a diet of oak leaves richly colonised by fungi. Therefore, it might be expected that shredders would be most abundant in streams where leaf-litter is associated with a particularly rich microflora. Decomposition processes have often been shown to be slower in acidified streams (e.g. Minshall & Minshall, 1978; Friberg, Otto & Svensson, 1980; Hildrew *et al.*, 1984; Mackay & Kersey, 1985). Although there are few direct estimates of microbial populations in streams of contrasting pH, recent work in the River Duddon in the English Lake District does suggest that microbial populations, particularly bacteria, are reduced in acid tributaries (Dr A.C. Chamier, personal communication). It is, therefore, somewhat anomalous that Townsend *et al.* (1983) found shredders were most prominent in acid streams and Sutcliffe & Carrick (1973a) reported a basically detritivorous community in acid sections of the Duddon system. Indeed, Townsend & Hildrew (1985) found a negative correlation between the number of shredder individuals in invertebrate samples and the rate of cellulose decomposition among a number of contrasting sites.

Townsend & Hildrew (1985) speculated that shredders may be at a disadvantage in streams with a rapid rate of litter decomposition because their food, which in temperate areas enters streams in an autumn pulse, might therefore not persist through the year. In acid streams the food, although it may be of lower quality, is at least always available. Such speculations await experimental test and, in general, research

Figure 6. Common stream invertebrates were fed upon surface layers on stones drawn from five stream sites and one sterile 'control' stone (organic material removed by ashing). The sites are arranged (left to right) in order of increasing organic fraction of grazed material. Old Lodge is the most acid stream (mean pH about 4.3). Broadstone is also acid (mean pH about 5.3 but the stones are covered in iron bacteria. Nutley Bridge (pH 6.0) and Withyham (pH 6.6) have abundant algae. Grazing rates are means ($n = 5$) ± 1 SE. Key to species; ■ *Leuctra nigra*, ○; *Nemurella picteti*, ●; *Baetis* spp., ▲; *Ecdyonurus dispar*, □; *Sericostoma personatum*, Δ; *Gammarus pulex*. Redrawn from Winterbourn *et al.* (1985).

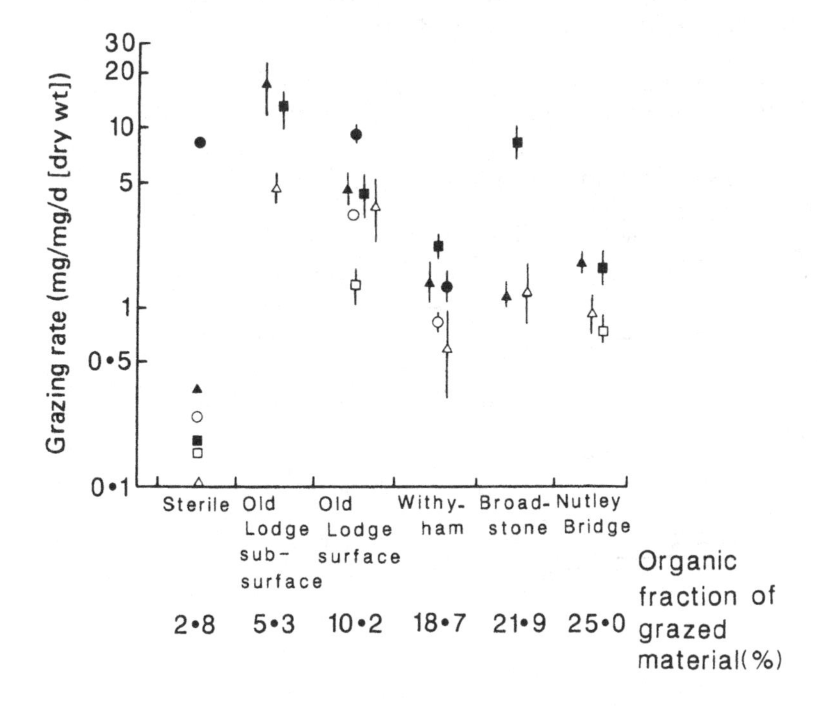

on the food quality and quantity of animals in acid streams, and particularly on the microbial flora, is urgently needed.

In one case, not usually considered in this context, the food hypothesis for community composition seems particularly well vindicated. There is a very clear correlation between the calcium concentration in British lakes and the species of predatory triclad flatworm present (Figure 7) (Reynoldson, 1983). However, there is no simple relationship between water chemistry and species occurrence. In this case each species is specialised on a particular prey taxon and the distribution of predators follows that of their prey. For triclads, calcium concentration is merely an indication of lake productivity, although calcium may have direct physiological effects on some prey organisms (e.g. *Gammarus* and *Asellus*).

The fish predation hypothesis

The composition of zooplankton communities in lakes is profoundly affected by fish predation (Zaret, 1980). Where predation is intense the larger planktonic species are absent or rare. Thus, in waters too acid for fish populations to persist we might expect larger invertebrates to flourish, as long as they can tolerate the chemical conditions. Eriksson *et al.* (1980) suggested that the planktonic phantom midge larva *Chaoborus*, and other large surface dwelling predators, were more abundant in acidified Scandinavian lakes without fish. Yan *et al.* (1985) however, found little relationship between *Chaoborus* abundance and acidity in central Canadian lakes. They suggested that *Chaoborus* populations are limited by the low productivity of acidified lakes rather than by predation.

Figure 7. The distribution of four species of predatory triclad flatworms in British lakes, related to the calcium concentration. Redrawn from Reynoldson (1983).

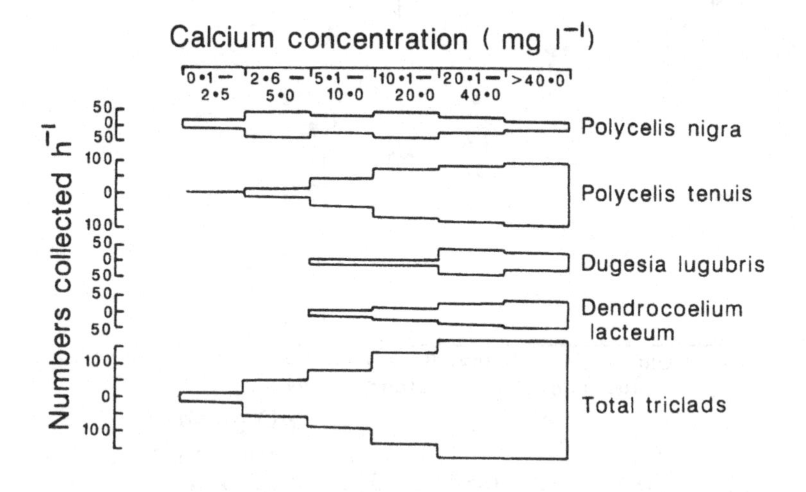

In streams it has proved much more difficult to demonstrate experimentally a strong effect of fish predation upon the benthos (Allan, 1982), so it seems less likely that the removal of fish by acidification would itself have much effect upon invertebrate communities. Nevertheless, there is some circumstantial evidence for it. Hildrew, Townsend & Francis (1984) divided the predatory insects in a number of stream communities from southern English streams into 'large' and 'small' species and found that the numerical proportion which was 'large' declined strongly with the abundance of fish (Figure 8). This relationship rested mainly upon the distribution of the large net-spinning caddis *Plectrocnemia* which may be particularly susceptible to predation by fish and is abundant in acid streams. Experimental field manipulations of brown trout in acid streams so far lend some support to the predation hypothesis, at least for *Plectrocnemia* (C.L. Schofield, unpublished data). However, it seems unlikely that the wholesale community differences between acid and circumneutral streams could arise from the pattern of fish predation alone.

Why are acid streams inpoverished?

It is hardly surprising that populations of those species unable to tolerate the chemical conditions of acidified waters are not found in acid streams. This is certainly the case for many common invertebrates, particularly molluscs, crustacea and some insects, although the physiological basis of why some species survive and others do not requires further investigation. Communities in acid streams might simply be chance assemblages drawn from a pool of a few tolerant species. This leaves several

Figure 8. Invertebrates and fish were sampled in 28 stream sites in southern England. The figure shows the relationship between the proportion of all predaceous insects with a maximum dry body weight greater than about 6 mg ('large') and the numbers of fish caught. Redrawn from Hildrew *et al.* (1984).

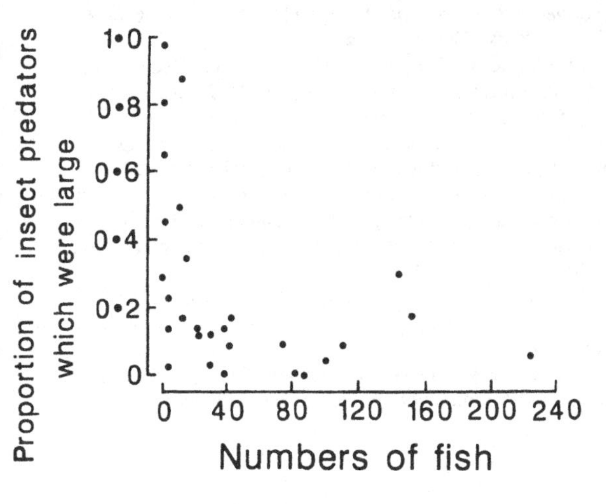

questions unanswered, however. For instance, we can see why acid streams are intolerable to many animals characteristic of neutral waters but it is less easy to explain what features of circumneutral streams might exclude animals more characteristic of acid waters. Such distribution patterns are common, however, and probably require explanation beyond physiological responses to water chemistry.

Even if some species could survive in acid streams they would find little food. For instance Willoughby & Sutcliffe (1976) showed that, whilst *Gammarus pulex* grew rapidly on decaying oak leaves, it did not grow when fed on a mixed diet of moorland grasses, bryophytes, algae and fungi from the acidic (and treeless) tributaries of the Duddon. These differences reflect the distribution of *Gammarus*; it is absent from acidic tributaries and common in the lower Duddon, where autumn-shed oak leaves are an important food source. Equally, however, survival and growth were poor in water from an acid tributary, even when *Gammarus* was provided with a rich diet.

It seems likely, therefore, that both food and water chemistry are important elements of the proximate explanation for community structure in acid streams. A low species diversity in acidified waters could ultimately be explained because the meagre resources would not lead to natural selection in favour of colonists. Finally, for almost all groups of animals and plants there is a strong species–area effect, species richness increasing with area (Williamson, 1981; Hildrew & Townsend, 1987). Perhaps the restricted area of acid waters, at least in Europe and North America, has limited the space and time available for colonisation and adaptation to a new environment.

Comparisons with other zones where acid waters may be naturally extensive, such as Amazonia (Sioli, 1985), could prove instructive.

References

Allan, J.D. (1982). Predator–prey relationships in streams. In *Stream Ecology: application and testing of general ecological theory*, ed. J.R. Barnes & G.W. Minshall, pp. 191–229. New York: Plenum Press.

Aston, R.J., Sadler, K., Milner, A.G.P. & Lynam, S. (1985). The effects of pH and related factors on stream invertebrates. CERL Report TPRD/L/2792/N84. Leatherhead, Surrey, UK: Central Electricity Research Laboratories.

Burns, J.C., Coy, J.S., Tervet, D.J., Harriman, R., Morrison, B.R.S. & Quine, C.P. (1984).The Loch Dee project: a study of the ecological effects of acid precipitation and forest management on an upland catchment in south-west Scotland. 1. Preliminary investigations. *Fisheries Management*, **15**, 147–67.

Burton, R.F. (1973). The significance of ionic concentrations in the internal media of animals. *Biological Reviews*, **48**, 195–231.

Burton, T.M., Stanford, R.M. & Allan, J.W. (1985). Acidification effects on stream biota and organic matter processing. *Canadian Journal of Fisheries and Aquatic Sciences*, **42**, 669–75.

Cummins, K.W. & Merritt, R.W. (Eds) (1984). *An Introduction to the Aquatic Insects of North America*, 2nd Edition. Dubuque, Iowa: Kenall/Hunt Publishing Co.

Engblom, E. & Lingdell, P.-A. (1984). The mapping of short-term acidification with the help of biological pH indicators. *Report of the Institute of Freshwater Research, Drottningholm*, **61**.

Eriksson, M.O.G., Henrikson, L., Nilsson, B.-I., Nyman, G., Oscarson, H.G. & Stenson, A.E.(1980). Predator–prey relations important for the biotic changes in acidified lakes. *Ambio*, **9**, 248–9.

Fiance, S.B. (1978). Effects of pH on the biology and distribution of *Ephemerella funeralis* (Ephemeroptera). *Oikos*, **31**, 332–9.

Friberg, F., Otto, C. & Svensson, B.S. (1980). Effects of acidification on the dynamics of allochthonous leaf material and benthic invertebrate communities in running waters. In *Proceedings of an international conference on Ecological Impact of Acid Precipitation* (ed. D. Drabløs & A. Tollan), pp. 304 –5. Oslo Ås: SNSF Project.

Hall, R.J., Likens, G.E., Fiance, S.B. & Henry, G.R. (1980). Experimental acidification of a stream in the Hubbard Brook Experimental Forest, New Hampshire. *Ecology*, **61**, 976–89.

Hall, R.J., Driscoll, C.T., Likens, G.E. & Pratt, J.M. (1985). Physical, chemical and biological consequences of episodic aluminium additions to a stream. *Limnology and Oceanography*, **30**, 212–20.

Harriman, R. & Morrison, B.R.S. (1982). Ecology of streams draining forested and non-forested catchments in an area of central Scotland subject to acid precipitation. *Hydrobiologia*, **88**, 251– 63.

Henriksen, A. & Seip, H.M. (1980). Strong and weak acids in surface waters of Southern Scotland. *Water Research*, **41**, 809 –13.

Hildrew, A.G. & Townsend, C.R. (1987). Organisation in freshwater benthic communities. In *Organisation of Communities: past and present*, Symposia of the British Ecological Society, ed. J.H.R. Gee & P.S. Giller, pp. 317–41. Oxford: Blackwell Scientific Publications.

Hildrew, A.G., Townsend, C.R. & Francis, J. (1984). Community structure in some southern English streams: the influence of species interactions. *Freshwater Biology*, **14**, 297–310.

Hildrew, A.G., Townsend, C.R., Francis, J. & Finch, K. (1984). Cellulolytic decomposition in streams of contrasting pH and its relationship with invertebrate community structure. *Freshwater Biology*, **14**, 323–8.

Houlihan, D.F., Rankin, J.C. & Shuttleworth, T.J. (1982). *Gills*, Society for Experimental Biology, Seminar Series, volume 16. Cambridge: Cambridge University Press.

Lavandier, P. (1979). Ecologie d'un torrent Pyrenéen de haute montagne: L'Estaragne. D.Sc. Thesis. Université Paul Sabatier de Toulouse (Sciences).

Lock, M.A., Wallace, R.R., Costerton, J.W., Ventullo, R.M. & Charlton, S.T. (1984). River epilithon: toward a structural functional model. *Oikos*, **42**, 10 –22.

Lucas, W.J. & Berry, J.A. (eds) (1985). *Inorganic Carbon Uptake by Aquatic Photosynthetic Organisms*. American Society of Plant Physiologists.

Mackay, R.J. & Kersey, K. (1985). A preliminary study of aquatic insect communities in acid streams near Dorset, Ontario. *Hydrobiologia*, **122**, 3 –11.

Mackereth, F.J.H., Heron, J. & Talling, J.F. (1978). *Water Analysis: some revised methods for limnologists*. Freshwater Biological Association Scientific Publication no. 36.

Minshall, G.W. & Minshall, J.D. (1978). Further evidence on the role of chemical factors in determining distribution of benthic invertebrates in the River Duddon. *Archiv für Hydrobiologie*, **83**, 324 –55.

Norton, S.A. & Henriksen, A. (1983). The importance of CO_2 in evaluation of effects of acidic deposition. *Vatten*, **39**, 346 –54.

Otto, C. & Svensson, B.S. (1983). Properties of acid brown water streams in South Sweden. *Archiv für Hydrobiologie*, **99**, 15–36.

Reynoldson, T.B. (1983). The population biology of *Turbellaria* with special reference to the freshwater triclads of the British Isles. *Advances in Ecological Research*, **13**, 236–316.

Sadler, K. & Lynam, S. (1984). Some chemical and physical characteristics of upland streams in north and mid Wales and the Peak District. CERL Report TPRD/L/2610/N83. Leatherhead, Surrey, UK: Central Electricity Research Laboratories.

Shaw, J. (1959). Solute and water balance in the muscle fibres of the East African freshwater crab, *Potamon niloticus* (M. Edw.). *Journal of Experimental Biology*, **36**, 145–56.

Simpson, K.W., Bode, R.W. & Colquhoun, J.R. (1985). The macroinvertebrate fauna of an acid-stressed headwater stream system in the Adirondack Mountains, New York. *Freshwater Biology*, **15**, 671–81.

Sioli, H. (ed.) (1985). *The Amazon: limnology and landscape ecology of a mighty tropical river and its basin*, Monographiae Biologicae, 56. Dordrecht: Dr W. Junk, Publishers.

Skeffington, R.A. (1985). The transport of acidity through ecosystems. Central Electricity Research Laboratories, Report TPRD/L/2907/N85. Also in *Pollutant Transport and Fate in Ecosystems*, British Ecological Society Special Publications, ed. P.J. Coughtrey, M.H. Martin & M. Unsworth (1986). Oxford: Blackwell.

Stobbart, R.H. (1967). The effect of some anions and cations upon the fluxes and net uptake of chloride in the larva of *Aëdes aegypti* (L.). and the nature of the uptake mechanisms for sodium and chloride. *Journal of Experimental Biology*, **47**, 35–57.

Stobbart, R.H. (1974). Electrical potential differences and ionic transport in the larva of the mosquito *Aëdes aegypti* (L.). *Journal of Experimental Biology*, **60**, 493–533.

Stoner, J.H., Gee, A.S. & Wade, K.R. (1984). The effects of acidification on the ecology of streams in the upper Tywi catchment in West Wales. *Experimental Pollution, Series A*, **35**, 125–57.

Sutcliffe, D.W. (1962). Studies on salt and water balance in caddis larvae (Trichoptera). III. Drinking and excretion. *Journal of Experimental Biology*, **39**, 141– 60.

Sutcliffe, D.W. (1967). Sodium regulation in the fresh-water amphipod, *Gammarus pulex* (L.). *Journal of Experimental Biology*, **46**, 499–518.

Sutcliffe, D.W. (1971). Regulation of water and some ions in gammarids (Amphipoda). II. *Gammarus pulex* (L.). *Journal of Experimental Biology*, **55**, 345–55.

Sutcliffe, D.W. (1984). Quantitative aspects of oxygen uptake in *Gammarus* (Crustacea, Amphipoda): a critical review. *Freshwater Biology*, **14**, 443–89.

Sutcliffe, D.W. & Carrick, T.R. (1973a). Studies on mountain streams in the English Lake District. I. pH, calcium and the distribution of invertebrates in the River Duddon. *Freshwater Biology*, **3**, 437– 62.

Sutcliffe, D.W. & Carrick, T.R. (1973b). Studies on mountain streams in the English Lake District. II. Aspects of water chemistry in the River Duddon. *Freshwater Biology*, **3**, 543 – 60.

Sutcliffe, D.W. & Carrick, T.R. (1973c). Studies on mountain streams in the English Lake District. III. Aspects of water chemistry in Brownrigg Well, Whelpside Ghyll. *Freshwater Biology*, **3**, 561–8.

Sutcliffe, D.W. & Carrick, T.R. (1975). Respiration in relation to ion uptake in the crayfish *Austropotamobius pallipes*(Lereboullet). *Journal of Experimental Biology*, **63**, 689–99.

Sutcliffe, D.W., Carrick, T.R., Heron, J., Rigg, E., Talling, J.F., Woof, C. & Lund, J.W.G. (1982). Long-term and seasonal changes in the chemical composition of precipitation and surface waters of lakes and tarns in the English Lake District. *Freshwater Biology*, **12**, 451–506.

Townsend, C.R. Hildrew, A.G. & Francis, J.E. (1983). Community structure in some southern English streams: the influence of physicochemical factors. *Freshwater Biology*, **13**, 521– 44.

Wiederholm, T. (1984). Responses of aquatic insects to environmental pollution. In *The Ecology of Aquatic Insects* (ed. V.H. Resh & D.M. Rosenberg), pp. 508–57. New York: Praeger Publishers.

Williamson, M.H. (1981). *Island Populations*. Oxford: Oxford University Press.

Willoughby, L.G. & Sutcliffe, D.W. (1976). Experiments on feeding and growth of the amphipod *Gammarus pulex* (L.) related to its distribution in the River Duddon. *Freshwater Biology*, **6**, 577–86.

Winterbourn, M.J., Hildrew, A.G. & Box, A. (1985). Structure and grazing of stone surface organic layers in some acid streams of southern England. *Freshwater Biology*, **15**, 363–74.

Yan, N.D., Nero, R.W., Keller, W. & Lasenby, D.C. (1985). Are *Chaoborus* larvae more abundant in acidified than in non-acidified lakes in central Canada? *Holarctic Ecology*, **8**, 93–9.

Zaret, T.M. (1980). *Predation and Freshwater Communities*, New Haven, Connecticut: Yale University Press.

D.J.A. BROWN AND K. SADLER

Fish survival in acid waters

Introduction

This paper will review the field and laboratory data on the chemical factors which affect fish and fishery survival in acid waters, and an attempt will be made to relate the two sets of data where this is possible. In general, the field data being considered are predominantly from Norway where the fish species is mainly brown trout, but, where relevant, data from North America will be described. Data from the UK are dealt with by Turnpenny (this volume).

The main chemical factors, other than pH (hydrogen ion concentration) that will be discussed are calcium (hardness) and aluminium concentration. The aquatic chemistry of aluminium, which is generally present in higher concentrations in more acid waters, is complex and it is necessary at this stage to summarise the details on the subject given by Freeman & Everhard (1977), Burrows (1977), Hunter *et al.* (1980), Spry *et al.* (1981) and O'Donnell *et al.* (1983).

Solubility of aluminium is a direct function of ambient pH, being at a minimum at around pH 5.5, and increasing towards both extremes of the pH scale. Soluble cationic species e.g. Al^{3+}, $AlOH^{2+}$ and $Al(OH)_2^+$, are formed at pH levels less than 5.5, and soluble aluminate species, e.g. $Al(OH)_4^-$, predominate at pH levels greater than 5.5. In natural waters, aluminium has a strong tendency to form complexes with other anions capable of forming coordinate bonds – for example, six different fluoride complexes are known (Burrows, 1977). Important also is the ability of aluminium to complex with organic molecules. Thus, three main fractions of aluminium in natural waters have been described by Driscoll *et al.* (1980), separated according to the following procedure.

The first division is into monomeric (rapidly reactive) and acid-soluble forms. The latter fraction consists of polymeric, colloidal and stable organic complexes. The monomeric fraction is then further subdivided on the basis of whether it is extracted by cation exchange resin. Thus the labile (inorganic) monomeric aluminium is retained (and this includes free aluminium, soluble hydroxy complexes, and complexes with fluoride and sulphate) and can be distinguished from the non-labile (organic) monomeric aluminium, which consists mainly of organically chelated species.

Against the background of this complex chemistry the field data will first be considered, followed by the field toxicity testing data and then the laboratory toxicity data. Finally, the overall results will be discussed and conclusions concerning future research requirements will be made.

Field data

Presence/absence data

(i) *pH and calcium concentration* Extensive chemical monitoring of acid lakes has been carried out in Norway together with estimates of their fishery status, in terms of their ability to support a sparse or a good fishery (usually of brown trout) or their being fishless (Wright & Snekvik, 1978). The data for 470 high altitude lakes demonstrated that fishery status was largely independent of pH for the 75% of the lakes with a pH below 5.1 with about 50% of the lakes being without fish and that only at pHs greater than 5.1 was there a marked improvement of fishery status with increasing pH, (Brown & Sadler, 1981). Wright & Snekvik (1978) found, from stepwise multiple regression techniques, that log [Ca^{2+}] and pH (in that order) emerged as the two most important chemical variables for the determination of fish status in southern Norwegian lakes. Brown (1982a) suggested isopleths of percentage of fishless lakes over the range of H^+ and Ca^{2+} concentrations and more recently, Chester (1984) has shown that the majority of fishless lakes are those with a Ca^{2+}/H^+ ratio (expressed as μequiv l^{-1}) of less than 3, whereas most of the lakes containing good fisheries have a Ca^{2+}/H^+ ratio greater than 4 (see Mason, this volume, Figure 4).

(ii) *pH and aluminium* Wright & Snekvik (1978) found no evidence from multiple regression analysis that total aluminium concentration was a factor determining fishery status of the southern Norwegian lakes. In fact the relationship between pH, total aluminium concentration, and the percentage of lakes which support a fishery might even be taken to suggest that at a given pH, there is a trend for the fishery status to improve with increasing aluminium (Brown, 1983a). This is most unlikely to be cause and effect, but may have arisen because at a given pH, higher aluminium concentrations are cross correlated with higher calcium concentrations. The improved fishery status with increasing aluminium implies that the beneficial calcium effect is dominating over the detrimental effect of aluminium. This tends to be confirmed by the plot in Figure 1 of fishery status against aluminium and calcium concentrations for a restricted range of hydrogen ion concentration (< 10 μequiv l^{-1}). Most of the lakes with a calcium concentration < 50 μequiv l^{-1} are fishless whereas the majority of the lakes with a calcium concentration > 50 μequiv l^{-1} contain fish, despite the fact that these same lakes have the higher aluminium concentrations.

Thus, because of the extensive cross correlation between the variables, the data of Wright & Snekvik (1978) do not give any clear indications as to the effect of aluminium on fisheries. Also, it must be noted that a single water sample taken from a

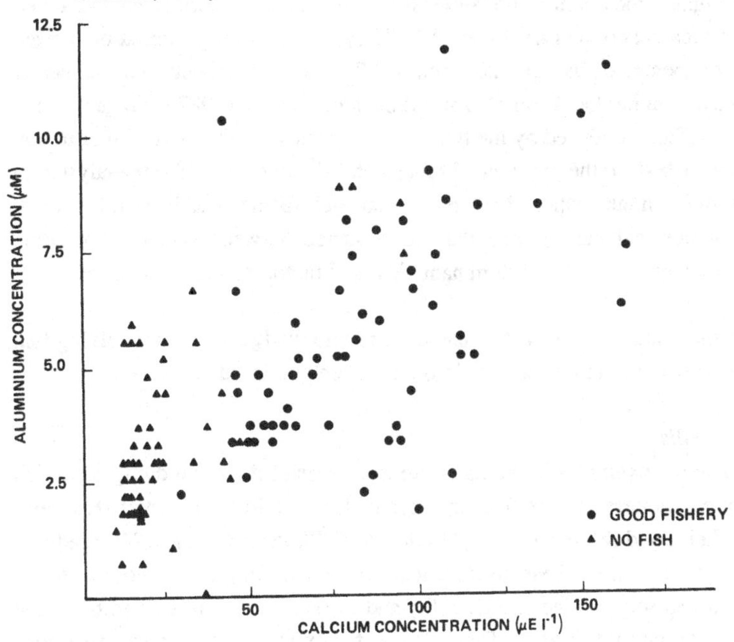

Figure 1. The fishery status of southern Norwegian lakes with a pH greater than 5.0 in relation to their calcium and aluminium concentrations, (data of Wright & Svekvik, 1978).

lake during the ice-free season may not adequately represent the conditions which are critical to the survival of fish populations, at, for example, snow melt events.

There are, however, some examples in the literature which indicate the toxic effects of total aluminium as measured in single samples of lake water. For example, in the Adirondack region of America, Schofield & Trojnar (1980) found that aluminium was a better determinator of the success/failure of attempts to stock brook trout (*Salvelimus fontinalis*) in 53 lakes, than pH or calcium. The mean total aluminium concentration in the lakes where stocked trout survived was 4.1 µM compared with 10.6 µM in lakes where they did not. Also, Bendell *et al.* (1983) found that the most striking difference between northern Ontario lakes which were fishless, from a total of 37 lakes, was their tendency to have significantly higher aluminium concentrations than the lakes which support fish when the data were adjusted for pH differences (analysis of covariance, $p < 0.01$).

Fish community data

(i) *pH and calcium concentration* Whereas most of the discussion in the previous section has been concerned with a single species, a variety of species is often found in North America and Swedish lakes. Rahel & Magnuson (1983) studied the distribution of 31 species in 138 naturally acidic lakes in northern Wisconsin and found a correlation between the rankings of species occurrence with minimum pH and rankings based on survival during laboratory exposure to low pH. Harvey (1975) investigated the relationship between fish diversity and various chemical and morphometric features of 66 La Cloche lakes. They found a strong correlation betwen the number of species of fish in a lake and pH. The correlation between number of species and calcium hardness was also significant and Brown (1982a) suggested that the pattern of effects produced by the two ions on species number was similar to that previously described for the presence/absence data. Henderson (1985) reanalysed La Cloche and Swedish lake data and suggested that each species was limited by one or two physico-chemical factors, and that these varied between species. For some species, calcium appeared to be a dominant chemical factor, and for others, pH.

(ii) *pH and aluminium* To the best of the author's knowledge, no data involving fish communities and aluminium concentrations have been published.

Fish kills

There are examples in the literature of reports of dead and dying adult fish having been encountered in the field in areas of low conductivity water (Huitfeldt Kaas, 1922; Leivestad & Muniz, 1976; Grahn, 1980; Skogheim *et al.*1984; Reader & Dempsey, this volume). These mortalities have been frequently thought to be associated with episodes of increased acidity and associated chemical conditions that occur in low conductivity waters, for example, following a rapid snowmelt. Proof that this is the case is often difficult because of lack of chemical data at the time of the kill, but there are a few instances of when contemporaneous water quality data are available or reconstructions have been possible, and these frequently involve aluminium. For example, Grahn (1980) hypothesised that the kill of adult ciscoe (*Coregonus albula*) that occurred in successive years in two Swedish lakes was the result of increased phytoplankton productivity which raised the pH of the lakes by 0.5 units (to 5.4 and 6.0) and hence the aluminium had become supersaturated and it was this that caused the fish kill. As evidence for this, the aluminium concentrations found in the gills of the dead fish were six to seven times higher than those in reference fish. A similar example of where aluminium-rich water at pH levels above 5.0 is thought to have resulted in a fish kill is given by Skogheim *et al.* (1984), for the River Ogna in SW Norway. Three kilometres from the mouth of the R. Ogna a hydroelectric station discharges acid, aluminium-rich water into the main river. During periods of low neutral pH river flow, the contribution from the hydroelectric station can increase, and during one such occasion in August 1982 when this happened, more than 50 Atlantic Salmon (3–10 kg) were found dead or dying in the river. In this case, there was no evidence of mucus clogging of the gills despite the supersaturation of aluminium, but the concentration of aluminium on the gills increased by an order of 10 (Skogheim *et al.*, 1984).

Field toxicity testing data

There are a few sets of data in the literature from experiments where fish have been kept in cages in natural waters, or the water from a river or stream has been pumped into tanks and observations of fish mortalities have been made. It was when the result of one such experiment involving white sucker fry was compared with survival in synthetic solutions of otherwise similar chemistry that it was first demonstrated by Driscoll *et al.* (1980) that not all the aluminium fractions encountered in the field were equally toxic. Median periods of survival in natural waters appeared to be determined by the levels of labile inorganic aluminium present and thus total aluminium values can overestimate the toxic potential. Further evidence of the detoxification of aluminium provided by, for example, organic complexation is given by the work of Hulsman *et al.* (1983). They investigated rainbow trout yolk sac fry survival in two La Cloche Mountain streams with similar pH (5.4–5.5) and total aluminium concentrations (5–6 $\mu M\ l^{-1}$) but differing in their organic (humic) content – the colour in one stream was 2 Hazen units compared with 13 in the other. After 10 d there was 95% survival in the coloured lake compared with only 5% in the clear lake.

Other examples of field toxicity studies implicating the toxicity of labile inorganic aluminium include that of Gunn & Keller (1984) who incubated lake trout sac fry in a lake in Ontario with a history of lake trout recruitment failure, during four short episodes of substantial pH depression. The mortalities (18%) occurred primarily during the longest depression (five days at pH 4.5–5.0) and highest concentrations of inorganic aluminium (*c.* 2 μM), but the authors make the point that the high survivals (82%) demonstrated that most sac fry could tolerate these conditions. They also give data which shows that substantially higher concentrations of inorganic aluminium (*c.*3 μM) were observed in the interstitial waters of the spawning rubble than in ambient waters, and thus fry within this spawning substrate may be subjected to more toxic conditions.

A similar partial mortality of salmonids has been reported by Henriksen *et al.* (1984). During a 20 d period that aluminium fractionation and speciation was being undertaken, the water was being pumped into tanks containing 0+ yr and 1+ yr (presmolt) Atlantic Salmon. The only fish that died were four out of 15 of the presmolts that died at times when the labile Al concentrations were, or had been *c.* 1.5 μM and pH levels below 5.5. No mortalities of 0+ yr fish were observed, or of either age group in the control tank.

Finally, several complete mortalities of brown trout were reported from cage experiments in four mountain streams during acid episodes in central Sweden during 1982 by Anderson & Nyberg (1984) but the authors were not able to distinguish whether the toxic metal at low pH was iron (10–22 μM) and/or manganese (1.5–3.3 μM) and/or aluminium (3.3–6 μM). The fact that the following year, exposed fish survived the beginning of the thaw despite maximum total concentrations of 930, 70 and 17 μM of iron, manganese and aluminium respectively further confuses the situation, but serves to reinforce the point that a much better resolution of variations in toxic metal levels and speciation in the field will be necessary to directly relate water chemistry and its ability or otherwise to support a fishery.

Laboratory data

Laboratory experimental work is most valuable in precisely defining the concentrations of elements, and their chemical speciation that are likely to be of concern and to establish cause effect relationships. This implies that studies within the range of conditions found in the field are of most relevance. Tests conducted under extreme conditions may well produce responses never normally encountered, and of no relevance to fish population losses caused by acidification. It is the intention of this review to concentrate on those studies conducted in conditions likely to be of most concern.

Egg survival

(i) *pH and calcium* There is a comparatively large body of literature concerning the survival of fish eggs at a low pH since the experimental systems required are relatively simple and compact so that large numbers of treatments may be made in confined laboratory space. Also when the effects of acid exposure have been investigated throughout the life cycle, eggs and larvae have been found to be particularly sensitive stages (Mount, 1973; Menedez, 1976; Craig & Baksi, 1977). Much of the early literature, however, failed to characterise adequately the water quality used or was inappropriate to oligotrophic waters susceptible to acidification. This has resulted in conflicting results in some cases, due, for example, to inadequate stripping of free CO_2 released by acidifying hard water, or from interactions with other toxic materials present in the water. A fair conclusion from the early literature would be that the freshly fertilised egg is an especially sensitive stage but that when the eggs have hardened, they are relatively insensitive until hatching.

The importance of calcium in determining survival of brown trout eggs at low pH has been demonstrated by Brown & Lynam (1981) and Brown (1982b). The relationship between the percentage survival of freshly fertilised eggs and a range of pH levels and calcium concentrations was described by Brown (1983a). Survival after eight days was 100% at pH 5.1 irrespective of calcium concentration down to 12 μequiv l^{-1}, and at 400 μequiv l^{-1} of calcium irrespective of pH, down to 4.2, whilst at low pH (4.2) and low calcium (50 μequiv l^{-1} and less) survival was nil (Figure 2).

(ii) *pH and aluminium* Relatively few experiments have been conducted on the effect of aluminium and low pH on the survival through to hatching of fish eggs, but those that have (e.g. Baker & Schofield, 1982) suggest that aluminium may have a beneficial effect on this stage of the life cycle.

Larval and adult survival

(i) *pH and calcium* The majority of experiments designed to measure the toxicity of low pH on older fish have been short term bioassays at less than pH 4.0. Such extreme conditions are not likely to be responsible for the elimination of fisheries in natural waters and only serve to indicate the relatively high resistance of adult fish compared with the eggs. A few longer term experiments have been conducted resulting in the survival rates indicated in Table 1. In terms of percentage survival rates, adult salmonids are able to tolerate pH levels down to 4.3–4.4 for long periods, even in low calcium concentrations.

Figure 2. The percentage survival of freshly fertilised brown trout eggs after 8 days in a range of pH and calcium concentrations.

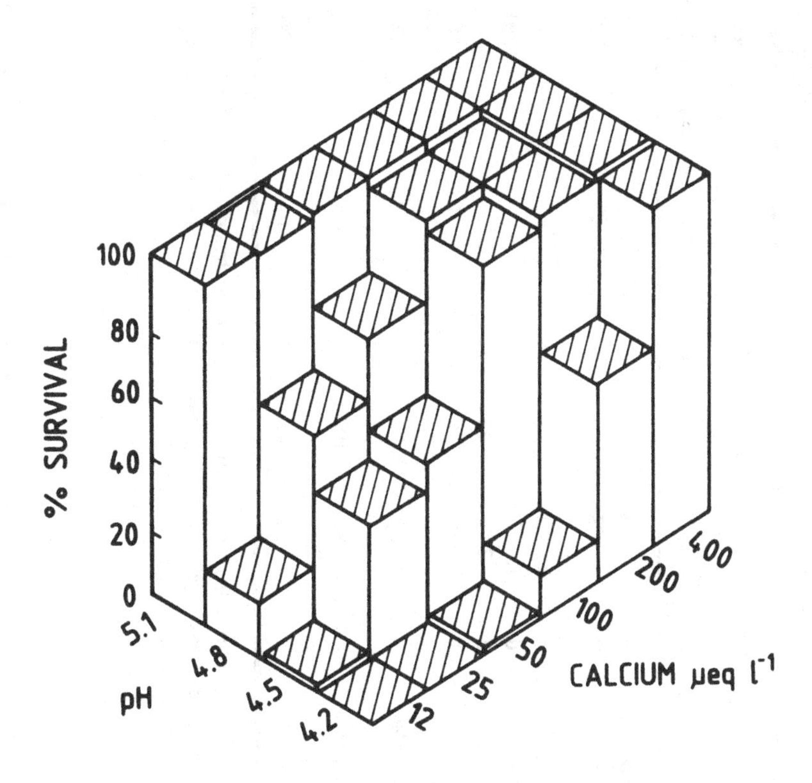

Table 1. *Percentage survival following exposure to low pH for yearling and older salmonids*

Species	Ref	Weight (g)	Duration (days)	pH							Water quality
				4.2–4.3	4.4–4.5	4.6–4.7	4.9–5.0	5.1–5.4	5.5–6.1	6.0+	
Salvelinus	a	100–170	100				90		100	100	River, 20 μS cm^{-1}[(a)]
alpinus	b	yearling	8	50	100		100		100	100	Tapwater, 13 μS cm^{-1}
Salvelinus	c	100–300	5	60–90						100	Tapwater
fontinalis	d	10–60	7	100			100	100	100		Tapwater, 30 μS cm^{-1}
	e	20	28					100		100	Synthetic, 125 μM Ca
	f	150–360	150		0		75			75–100	Tapwater, 120 μS cm^{-1}
Salmo gairdneri	a	200–300	100				93		90	97	River, 20 μS cm^{-1}
Salmo trutta	a	60–80	100				94		98	95	River, 20 μS cm^{-1}
	g	25	48				58	92		83	Tapwater, 30 μS cm^{-1}
	h	15–60	42	20	100	100	100	98	98	100	Synthetic, 25 μM Ca

[(a)] μS cm^{-1} is a measure of the conductivity (i.e. the total ionic strength) of a solution.

References: a Edwards & Hjeldnes, 1977; b. Jagoe *et al.*, 1984; c. Dively *et al.*, 1977; d. Daye & Garside, 1975; e. Rodgers, 1984; f. Menedez, 1976; g. Jacbsen, 1977; h. Sadler & Lynam, 1985.

(ii) *pH and aluminium* The toxicity of aluminium to fish has been the subject of reviews by Burrows (1977), and O'Donnell *et al.* (1983). However, as pointed out by Burrows (1977), most of the early studies were deficient in that toxicity was evaluated over a very short time period, pH was not controlled or measured in some cases, or water quality was not determined. Clearly pH control is vital in aluminium toxicity studies, not only because of the complex pH related speciation of aluminium, but also because hydrolysis of aluminium salts yields acid solutions in poorly buffered waters and this may compound apparent aluminium effects with low pH effects.

The majority of the recent published bioassay work involving pH and aluminium has used aluminium concentrations in excess of 9 μM and has reported maximum toxicity at pH 5.0–5.5 with beneficial effects at pH levels around 4.0 (Schofield, 1980; Schofield & Trojnar, 1980; Baker & Schofield, 1982; Brown, 1983b). Older fish also appear to be much more sensitive to aluminium than are eggs and fry, contrasting with the effects of pure acid stress (Brown, 1983a). There is also evidence that calcium interacts with both pH and aluminium concentration in modifying the results of 16 d bioassay experiments using brown trout fry (Brown, 1983b). The survival in acid solutions with aluminium concentrations of 9 and 18 μM was significantly higher if the calcium concentration was 50 μequiv l^{-1} or higher when compared with 25 μequiv l^{-1} or lower. However more recent experiments (Dalziel & Brown, 1984) suggest that the calcium effect is less marked with lower aluminium concentrations.

The amelioration of aluminium toxicity by other substances found occurring naturally in surface waters (e.g. organics) was noted in an earlier section, and has been confirmed in the laboratory by Driscoll *et al.* (1980). They found that citrate (which forms very strong complexes with aluminium) completely eliminated the toxicity of aluminium to brook trout fry.

Evidence is also emerging that not all labile monomeric aluminium species are equally toxic. Driscoll *et al.* (1980) reported that adding fluoride to aluminium solutions halved their toxicity (and aluminium fluoride complexes are often a major component of natural waters – see later). Similarly, Fivelstad & Leivestad (1984), in a study of trout survival in natural waters with added aluminium, also considered fluoride complexes to be relatively low in toxicity. They also dismissed the previously held belief that the insoluble polymeric $Al(OH)_3$ complexes were of greatest toxicity (aluminium has minimum solubility in its most toxic pH range), and suggested that the $Al(OH)^{2+}$ species was the most important to toxicity. (It is also of interest that Helliwell *et al.* (1983) reached similar conclusions implicating $Al(OH)_2^-$ in algal bioassay experiments).

Adult fish growth

(i) *pH and calcium* With regard to chronic (sublethal) effects of acid exposure the data are confused. Several workers report slower growth rates of fish around pH 5 compared with that at pH levels above 6 (Menedez, 1976; Edwards & Hjeldnes, 1977; Rogers, 1984) whereas Jacobsen (1977) found no effect on brown trout growth down to pH 5, and Sadler & Lynam (1984) found that good growth rates were maintained even down to pH 4.4. This latter study, however, did indicate an effect of pH on the rate of weight loss of starving fish, presumably indicating a change in metabolic rate of acid stressed fish. This may partly explain some of the differences in the findings between studies, in that high feeding rates are needed to compensate for the low pH stress, and if the feeding rate is not high enough, growth effects can be observed at low pH.

The importance of calcium in ameliorating any acid effect on growth rates is also confused. Rogers (1984) found that both pH and calcium had significant effects on brook trout growth (maximum 1.2% d^{-1}), whereas Sadler & Lynam (1984) found that brown trout could maintain a growth rate of 1.3% d^{-1} down to pH 4.4 and with a calcium concentration nearly 20 times lower than the minimum used by Rogers (1984). The difference between these studies must at this stage be attributed to species differences unless differences in experimental design produced some stress exacerbating the pH stress effects in the study of Rogers (1984).

(ii) *pH and aluminium* Sadler & Lynam (1986) have recently reported reductions in brown trout growth rates at pH levels below 5.5 with concentrations of total aluminium as low as 1 μM. At pH levels above 5.5 this effect was markedly reduced. When growth rates were correlated with the different aluminium species, it appears that $Al(OH)^{2+}$ species was again the one having the most deleterious effect.

Discussion and conclusions

Clearly, there are certain parallels between the field observations and the results of the laboratory studies, but relating the two is not always straightforward. Calcium is obviously very important for freshly fertilised fish egg survival at low pH and there is an undeniable relationship between pH and calcium concentration and the ability of a lake to support a fishery (see Mason, this volume, Figure 4). As far as aluminium is concerned, superficially, at least, there is also a connection between the bioassay results of Brown (1983b) and the field data in Figure 1, with improved fish survival and lakes more likely to have a fishery, if the calcium concentration is 50 μequiv l^{-1} or more, even though these lakes are the ones with higher aluminium concentrations. However this simplistic explanation does not seem to have been supported by the more recent results, and clearly a single field measurement of total aluminium is totally inadequate to characterise the ability of a lake to support a fishery. Not only are there the complications caused by differing toxicities of different aluminium fractions and species described earlier, but also those caused by seasonal variations in the concentrations of these fractions and species in natural waters. Several researchers have investigated these variations, along with pH, for various time periods ranging from daily for three weeks to monthly for a year (Driscoll, 1980; La Zerte, 1984; Seip *et al.*, 1984; Henriksen *et al.*, 1984). The common findings are that total aluminium levels increase during low pH conditions and that most of this increase is accounted for by the labile inorganic fraction. In fact, even highly coloured streams can switch from primarily organically complexed to inorganically complexed monomeric aluminium during peak flow (La Zerte, 1984). A further common finding is that the dominant component of this monomeric fraction is the fluoride complexes which constitute between 40 and 50% in the Norwegian examples (Seip *et al.*, 1984; Henriksen *et al.*, 1984); 50–60% in the American (Adriondack) data (Driscoll, 1983) and 70–90% in the Canadian examples (La Zerte, 1984).

In conclusion, a substantial amount of work is still required to evaluate the toxic effects of various species and complexes of aluminium and their interactions with various other water quality parameters. In particular, the effects of chronic levels of exposure require more study in view of the findings that very low concentrations may affect adult growth (Sadler & Lynam, 1984). These studies should proceed along with field studies designed to get a better characterisation of the types of aluminium present in natural waters, and to evaluate the relative importance of episodic rather than long term exposure.

Acknowledgements

This paper is published with the permission of the Central Electricity Generating Board.

References

Almer, B. & Hanson, M. (1980). Forsurninsseffeker i vastkustsjoar. Information from the Freshwater Lab., Drottningholm, pp. 44.

Andersson, P. & Nyberg, P. (1984). Experiments with brown trout (*Salmo trutta* L.) during spring in mountain streams at low pH and elevated levels of iron, manganese and aluminium. *Rep. Inst. Freshwater Res. Drottningholm*, **61**, 34–47.

Baker, J.D. & Schofield, C.L. (1982). Aluminium toxicity to fish in acidic waters. *Water Air Soil Pollut.*, **18**, 289–309.

Bendell, B.E., McNicol, D.K. & Ross, R.K. (1983). Effects of acidic precipitation on waterfowl poulations in northern Ontario. II. Fish community associations in small lakes in the Ranger Lake Area, their relationships to chemical and physiological variables, and their implications for waterfowl productivity. *Can. Wildl. Serv. LRTAP Program*, May 1983.

Brown, D.J.A. (1982a). The effect of pH and calcium on fish and fisheries. *Water Air Soil Pollut.*, **18**, 343–51.

Brown, D.J.A. (1982b). The influence of calcium on the survival of eggs and fry of brown trout (*Salmo trutta*) at pH 4.5. *Bull. Envir. Contam. Toxicol.*, **28**, 664–8.

Brown, D.J.A. (1983a). The relationship between surface water acidification and loss of fisheries. Kolloquium Lindau 1983, VDI–Berichte 500, Düsseldorf.

Brown, D.J.A. (1983b). Effect of calcium and aluminium concentrations on the survival of brown trout (*Salmo trutta*) at low pH. *Bull. Envir. Contam. Toxicol.*, **30**, 582–7.

Brown, D.J.A. & Lynam, S. (1981). The effect of sodium and calcium concentrations on the survival of the yolk sac fry of brown trout (*Salmo trutta* L.) at low pH. *J. Fish Biol.*, **19**, 205–11.

Brown, D.J.A. & Sadler, K. (1981). The chemistry and fishery status of lakes in Norway and their relationship to European sulphur emissions. *J. Appl. Ecol.*, **18**, 433–41.

Burrows, W.D. (1977). Aquatic aluminium: chemistry, toxicology and environmental prevalence. *Critical Reviews in Environmental Control*, **7**, 167–216.

Chester, P.F. (1984). Chemical balance in lakes in Sorlandet. *Proc. R. Soc. Lond. B.*, **305**, 564–65.

Craig, G.R. & Baksi, W.F. (1977). The effects of depressed pH on flagfish reproduction, growth and survival. *Water Res.*, **11**, 621–6.

Dalziel, T.R.K. & Brown, D.J.A. (1984). Survival of brown trout (*Salmo trutta* L.) fry in solutions of aluminium and calcium at low pH. CERL Note TPRD/L/2562/N83. Leatherhead, Surrey, UK: Central Electricity Research Laboratories.

Daye, P.G. & Garside, E.T. (1975). Lower lethal levels of pH for embryos and alevins of Atlantic salmon, *Can. J. Zool.*, **55**, 1504 –8.

Dively, J.L., Mudge, J.E., Neff, W.H. & Anthony, A. (1977). Blood pO, pCO_2 and pH changes in brook trout (*Salvelinus fontinalis*) exposed to sublethal levels of acidity. *Comp. Biochem. Physiol.*, **57A**, 347–51.

Driscoll, C.T., Baker, J.P., Bisogni, J.J. & Schofield, C.L. (1980). Effects of aluminium speciation on fish in dilute acidified water. *Nature (Lond.)*, **284**, 161–4.

Driscoll, C.T., Baker, J.P., Bisogni, J.J. & Schofield, C.L. (1983). Aluminium speciation and equilibria in dilute acidic surface waters of the Adirondack Region of New York State. In *Geological aspects of acid deposition*, ed. O.P. Bricker, pp. 55–7. Ann Arbor Science: Butterworth.

Edwards, D.J. & Hjeldnes, S. (1977). Growth and survival of salmonids in water of different pH. SNSF Research Report FR 10/77, 1432 As–NLH, Norway, 27 pp.

Fivelstad, S. & Lievestad, H. (1984). Aluminium toxicity of Atlantic salmon (*Salmo salar* L.) and brown trout (*Salmo trutta* L.): Mortality and physiological response. *Rep. Inst. Freshwater Res., Drottningholm*, **61**, 69–77.

Freeman, R.A. & Everhard, W.H. (1977). Toxicity of aluminium hydroxide complexes in neutral and basic media to rainbow trout. *Trans. Am. Fish. Soc.*, **100**, 644 –58.

Grahn, O. (1980). Fishkills in two moderately acid lakes due to high aluminium concentration. In *Proc. Int. Conf. Ecological Impact of Acid Precipitation*, ed. D. Drabløs & A. Tollan. Oslo–Ås: SNSF project.

Gunn, J.M. & Keller, W. (1984). Spawning site water chemistry and lake trout (*Salvelinus namaycush*) sac fry survival during snowmelt. *Can. J. Fish. Aquat. Sci.*, **41**, 319–29.

Harvey, H.H. (1975). Fish populations in a large group of acid stressed lakes. *Verh. Int. Ver. Limnol.*, **19**, 2406 –17.

Henderson, P.A. (1985). An approach to the prediction of temperate freshwater fish communities. *J. Fish Biol.*, **27** *Suppl. A*, 279–91.

Helliwell, S., Batley, G.E., Florence, T.M. & Lumsden, B.G. (1983). Speciation and toxicity of aluminium in a model fresh water. *Envir. Technol. Letters*, **4**, 141–4.

Henriksen, A., Skogheim, O.K. & Rosseland, B.O. (1984). Episodic changes in pH and aluminium speciation kill fish in a Norwegian salmon river. *Vatten*, **40**, 255–60.

Huitfield-Kaas, H. (1922). Om aarsaken til massedod av laks os orret i Frafjordelven, Helleelven os Dirdalselven i Ryfylke hosten 1920. *Norsk Jaeger Fiskefor. Tidskr.*, **1/2**, 37–44.

Hulsman, P.F., Powles, P.M. & Gunn, J.M. (1983). Mortality of walleye egges and rainbow trout yolk-sac larvae in low pH waters of the LaCloche Mountain area, Ontario. *Trans. Am. Fish. Soc.*, **112**, 580–8.

Hunter, J.B., Ross, S.L. & Tannahill, J. (1980). Aluminium pollution and fish toxicity. *Water Pollut. Control, 1980*, 413–20.

Jacobsen, O.J. (1977). Brown trout (*Salmo trutta* L.) growth at reduced pH. *Aquaculture*, **11**, 81–4.

Jagoe, C.H., Haines, T.A. & Kircheis, F.W. (1984). Effects of reduced pH on three life stages of sunapee char *Salvelinus alpinus*. *Bull. Envir. Contam. Toxicol.*, **33**, 430–8.

La Zerte, B.D. (1984). Forms of aqueous aluminium in acidified catchments of central Ontario: a methodological analysis. *Can. J. Fish. Aquat. Sci.*, **41**, 766–76.

Leivestad, H. & Muniz, I.P. (1976). Fish kill at low pH in a Norwegian river. *Nature (Lond.)*, **259**, 391–2.

Menendez, R. (1976). Chronic effects of reduced pH on brook trout (*Salvelinus fontinalis*). *J. Fish. Res. Bd. Can.*, **33**, 118–23.

Mount, D.I. (1973). Chronic effects of low pH on fathead minnow survival, growth and reproduction. *Water Res.*, **7**, 987–93.

O'Donnell, A.R., Mance, G.E. & Norton, R. (1983). A review of the toxicity of aluminium in freshwater. WRC Report 541–M.

Rahel, F.J. & Magnuson, J.J. (1983). Low pH and the absence of fish species in naturally acidic Wisconsin lakes: inferences for cultural acidification. *Can. J. Fish. Aquat. Sci.*, **40**, 3–9.

Rogers, D.W. (1984). Ambient pH and calcium concentration as maodifiers of growth and calcium dynamics of brook trout, *Salvelinus fontinalis*. *Can. J. Fish. Aquat. Sci.*, **41**, 1774–80.

Sadler, K. & Lynam, S. (1984). Some effects of low pH and calcium on the growth and tissue mineral content of yearling brown trout (*Salmo trutta*). CERL Note TPRD/L/2789/84. Leatherhead, Surrey, UK: Central Electricity Research Laboratories.

Sadler, K. & Lynam, S. (1986). Some effects on the growth of brown trout from exposure to aluminium at various pH levels. CERL Note TPRD/L/2982/R86. Leatherhead, Surrey, UK: Central Electricity Research Laboratories.

Schofield, C.L. (1980). Processes limiting fish populations in acidified lakes, In *Atmospheric Sulphur Deposition: Environmental Impact and Health Effects*, ed. D.S. Shriner, C.R. Richmond & S.E. Lindberg, pp. 345–56. Ann Arbor Science: Butterworth.

Schofield, C.L. & Trojnar, J.R. (1980). Aluminium toxicity to brook trout (*Salvelinus fontinalis*) in acidified waters. In *Polluted Rain*, ed. T.Y. Toribara, M.W. Miller & P.E. Morrow, pp. 341–465. New York: Plenum Press.

Seip, H.M., Muller, L. & Naas, A. (1984). Aluminium speciation: Comparison of two spectrophotometric analytical methods and observed concentrations in some acidic aquatic systems in southern Norway. *Water Air Soil Pollut.*, **23**, 81–95.

Sevaldrud, I.H. & Muniz, I.P. (1980). Sure vatn og innlandsfisket i Norge. Resultater fra intervjuundersokelsene 1974–1979. SNSF Research report IR 77/80, 1432 As–NLH, Norway, pp. 92.

Skogheim, O.K., Rosseland, B.O. & Sevaldrud, I.H. (1984). Deaths of spawners of Atlantic salmon (*Salmo salar* L.) in River Ogna, SW Norway, caused by acidified aluminium-rich water. *Rep. Inst. Freshwater Res., Drottningholm*, **61**, 195–202.

Spry, D.J., Wood, C.M. & Hodson, P.V. (1981). The effects of environmental acid on freshwater fish with particular reference to the soft water lakes in Ontario and the modifying effects of heavy metals. A review. *Can. Tech. Rep. Fish. Aquat. Sci.*, **999**, pp. 145.

Wright, R.F. & Snekvik, E. (1978). Acid precipitation – chemistry and fish populations in 70 lakes in southernmost Norway. *Verh. fur Limnol.*, **20**, 765–75.

A.W.H. TURNPENNY

Field studies on fisheries in acid waters in the United Kingdom

Introduction

The term 'acid water' is variously defined in the literature, but it is usual to adopt an upper limit of pH 5.6, equivalent to the pH of pure water in equilibrium with atmospheric carbon dioxide. The UK Acid Waters Review Group (1986) thus defines three categories of water with respect to acidity:

(i) 'permanently acid': pH usually <5.6, alkalinity zero or close to zero;

(ii) 'occasionally acid': pH occasionally <5.6, low alkalinity;

(iii) 'never acid': pH never <5.6, well buffered.

Although no survey of acid water distribution in the UK has ever been undertaken, the likely distribution can be inferred from the combination of solid and drift geology and soil type. Maps of these characteristics are given by the UK Acid Waters Review Group (1986), and a composite map is shown in Figure 1. This indicates a probable distribution of acid waters in the UK, predominantly to the west and north.

As far as fish are concerned, the definition of an acid water is arbitrary. Fish have been shown to exist in a wide range of pH levels, from 3.5 to 11; also, their response to acid conditions is determined by a complex of other factors such as related increases in toxic metal levels or indirect effects upon food organisms (Alabaster & Lloyd, 1980; Howells, 1983). The purpose of field studies is to establish the net effect of these combining or competing factors on fishery status.

The present paper draws together information, firstly on fishery problems which appear to have been associated with water acidity, and secondly observations on fishery status in acid waters recorded by various regional groups in the UK. The significance of factors other than acidity will be discussed.

Fishery problems associated with acid waters in the UK

Two types of fishery problem associated with acid waters are noted in the literature:

(i) kills of extant wild fish populations during or following acid episodes;

(ii) decline of fish populations in waters known to be acid.

Evidence of fish kills associated with acid episodes

Directly observed fish kills which can be unequivocally blamed on acid episodes have seldom been reported. The main difficulty is that pulses of acidity tend to be brief, usually following heavy rainfall or the first flush of a snowmelt (Dempsey, 1985), whilst mortalities of fish may be delayed or go unnoticed until some time after the event. Also, at the time of a spate, rivers are often inaccessible for objective assessments to be made of the extent of a fish kill. Incidents which have been reported relate only to salmonid species, though it is not clear whether other species are unaffected or just unnoticed.

Figure 1. Acid waters in the United Kingdom. The maps shows a likely distribution of permanently acid (stippled) and occasionally acid (vertical hatching) surface waters, based on information about lithology and soil types. Numbers indicate regions where fishery status of upland streams has been investigated (see Table 5). (Source of data: UK Acid Waters Review Group, 1986.)

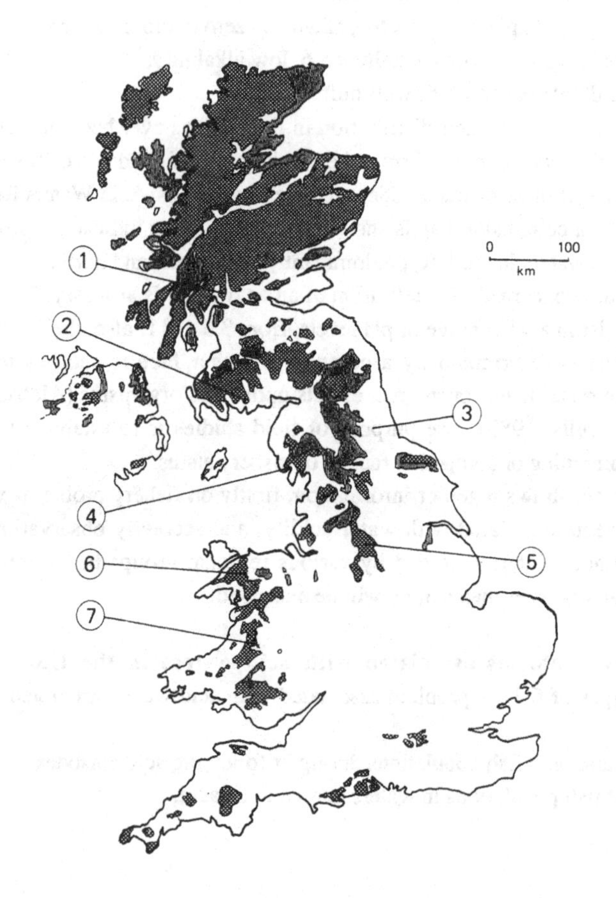

In Cumbria, a number of fish kills have been attributed to acidity, notably in the Esk and Duddon catchments (Prigg, 1983). Both of these rivers are underlain by igneous rocks of the Borrowdale Volcanic series. The first major fish kill associated with acidity was detected on the R. Esk in June 1980, when a severe mortality of fresh run adult and juvenile sea trout (*Salmo trutta* L.) and a salmon (*Salmo salar* L.) was reported after heavy rainfall, following a dry period. Such conditions have been shown elsewhere to increase the acidity of runoff as a result of 'occult' (i.e. deposited via cloud, fog or mist) or dry-deposited acidity accumulating on catchment vegetation (Miller & Miller 1980; Dollard, Unsworth & Harve, 1983). pH values measured after the kill were depressed to almost 5.0 in the middle reaches of the main river, but dead fish had been seen two days earlier, suggesting that the peak pulse of acidity had already passed. Over the same period a smaller, and at the time unreported, mortality occurred in the adjacent Duddon catchment.

In the R. Esk case, efforts were made to identify other possible causes of the fish kill. As an upland rural catchment, pesticides offered the most likely alternative to acidity, but no significant point sources of material nor any insecticidal effect on bottom fauna were found. The subsequent installation of continuous recording pH meters at flow gauging stations in the lower reaches of Rivers Esk and Duddon had revealed a marked incidence of low pH values related to high river flows for both rivers (Figure 2).

Figure 2. Relationships between daily maximum flow and daily minimum pH for the Rivers Esk and Duddon, August 1983–February 1984. (Source of data: UK Acid Waters Review Group, 1986.)

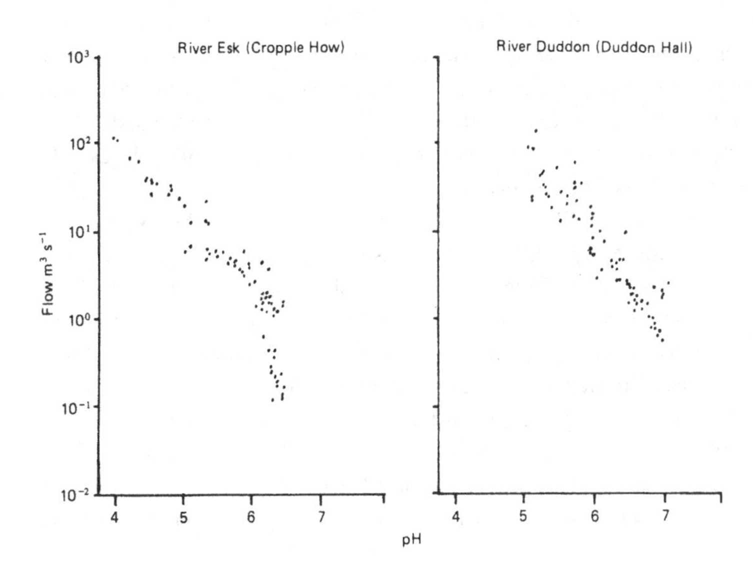

Elsewhere in Cumbria, Prigg (1985) reports an observed kill of brown trout (*S. trutta* L.) at a later date (18 June 1985) on the River Glenderamackin. When the event was investigated, some days after the kill, pH in the streams was 7.2 and alkalinity 2.0 mg l^{-1} $CaCO_3$. However, it was noted that a spate had occurred five days earlier (13 June), and that the invertebrate fauna showed some signs of acid restriction (notably lack of Ephemeroptera). The possibility of an insecticide incident was again considered and discarded, whilst subsequent chemical sampling indicated occasional pH values falling to 5.3–5.4, in conjunction with high total aluminium levels (up to 170 μg l^{-1}) and low calcium concentrations (0.4–0.6 mg l^{-1}). Laboratory toxicity data (Dalziel & Brown, 1984) indicate that this combination of chemical conditions is potentially acutely lethal to brown trout.

Pathological studies on fish can be useful in identifying the cause of a fish kill. For example, an incident on the River Glaslyn in north Wales killed 108 adult sea trout and nine adult salmon, following spate flows in September 1984 (Milner & Hemsworth, unpublished data, cited in report of UK Acid Waters Review Group, 1986). Blood samples taken from moribund fish exhibited low plasma ionic concentration, characteristic of exposure to low pH.

Decline of fish populations in acid waters

There are few instances in the UK where loss or decline of fish populations due to acidification can be validated. The best examples are to be found in southwest Scotland. In 1978, the Department of Agriculture and Fisheries for Scotland undertook a survey of some 22 upland lochs and associated streams in the southwest of Scotland (DAFS Freshwater Fisheries Laboratory, Triennial Review, 1979–81). In addition, information on earlier fishery status was obtained from the literature, angling records and by interview with commercial fishermen. The results indicated a decline of fish in a number of lochs and streams overlying granite geology. Evidence from water sampling and analysis of diatom assemblages in lake sediment cores (Battarbee *et al.*, 1985) is consistent with a decline in pH of between 0.5 and 1.2 pH units within the last 150 years in a number of lochs (L. Enoch, L. Valley, L. Dee, L. Grannoch and Round Loch of Glenhead).

Records for one loch, L. Fleet, show a progressive decline in the rod catch of brown trout from the early 1950s, and no fish caught after 1972 (Figure 3). Subsequent surveys in 1984–5 failed to find any fish. Survival studies carried out over the same period using artificially implanted brown trout eggs and captive fry demonstrated that the waters of the Loch and its main afferent spawning stream were acutely toxic to these life stages owing to low pH and calcium and high aluminium levels (pH 4.0–4.5, Al 200 μg l^{-1}, Ca 0.5–1.2 mg l^{-1}; Turnpenny, Dempsey, Davis & Fleming, 1988).

There is also evidence of a more widespread decline of salmon fisheries in Scotland associated with areas of heavy coniferous afforestation (Egglishaw,

Gardiner & Foster, 1986). Such forests are known to increase acidity of drainage waters (Harriman & Morrison, 1982).

Stoner & Gee (1985) have reported the decline over the last decade of some upland lake and river fisheries in central and west Wales. One case concerns the headwaters of the R. Tywi, above the Llyn Brianne regulating reservoir. Since 1969, fish traps have been operated on the inlets and outlet of the reservoir, as part of a 'trucking and trapping' scheme devised to maintain fish access to the headwaters. Despite regular stocking of the headwaters, trap catches of salmon and sea trout have declined sharply in recent years (Table 1). Surveys of the streams have revealed low densities of salmon (0.014–0.043 m^{-2}) and trout (0.066–0.070 m^{-2}), implying high juvenile mortality rates. Water samples collected in 1981–2 (Table 2) show conditions of pH, calcium and aluminium which are potentially lethal to trout (Dalziel & Brown, 1984), whilst salmon have been shown elsewhere to be more sensitive than trout with respect to acidity (Grande, Muniz & Anderson, 1973).

More direct evidence of low pH/aluminium toxicity in these streams has come from results of caged fish tests with trout, in which acute mortality was demonstrated, the rate of mortality ($1/LT_{50}$) being highly correlated with the concentration of aluminium present in the water ($r = 0.98$, $p < 0.01$; Figure 4). Stoner & Gee point out that a number of sub-catchments have been planted with coniferous forest since 1957, and that streams associated with these exhibit toxicity towards fish.

Stoner & Gee (1985) also describe the decline of fish populations in three upland lakes in central Wales, known as Syfydrin, Blaenmelindwr and Pendam. All three of these catchments overly hard bedrock and have been planted with conifers since

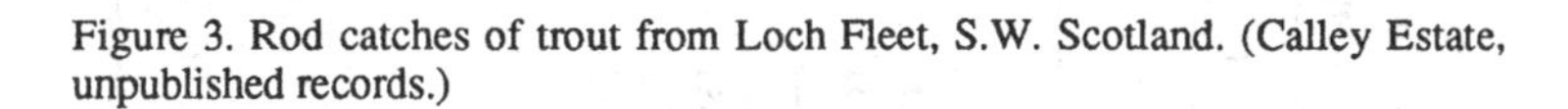

Figure 3. Rod catches of trout from Loch Fleet, S.W. Scotland. (Calley Estate, unpublished records.)

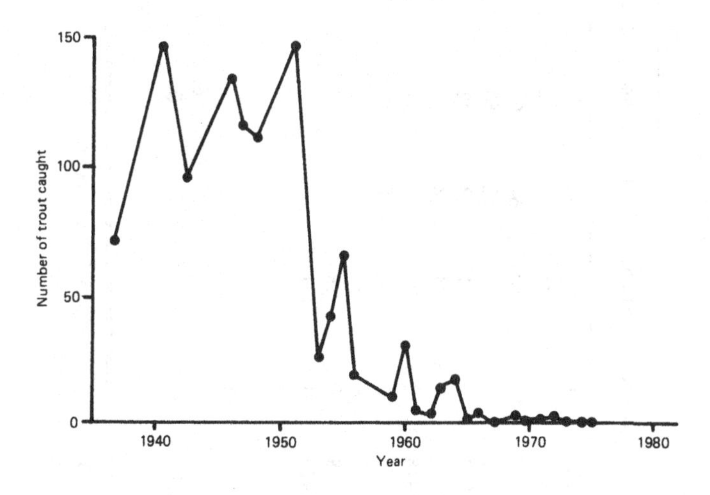

Table 1. *Trap catches and stocking rates for the Llyn Brianne Fisheries Protection Scheme*

Year	Number of adult fish trapped			Number of smolts trapped						Number of fish stocked (as unfed fry)			
				Camddwr			Tywi			Camddwr		Tywi	
	Salmon	Sea trout	Total	Salmon	Sea trout	Total	Salmon	Sea trout	Total	Salmon	Sea trout	Salmon	Sea trout
1969	2	28	30	–	–	–	–	–	–	–	–	–	–
1970	22	74	96	–	–	–	–	–	–	–	60,000	21,000**	70,000
1971	14	50	64	67	406	473	2	0	2	26,000	54,000	–	23,000
1972	0	12	12	60	331	391	2	72	74	28,000	28,000	–	–
1973	2	21	23	42	26	68	0	5	5	22,000	35,000	–	30,000
1974	0	44	44	89	128	217	0	5	5	–	22,000	–	12,00
1975	2	91	93	188	651	839	0	1	1	1,000*	77,000	–	–
1976	0	58	58	69	1216	1,285	0	174	174	103,000*	–	–	65,000
1977	0	32	32	3	132	135	0	161	161	–	–	–	123,00
1978	0	18	18	0	6	6	0	8	8	40,000***	–	–	–
1979	0	137	137	1	1	2	0	0	0	–	–	–	–
1980	0	27	27	0	18	18	1	0	1	1,038*	–	1,000*	–
							(Dead)						
1981	0	25	25	7	16	23	1	3	4				
				(incl. 2 dead)	(incl. 2 dead)			(incl. 1 dead)					
1982				0	23	23	0	0	0				
					(incl. 4 dead)								

* = 1+ parr; ** = O+ parr; *** = Eyed ova

From Stoner & Gee, 1985.

Table 2. *Means and ranges of pH, calcium and aluminium levels in River Tywi Headwaters, April 1981-March 1982.*

Stream	pH	Ca ($mg\ l^{-1}$)	Al ($\mu g\ l^{-1}$)
Tywi above Brianne	4.8	1.38	234
	3.9–6.1	0.60–2.60	99–801
Camddwr above Brianne	5.3	1.36	126
	4.4–6.7	0.20–3.00	99–504

From Stoner & Gee, 1985.

1940. Angling catches of brown trout have dropped off sharply in the last ten years compared with the neighbouring unafforested catchments of Rhosgoch and Craig y Pistyll on comparable geologies (Table 3).

Although there is no historical record of water quality for these lakes, levels of pH, calcium and aluminium measured in recent years (Table 4) fall within ranges likely to be toxic to brown trout (Dalziel & Brown, 1984). Unfortunately, no accompanying caged fish tests are reported to confirm this finding.

Figure 4. Mortality of caged trout in relation to mean aluminium concentrations in River Tywi headwater streams. ■ = unafforested, × = afforested.) (Source of data: Stoner & Gee, 1985.)

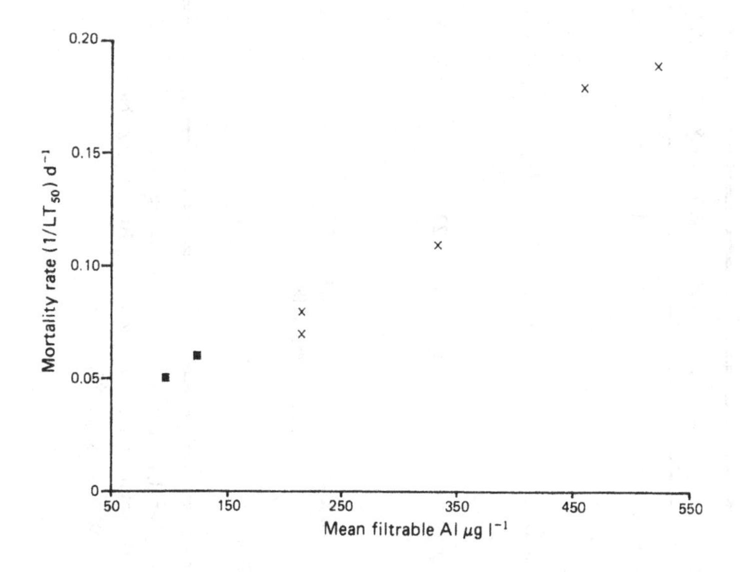

Table 3. *Catches of brown trout (except where indicated) declared by anglers for the period 1965–82 in five Welsh lakes (the numbers of fish stocked into the lakes are given in parentheses)*

Year	1964	1965	1966	1967	1968	1969	1970	1971	1972	1973	1974	1975	1976	1977	1978	1979	1980	1981	1982
Craig y Pistyll		0	0	1	2	75	31	26	15	83 (100 RT) (100 RT)	9	28	57	33	30	19	17	20	41
Sydydrin		84	111	224	96	104	45	74	31	12 (8)	16	40	2	6	7	1	1	2	6
Rhosgoch		20	16	22	22	59	30	68BT 20RT	37	44 (100BT) (50RT)	67 (100)	33 (100)	25 (100)	51 (100)	58 (100)	30 (100)	49 (100)	36 (100)	24 (100)
Blaenmelindwr	(1194)	1 (400)	83 (700)	20	10	35	2	37	6	0 (300)	1	4	0	2	0	0	158 (500BC)	55 (200BC)	(a)
Pendam		19	2	12	34	16	4	50	17	4 (100)	14	0	–	9	15	1	1	0	0

(a) Undeclared, but known to include 0+ brook charr.
BT = Brown trout; RT = Rainbow trout; BC = Brook charr.
From Stoner & Gee, 1985.

Table 4. *Means and ranges of pH, calcium and aluminium levels from upland lakes in Central Wales, April 1981–March 1982*

	pH mg l^{-1}	Ca	Al $\mu g\ l^{-1}$
Craig y Pistyll	5.0	1.08	108
	4.1–6.1	0.20–2.0	30–144
Syfydrin	4.7	0.94	180
	4.0–5.1	0.20–2.2	108–243
Rhosgoch	5.4	1.40	78
	4.0–5.1	0.40–3.4	40–200
Blaenmelindwr	4.9	1.64	300
	4.4–5.7	0.70–2.2	90–520
Pendam	5.1	2.0	370
	4.6–5.8	0.90–2.9	90–860

From Stonor & Gee, 1985

Synopsis of data on fishery status and water quality of UK upland streams

Fishery status can be measured in terms of species presence/absence, standing crop or in terms of production or any or its component processes (birth, growth and death). These measures can then be related to water quality parameters in a variety of ways.

The bulk of the observations of this type relate to upland streams, the main spawning habitat for salmonids. Upland streams are important in the acid water context, firstly, because they are subject to the greatest extremes of acidity, and secondly since an extensive literature has identified spawning or recruitment failure as the cause of fishery decline in many acid water areas (see Howells, 1983). Streams also offer sampling advantage over lake studies, since fish samples and population parameters can be more easily, cheaply and precisely obtained.

This section provides a synopsis of field survey data from the United Kingdom which demonstrate the overall effects of acidity and associated factors on fishery status.

Sources of data

The sources of data sets used are listed in Table 5. A total of 181 upland stream sites is represented, covering areas of central and southwest Scotland, Cumbria, Westmorland, the south Pennines and central and northern Wales. All but one of these areas contain acid streams. The exception is Westmorland, for which

Table 5. *Fishery status in UK upland streams: regional coverage and sources of data*

Region no.[a]	Country/Region	Parent catchment	No. of streams in sample	Reference
	Scotland			
1	Trossachs	Loch Ard	10	Harriman & Morrison, 1982[b]
2	Galloway	Loch Dee	3	Burns *et al.*, 1984[b]
	England			
3	Westmorland	River Tees	12	Crisp *et al.*, 1975
4	Cumbria	Rivers Esk, Duddon & Others	74	Prigg, 1983 *et seq.* [b]
5	S. Pennines	Rivers Don, Weaver, Mersey & Trent	21	Turnpenny, 1985
	Wales			
6	Central	Rivers Wye, Severn, Tywi & Teifi	34	Stoner *et al.*, 1984[b]
7	Northern	Various	19	
		Total	181	

[a] See Figure 1.
[b] Additional unpublished chemical or biological data obtained from the authors.

Table 6. *UK upland streams: chemistry of five stream groups categorised by pH (values as means, with ranges in parentheses)*[a]

	Group 1 pH < 5.0	Group 2 pH 5.01–5.50	Group 3 pH 5.51–6.00	Group 4 pH 6.01–6.5	Group 5 pH> 6.50
No of sites in sample	25	36	30	48	42
Mean pH	4.63 (4.16–5.00)	5.26 (5.02–5.50)	5.72 (5.53–6.0)	6.23 (6.01–6.50)	6.96 (6.54–7.56)
Min pH	4.21 (3.70–4.70)	4.61 (4.00–5.00)	4.78 (3.70–5.70)	5.51 (3.82–6.00)	6.50 (5.80–7.18)
Calcium mg l^{-1}	1.5 (0.74–3.2)	2.2 (0.80–6.4)	2.6 (1.0–6.2)	3.6 (0.90–14)	7.9 (2.1–27)
Total aluminium µg l^{-1}	231 (81–550)	137 (80–280)	131 (17–530)	90 (13–380)	79 (10–300)
Metox[b]	–	0.49 (0.27–0.71)	0.45 (0.31–0.86)	0.40 (0.23–0.66)	0.26 (0.11–0.51)

[a] Values given only for Welsh and Pennine streams listed by Turnpenny (1985).
[b] Zn/Cu/Pb toxicity as a fraction of the predicted 48 h LC_{50} to rainbow trout (Brown, 1968).

data from a number of more alkaline sites is included for comparison. Otherwise, the following criteria were adopted in selecting data sets:

(i) fishery status: for salmonids and eel (*Anguilla anguilla* L.), quantitative estimates of summer population biomass and/or density; for other species, presence/absence only;

(ii) water quality: chemical analyses of pH, calcium and total aluminium levels at least five times a year for one or more years;

(iii) concurrent chemical and biological sampling.

Stream chemistry

To simplify interpretation of acid-associated trends, streams were allocated to one of five groups according to mean pH value. In Table 6 the number of streams in each group is shown, together with details of mean and minimum pH values, mean calcium and total aluminium concentrations, and their respective ranges. In the case of the Westmorland streams (Crips, Mann & McCormack, 1975), pH and aluminium data were not available. However, it is clear from water hardness values (16–30 mg l^{-1} $CaCO_3$) that they belong in the circumneutral stream group (pH >6.5). In addition, for the 61 Welsh and Pennine streams surveyed by Turnpenny (1985), concentrations of the heavy metals zinc, copper and lead are shown, expressed as a combined index of toxicity to rainbow trout, *Salmo gairdneri* Richardson (Brown, 1968). This method is commonly used to provide an index of heavy metal toxicity to fish of river waters (e.g. Howells, Howells & Alabaster, 1983).

The overall pattern of water chemistry indicates trends of decreasing calcium concentration and increasing levels of total aluminium concentration and heavy metal toxicity with decreasing pH values. These are consistent with trends in surface waters found elsewhere in Europe and North America (see Howells, 1983).

Fishery status

Figure 5 shows the percentage frequency of occurrence in different pH groups of fish species recorded. The total numbers of sites considered for each species varied. In the case of eels and salmon, south Pennine sites were excluded, since absence of these species from all sites in the region is considered to be due to barriers to upstream movement rather than water quality (Turnpenny, 1985). Counts of bullhead (*Cottus gobio* (L.)) were limited to English and Welsh sites, since this species is scarcely present in Scotland (Maitland, 1972).

Figures 6 and 7 show standing crop population estimates of density and biomass for salmonids and eels respectively. Values shown represent group geometric means of only those sites in each pH class where the species was present. Salmonid data are

Figure 5. Percentage of streams in different pH categories in which fish species are present.

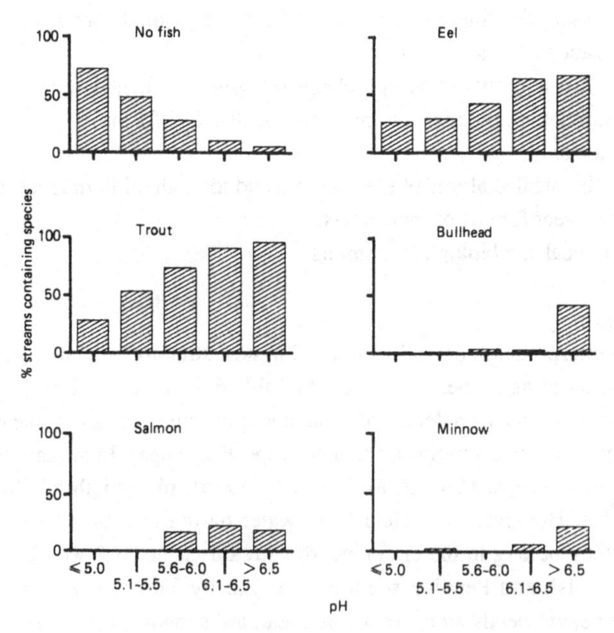

presented as combined figures for salmon and trout of I+ age class and older. Insufficient data were available to present similar comparisons for other species.

Salmonids Trout was the predominant species in all pH groups (Figure 5) and was present at all sites where fish were found, except for two streams in the Galloway area, where only eels were present (Turnpenny, Fleming & Davis, unpublished data). Salmon were much more restricted in their distribution generally, undoubtedly in part because of problems of access to low order streams. For both species, a lower

Figure 6. Salmonid population density and biomass values in different pH categories (Means and 95% confidence intervals; density values exclude data for O-group fish).

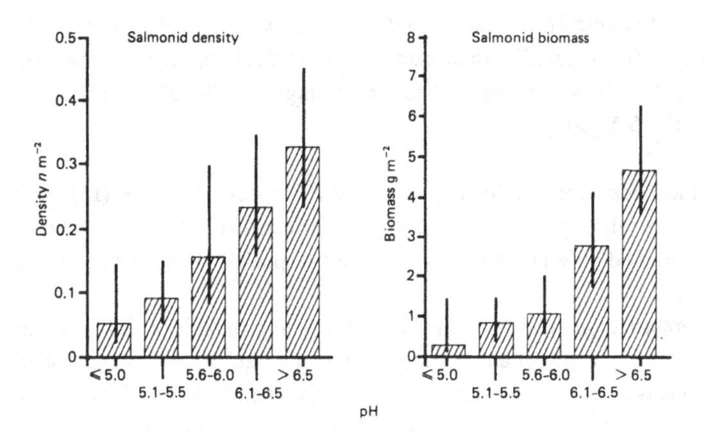

Figure 7. Eel population density and biomass values in different pH categories (means and 95% confidence intervals).

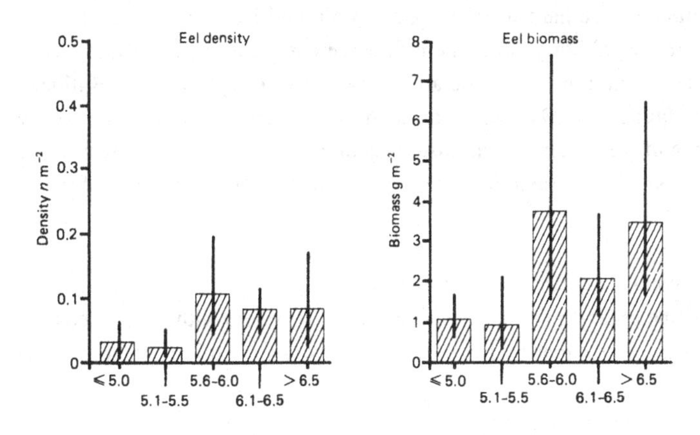

frequency of occurrence was found in more acid streams, only 28% of streams in the most acid category (pH <5.0) containing trout, compared with 95% in the circumneutral (pH >6.5) category, whilst no streams of pH <5.50 contained salmon.

The standing crop of salmonids was also much lower in more acid streams. Geometric mean population density in the most acid group (pH <5) was reduced to 16.1% of that in the circumneutral group (pH >6.5), whilst biomass was reduced to 6.2%.

Eels Like trout, eels were widely distributed with respect to acidity. This result is consistent with observations reported by Almer *et al.* (1978) for a set of 100 Swedish lakes which ranged in pH from 7.5 to 4.3. All but six of these lakes (three of which had pH 7.5) contained eels.

Despite this relative tolerance of acid conditions, eels were twice as common in streams with pH >6.0 compared with streams of pH <5.5. Also, the standing crop biomass and density values were three to four times higher in the pH >6.0 groups compared with pH <5.5 groups.

Minnow and bullhead Observations of minnow *Phoxinus phoxinus* (L.), and bullhead were sparse throughout. Of circumneutral streams (pH >6.5) only 41% contained bullhead, and 22% contained minnow. Both species were scarce in more acid stream groups.

Their poor representation in low order upland streams in general is probably due to lack of access, since small species are less able to ascend waterfalls and cascades than salmonid fish. This is true of more alkaline streams also, where bullhead and minnow seldom penetrate as far upstream as the trout (e.g. Crisp *et al.*, 1974). The absence of minnows from the more acid streams is consistent with distribution patterns found in Swedish lakes (Almer *et al.*, 1978).

There is practically no information concerning the water quality preferences of bullhead. Mann (1971) compared fish populations in southern England between one soft water stream and three hard water streams, and found bullhead present only in the hard water streams, although elsewhere they are known to occur in soft waters (Smyly, 1957). Diet could, however, be a limiting factor in acid streams. Bullhead feed on bottom fauna, chiefly crustaceans, molluscs and insects, particularly ephemeropteran and plecopteran nymphs. (Wheeler, 1969; Hyslop, 1982); all of these groups are found to be restricted in acid streams (Aston *et al.*, 1984; Stoner, Gee & Wade, 1984).

Factors controlling fishery status in acid streams

Clearly, the observation that fewer fish occur in waters with lower mean pH does not necessarily imply that hydrogen ion toxicity is the limiting factor. Some

Table 7. *Welsh and Pennine streams: Blegvad Index of stomach fullness for trout in different pH categories*

	Stomach fullness index			
	pH < 5.5	pH 5.51–6.00	pH 6.01–6.50	pH > 6.50
Mean	131	137	138	138
SD	51	72	46	58
n	3	11	9	12

From A.G.P. Milner, unpublished data.

other physical, chemical or biological correlate could equally well be the underlying cause. This possibility has already been indicated in the case of minnow and bullhead, but salmonids have been more fully investigated and data on trout offer better opportunities to explore these questions.

Physical factors The physical properties of a stream catchment can determine fishery status in a number of ways.

Streams which are susceptible to acidity are usually hydrologically unstable, with a high risk of drought or spate, either of which can displace fish (Deacon, 1961; Cowx, Young & Hellawell, 1984). During the intra-gravel phase of salmonids (typically four months in upland streams in the UK) there is a further risk of washout of eggs in spate flows (Elliott, 1976; E.M. Ottoway, A. Clarke & D.R. Forrest, personal communication).

It is clear from the literature that these risks affect circumneutral as well as acidic upland streams, and may account for recruitment variability (Crisp *et al.*, 1974) and the absence of fish from some circumneutral streams (Figure 5). Turnpenny (1985) examined the incidence of missing age classes in 45 trout-containing streams in Wales and the Pennines and found a high incidence (40–58%) of missing 0 and I group throughout, irrespective of pH (Table 7).

Streams draining coniferous forests tend to be more acidic (Harriman & Morrison, 1982) and these forests may have other effects upon fisheries unassociated with this acidity. Examples are effects of light-shading on primary productivity and stream temperature (Smith, 1980), lowering of the water table through evapo-transpiration (Gash *et al.*, 1978) and increased soil erosion due to drainage practices (Stoner & Gee, 1985). The latter can lead to streambed sedimentation, and thus limit benthic productivity and spawning area (Cordone & Kelley, 1961).

Table 8. *Welsh and Pennine streams: % of streams with missing O+ and I+ trout age classes, classified according to mean pH*

	% of streams with missing age class			
	pH < 5.5	pH 5.51–6.00	pH 6.01–6.50	pH > 6.50
Age class				
O+	40	40	58	53
I+	40	50	50	50

From Turnpenny, 1985.

Biological factors Although acidity could affect fish indirectly through a variety of biological pathways (e.g. food supply, changes in fish community, exposure to pathogens and parasites, etc.), the effect upon food supply is likely to be the most important, and is the only one for which any data are available from the UK.

For the Welsh and Pennine streams surveyed for fish by Turnpenny (1985), benthic macrofauna (Aston *et al.*, 1984) and gut contents of trout (A.G.P. Milner, unpublished data) have been examined. Despite a restriction of macrofaunal diversity in beds of the more acid streams, the gut fullness index of trout (Blegvad, 1917) was similar in all pH groups (Table 8). A high incidence of surface and terrestrial insects was found, indicating that deficiencies in the streambed fauna are offset by allochthonous inputs. Turnpenny (1985) also showed that growth rates of O-group trout were unrelated to stream pH in these streams.

Water chemistry It is seen from Table 6 that other water quality changes accompany differences in the mean pH of stream waters. Laboratory studies (Brown, 1983; Dalziel & Brown, 1984) demonstrate that when aluminium is present, its toxicity to trout is determined by the pH and calcium content of the water. Thus toxic effects of stream waters on trout are likely to be related to all three of these characteristics, as well as any other toxicants (e.g. heavy metals).

Turnpenny (1985) showed that, for Welsh and Pennine sites, absence or sparsity of trout was related to high levels of total aluminium in Pennine treams and high levels of heavy metal toxicity (Cu/Pb/Zn) in Welsh streams. This is illustrated in Figure 8. The results indicate threshold values for loss of fish at mean values of about 40 μg Al l^{-1} or 0.6–0.8 of the 48 h LC_{50} to rainbow trout for Cu/Pb/Zn. The latter value corresponds reasonably well with studies on the R. Mawddach in Wales reported by Howells (1983), who found the extinction point for brown trout to be 0.56 of the 48 h LC_{50} to rainbow trout, whilst toxic effects (growth reduction) of inorganic aluminium on brown trout have been demonstrated in the laboratory at levels down to 20 µg l^{-1} (Sadler & Lynam, 1986a). The importance of aluminium in the field has also been demonstrated by Stoner *et al.* (1984) (see Figure 4).

It is quite clear from North American studies (e.g. Driscoll *et al.* (1980)) that not all of the 'total' aluminium measured will be in toxic form, but that toxicity is largely attributable to inorganic monomeric forms. Sadler & Lynam (1986b) have more recently carried out fractionation of aluminium in the same set of Welsh and Pennine streams, and their findings indicate that, on average, some 45% of the total aluminium was in inorganic monomeric form, though this fraction varied considerably from site to site. Other recent findings from the L. Fleet catchment, Galloway, show that this fraction can vary from 20 to 100% in streams and lakes (Turnpenny, Davis & Fleming, unpublished data), and hence no useful generalisation can be made.

Figure 8. Biomass of salmonids in Welsh and Pennine streams in relation to levels of labile monomeric aluminium and Cu/Zn/Pb toxicity. Biomass: solid symbols, >5 gm^{-2}; semi-solid, >0.5 gm^{-2}; open symbols, no fish.) (Source of data: Turnpenny, 1985.)

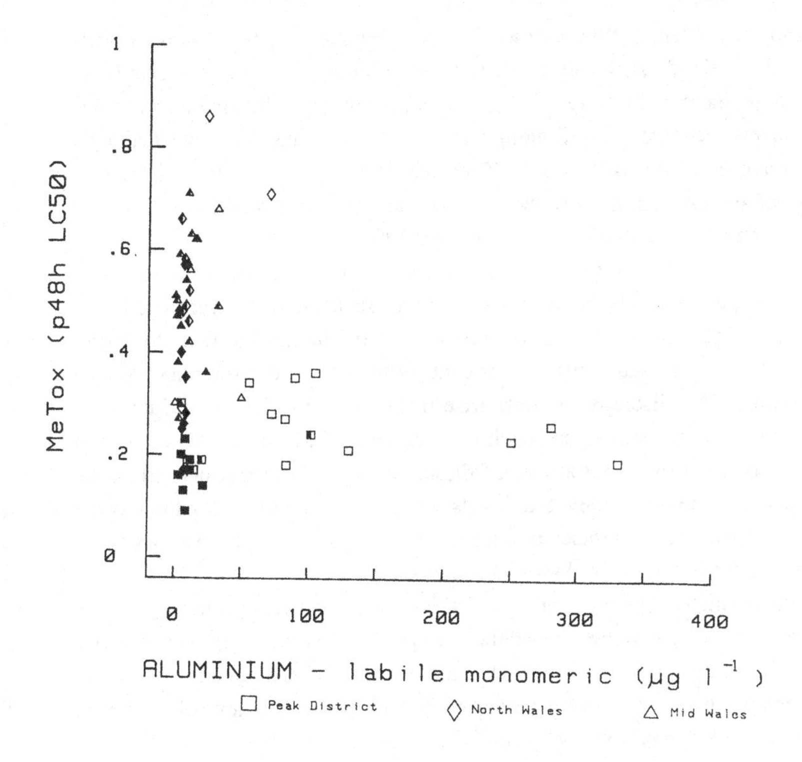

Conclusions

Fish have long been known to occur in acid waters in the UK (Frost & Brown, 1967). The fact that a number of recent studies reviewed here have shown impoverished fish fauna in acid streams in the UK does not in itself imply that a problem has suddenly arisen, but rather reflects the current global interest in surface water acidification (Sutcliffe, 1983). Indeed, the findings of the UK Acid Waters Review Group (1986) indicate that very little firm evidence from water quality records of recent change in pH status of UK waters can be found, other than in some Scottish waters, although it is accepted that currently available data are inadequate to make a full assessment.

Reports of fish kills associated with acidity have only emerged in recent years. It is not clear whether this is because the incidence of such events has become more common, or merely that this possibility is now more frequently investigated. With the recent introduction of continuous recording pH meters on a number of acid-susceptible rivers of the UK (e.g. R. Esk and Duddon: Figure 2), and the use of pathological techniques (e.g. analysis of blood chemistry) for the identification of acid stress, it is probable that such incidents will be more frequently recognised and reported in the future.

There is strong evidence of links between fishery decline and afforestation in parts of Wales and Scotland. This may be due to a combination of direct and indirect effects. Of these, increased acidity of drainage waters together with elevation of toxic trace metal levels appear to be the limiting factor on fisheries since acutely lethal toxic conditions have been shown to result. The area covered by forests in Britain is expected to increase substantially in the next 50 years, and this could have important consequences for upland fisheries (Stoner & Gee, 1985).

The upland stream data reviewed here demonstrate that the fish fauna of these waters becomes progressively impoverished through successively lower pH classes. At values in the range pH 5–6.5, not normally considered harmful to fish (Alabaster & Lloyd, 1980), fishery status was significantly lower than in circumneutral (pH 6.5–7.5) streams. This discrepancy must arise from the fact that fish production in natural waters is dependent upon a whole complex of physical, chemical and biological factors, not mean pH alone. Of these, chemical factors appear to be the most important, especially elevated levels of toxic trace metals which often accompany low pH. The importance of trace metals in acid waters has been shown to vary regionally, depending on background geochemistry.

There is also evidence that the limiting conditions in upland streams may not relate to mean water chemistry values, but rather to episodic extremes. It is commonly found in small streams during such episodes, that toxic hydrogen ion and aluminium concentrations rise by an order of magnitude or more, whereas calcium, which has an ameliorating effect on their toxicity, declines in concentration (Dempsey, 1985). The net effect of this is a substantial transient increase in toxicity.

The effect of transient changes in water chemistry is currently the subject of extensive laboratory and field investigations in the UK.

Acknowledgements

I would like to thank Dr D. Crawshaw and Mr R.F. Prigg (North West Water), Dr A.S. Gee (Welsh Water), Mr B.R.S. Morrison (Freshwater Fisheries Laboratory, Pitlochry), Mr A.G.P. Milner, Dr K. Sadler and Mrs S. Lynam (CEGB) for the use of unpublished data. I am also grateful to Dr R.N. Bamber and Dr D.J.A. Brown for their helpful comments on the manuscript. This paper is published by permission of the Central Electricity Generating Board.

References

Alabaster, J.S. & Lloyd, R. (1980). *Water Quality Criteria for Freshwater Fish.* London: Butterworth.

Almer, B., Dickson, W., Ekstrom, C. & Hornstrom, E. (1978). Sulphur pollution and the aquatic ecosystems. In *Sulphur in the Environment, part II,* ed. J.O. Nriagu, pp. 271–311. New York: John Wiley.

Aston, R.J., Sadler, K., Milner, A.G.P. & Lynam, S. (1985). The effects of pH and related factors on stream invertebrates. CERL Report TPRD/L/2792/N84. Leatherhead, Surrey, UK: Central Electricity Research Laboratories.

Battarbee, R.W., Flower, R.J., Stevenson, A.C. & Rippey, B. (1985). Lake acidification in Galloway: a palaeoecological test of competing hypotheses. *Nature (Lond.)*, **314**, 350–2.

Blegvad, H. (1917). On the food of fish in the Danish waters within the Skaw. *Rep. Dan. Biol. Stn.*, **24**, 17–72.

Brown, V.M. (1968). The calculation of the acute toxicity of mixtures of poisons to rainbow trout. *Water Res.*, **2**, 723–33.

Brown, D.J.A. (1983). The effect of calcium and aluminium concentrations on the survival of brown trout (*Salmo trutta*), at low pH. *Bull. Envir. Contam. Toxicol.*, **30**, 582–7.

Burns, J.C., Coy, J.S., Tervet, D.J., Harriman, R., Morrison, B.R.S. & Quine, C.P. (1984). The Loch Dee Project: a study of the ecological effects of acid precipitation and forest management on an upland catchment in south-west Scotland. 1. Preliminary investigations. *Fish Mgmt.*, **15** (4), 145–67.

Cordone, A.J. & Kelley, D.W. (1961). The influences of inorganic sediment on the aquatic life of streams. *Calif. Fish Game*, **47**, 189–228.

Cowx, I.G., Young, W.O. & Hellawell, J.M. (1984). The influence of drought on the fish and invertebrate populations of an upland stream in Wales. *Freshwater Biol.*, **14**, 165–77.

Crisp, D.T., Mann, R.H.K. & McCormack, J.C. (1974). The populations of fish at Cow Green, Upper Teesdale, before impoundment. *J. appl. Ecol.*, **11**, 969–96.

Dalziel, T.R.K. & Brown, D.J.A. (1984). Survival of brown trout (*Salmo trutta* L.) fry in solutions of aluminium and calcium at low pH. CERL Report TPRD/L/2562/N85. Leatherhead, Surrey, UK: Central Electricity Research Laboratories.

Deacon, J.E. (1961). Fish populations following a drought, in the Neosho and Marais des Cygnes Rivers of Kansas. *Univ. Kans. Publs., Mus., Nat. Hist.*, **13** (9), 359–427.

Dempsey, C.H. (1985). A review of the effects on fish of changes in water chemistry during episodes of lowered pH. CERL Report TPRD/L/2869/N85. Leatherhead, Surrey, UK: Central Electricity Research Laboratories.

Dollard, G.J., Unsworth, M.H. & Harve, M.J. (1983). Pollutant transfer in upland regions by occult precipitation. *Nature (Lond.)*, **302**, 241.

Driscoll, C.D., Baker, J.P., Bisogni, J.J. & Schofield, C.L. (1980). Effects of aluminium speciation on fish in dilute waters. *Nature (Lond.)*, **284**, 161–64.

Egglishaw, H., Gardiner, R. & Foster, J. (1986). Salmon catch delcine and forestry in Scotland. *Scott. Geogr. Mag.*, **102** (1), 57–61.

Elliott, J.M. (1976). The energetics of feeding metabolism and growth of brown trout (*Salmo trutta* L.) in relation to body weight, water temperature and ration size. *J. Anim. Ecol.*, **45**, 923–48.

Frost, W. & Brown, M.E. (1967). The Trout. London: Collins.

Gash, J.H.C., Oliver, H.A., Shuttleworth, W.J. & Stewart, J.B. (1978). Evaporation from forests. *J. Inst. Wat. Engres. Sci.*, **32** (2), 104.

Grande, M., Muniz, I.P. & Sanderson, S. (1979). The relative tolerance of some salmonids to acid waters. *Verh. Int. Ver. Limnol.*, **20**, 2976.

Harriman, R. & Morrison, B.R.S. (1982). Ecology of streams draining forested and non-forested catchments in an area of central Scotland subject to acid precipitations. *Hydrobiologia*, **88**, 251–63.

Howells, G.D. (1983). Acid waters – the effect of low pH and acid associated factors on fisheries. *Adv. Appl. Biol.*, **9**, 143–255.

Howells, E.J., Howells, M.E. & Alabaster, J.S. (1983). A field investigation of water quality, fish and invertebrates in the Mawddach river system, Wales. *Fish Biol.*, **22**, 447–70.

Hyslop, E.J. (1982). The feeding habits of 0+ stoneloach. *Noemacheilus barbatulus* (L), and bullhead *Cottus gobio* L. . *Fish Biol.*, **21**, 187–96.

Maitland, P.S. (1972). *Key to British Freshwater Fishes*. Freshwater Biological Association, Scientific Publication no. 27.

Mann, R.H.K. (1971). The populations, growth and production of fish in four small streams in southern England. . *Anim. Ecol.*, **40**, 155–90.

Miller, H.G. & Miller, J.D. (1980). Collection and retention of atmospheric pollutants by vegetation. In *Proc. Int. Conf. Ecological Impact of Acid Precipitation*, ed. D. Drabløs & A. Tollan. Oslo–Ås: SNSF Project.

Prigg, R.F. (1983). Juvenile salmonid populations and biological quality of upland streams in Cumbria, with particular reference to low pH effects. North West Water, Report no. BN 77–2–83 (unpublished). Carlisle, UK: North West Water, Rivers Division.

Prigg, R.F. (1985a). Acid rain project biosurveys in the Wastwater catchment. North West Water, Technical Note NC 297(12/85) (unpublished). Carlisle, UK: North West Water, Rivers Division.

Prigg, R.F. (1985b). Faunal and chemical observations on the River Glenderamackin above Mungrisdale following a probable acid event. North West Water Technical Note NC 299(I2/85) (unpublished). Carlisle, UK: North West Water, Rivers Division.

Sadler, K. & Lynam, S. (1986a). Some effects on the growth of brown trout from exposure to aluminium at different pH levels. CERL Report TPRD/L/2982/R86. Leatherhead, Surrey, UK: Central Electricity Research Laboratories.

Sadler, K. & Lynam, S. (1986b). Water chemistry measurements, including inorganic aluminium and organic aluminium complexes in some Welsh and Penine streams. CERL Report TPRD/L/3015/R86.

Smith, B.D. (1980). The effects of afforestation on the trout of a small stream in southern Scotland. *Fish Mgmt*, **11**, 39.

Smyly, W.J.P. (1957). The life history of the bullhead or Millers, Thumb (*Cottus gobio* L.). *Proc. Zool. Soc. Lond.*, **128**, 431–53.

Stoner, J.H. & Gee, A.S. (1985). Effects of forestry on water quality and fish in Welsh rivers and lakes. *J. Inst. Wat. Engrs. Sci.*, **39** (1), 27–45.

Stoner, J.H., Gee, A.S. & Wade, K.R. (1984). The effects of acidification on the ecology of streams in the upper Tywi catchment, in west Wales. *Envir. Pollut. Ser. A.*, **35**, 125.

Sutcliffe, D.W. (1983). Acid precipitation and its effects on aquatic systems in the English Lake District (Cumbria). Freshwater Biological Association, 51st Ann. Rpt., 30–59.

Turnpenny, A.W.H. (1985). The fish populations and water quality of some Welsh and Pennine streams. CERL Report TPRD/L/2859/N85. Leatherhead, Surrey, UK: Central Electricity Research Laboratories.

Turnpenny, A.W.H., Dempsey, C.H., Davis, M.H. & Fleming, J.M. (1988, in press). Factors limiting fish populations in the Loch Fleet Systems, an acidic drainage system in S.W. Scotland. *J. Fish Biol.*, **32**.

United Kingdom Acid Water Review Group (UKAWRG) (1986). *Acidity in United Kingdom Fresh Waters: Interim Report* London: Dept. of the Environment.

Wheeler, A. (1969). *The Fishes of the British Isles and North-West Europe.* London: Macmillan.

J.P. READER AND C.H. DEMPSEY

Episodic changes in water quality and their effects on fish

Introduction

The areas of North America and Northern Europe from which decline of fish populations in acid waters (pH ≤5.6) has been reported are usually characterised by high levels of precipitation and by weathering-resistant bedrock, resulting in poor soils and surface waters very low in dissolved substances ([Ca] ≤100 μmol l^{-1}, 4 mg l^{-1}; conductivity ≤60 μS cm^{-1}) (reviews: Haines, 1981; Spry *et al.*, 1981; Harvey, 1982; Harvey & Lee, 1982; Howells, 1983; see also papers in this volume by Brown & Sadler; Mason; Turnpenny).

The streams and lakes in these predominantly mountainous areas are, consequently, vulnerable to rapid and large changes in their chemistry. The soil may frequently be waterlogged or frozen, and may be thin or absent altogether (at bedrock outcrops and scree slopes). Episodes of heavy rainfall or snowmelt may therefore result in more or less direct runoff into the water bodies. The low concentration of dissolved substances in the water bodies ensures that the ability to buffer changes in acidity is poor, and concentration changes caused by runoff episodes are likely to be relatively large. The pH of the water may fall dramatically during such episodes (Tables 1, 2).

From time to time, fishkills (particularly of Salmonidae) are observed in such waters (Table 2), occurring over short periods (hours or days) and coinciding with increased water discharge associated with heavy rain, snowmelt or release of artificially impounded water. Intensive mining, industrial or agricultural activities are usually absent from the immediate area, so localised introduction of toxic substances by human activity is unlikely to be the cause of death. The mechanism of these fishkills may have an important part to play in fish population decline in acid waters.

Episodic changes in water quality

Episodes of increasing discharge in low conductivity waters can be accompanied by some or all of the following changes; declining pH, increasing concentrations of potentially toxic trace metals and changes in their chemical speciation and changes in the concentrations of organic materials and of a variety of inorganic ions (e.g. Ca^{2+}, K^+, Mg^{2+}, Na^+, Cl^-, HCO^-_3, NO^-_3, SO^{2-}_4).

Table 1. *Low pH episodes in low conductivity waters (for further explanation see text)*[(a)]

Reference	Area	Water body	Weather	Sample interval
Bjärnborg (1983)	NW Sweden	stream	snowmelt	2–4 wk
Christopherson *et al.* (1982)	S Norway	stream	rain	1 d
Dale *et al.* (1974)	S Norway	stream	rain	1–2 d
Driscoll *et al.* (1980)	New York State	stream	rain	*c.* 4 wk
Driscoll *et al.* (1980)	New York State	stream	snowmelt	*c.* 4 wk
Hagan & Langeland (1973)	S Norway	lake surface	snowmelt	*c.* 4 wk
Henriksen & Wright (1977)	S Norway	stream	rain and snowmelt	<1 wk
Henriksen & Wright (1977)	S Norway	lake	rain and snowmelt	<1 wk
Hultberg (1977)	SW Sweden	lake surface	snowmelt	*c.* 4 wk
Jeffries *et al.* (1979)	Ontario	stream	snowmelt	*c.* 1 d
Jones *et al.* (1983)	N Carolina	stream	rain	*c.* 2 h
Langan (1987)	SW Scotland	stream	rain	1 h
Leibfried *et al.* (1984)	Pennsylvania	stream	rain and snowmelt	12–24 h
Schofield *et al.* (1985)	New York State	stream	snowmelt	12–24 h
Schofield *et al.* (1985)	New York State	lake surface	snowmelt	1 wk
Schofield & Trojnar (1980)	New York State	stream	snowmelt	< 1 wk
UKAWRG (1986)	mid-Wales	stream	?	continuous

[(a)]Some of the above data are derived from figures.
[(b)]Inorg: labile inorganic or monomeric inorganic aluminium.
[(c)]Recorded in same stream in subsequent year.

Table 1 (*contd.*)

Duration	Time to min. pH	Minimum pH	pH excursion	Maximum [Al] µmol l⁻¹ total	Maximum [Al] µmol l⁻¹ inorg(b)	Other metals elevated
8 wk	*c.* 2 wk	6.6	0.7	6.7		Fe Mn
c. 11 d	6 d	4.2	1.8	30		
4 wk	1 wk	4.2	1.0			
<8 wk	<4 wk	4.8	1.1	31	12	
<12 wk	<6 wk	4.7	1.3	20	11	
8 wk	*c.* 4 wk	3.9	2.2			Fe Mn Pb Zn
4–6 wk	≤1 wk	4.3	≤1.4			
4–6 wk	≤1 wk	4.5	≤0.9			
c. 6 wk	≤3 wk	3.8	2.2			
c. 25 d	13 d	4.5	0.3			
c. 24 h	6 h	4.3	1.1	108		Fe Mn
c. 24 h	11.5 d	4.6	0.9			
c. 6 d	3 d	4.5	1.4	26		
c. 8 wk	*c.* 3 wk	4.7	2.7	12(*c*)	≥5.2(*c*)	
c. 5 wk	*c.* 3 wk	<4.2	>1.8			
20 d	6 d	4.6	0.8	37		
c. 5 d	24 h	4.2	1.8			

Table 2. *Fishkills in low conductivity waters (for further explanation see text)*[a]

Reference	Area	Water body	Weather	Sample interval	Duration
Andersson & Nyberg (1984)	Central Sweden	stream	snowmelt	1 d[c]	6.5 wk
Andersson & Nyberg (1984)	Central Sweden	stream	snowmelt	1 d[c]	5.5 wk
Andersson & Nyberg (1984)	Central Sweden	stream	snowmelt	1 d[c]	>6.5 wk
Gunn & Keller (1984)	Ontario	lake margin	rain and snowmelt	1 d[c]	8 d
Henriksen *et al.*(1984)	SW Norway	river	rain and snowmelt	continuous[d]	*c.* 3 d
Henriksen *et al* (1984)	SW Norway	river	rain and snowmelt	continuous[d]	c. 3 wk?
Hultberg (1977)	SW Sweden	streams	snowmelt	2–14 d	2–14 d
Leivestad & Muniz (1976)	S. Norway	river	snowmelt	c. 1 d	>5 wk
Muniz & Leivestad (1979)	S Norway	stream	rain	≥ 3h	≥9 d
Muniz *et al.* (1979)	S Norway	river	snowmelt	1–3 d	?
Prigg (1983)	Cumbria	river	rain	–	?
Schofield & Trojnar (1980)	New York State	lake	snowmelt	< 1 wk	7 d
Skogheim *et al.* (1984a)	S Norway	river	rain	*c.* 12 h	3 wk
Skogheim *et al.* (1984b)	SW Norway	river	[e]	*c.* 1 d	≥2 d

[a]Some of the above data are derived from figures.
[b]Inorg: labile inorganic or monomeric inorganic aluminium.
[c]Sample interval for trace metals: *c.* 1 week or greater.
[d]Sample interval for aluminium: 1 day.
[e]Release of water impounded by hydroelectric scheme.
[f]Measurement before water was limed (see text).
[g]Exp: experimental exposure in the field.

Table 2 (*contd.*)

Time to min. pH	Minimum pH	pH excursion	Maximum [Al] μmol l^{-1}		Other metals elevated	Animals killed
			total	inorg[(b)]		
4 wk	4.5	1.7	5.6	<1.1	Fe Mn	*Salmo trutta* (exp.)[(g)]
3 wk	4.5	2.5	4.5	<1.1	Fe Mn	*Salmo trutta* (exp.)
3 wk	4.6	2.1	17	2.0	Fe Mn	*Salmo trutta* (exp.)
2 d	4.5	1.0	4.1	1.7	Mn	*Salvelinus namaycush* (yolk sac fry) (exp.)
1 d	5.1	0.8	2.9	1.5		*Salmo salar* (exp.)
1 d	5.3	0.5	2.0	1.4		*Salmo salar* (exp.)
?	5.3–4.3?	≤1.1				*Phoxinus phoxinus*, *Salmo trutta* (exp.)
c. 12 d	4.7	0.5				*Salmo trutta*
2.5 d	4.9	1.1	19[(f)]			*Salvelinus fontinalis* (exp.)
?	4.6?	?	8.9?			*Salmo trutta*
?	5.0?				Cu?	*Salmo salar*, *S. trutta*
≤3 d	5.9	1.0	12			*Salvelinus fontinalis* (exp.)
5 d	4.6	0.5	8.4	4.9		*Salmo salar* (exp.)
?	4.8	1.2	13	11		*Salmo salar*

It is difficult to characterise a 'typical' episode. The nature and sequence of events may be influenced by numerous variables; precipitation chemistry, snowpack changes, vegetation, agricultural practices, bacterial activity, chemistry of the soil, sediment, bedrock and water and patterns of water retention and flow. Furthermore, the majority of studies have involved rather long sampling intervals (daily or less frequently) (Tables 1, 2), and usually only a small selection of the numerous water quality variables have been examined.

Common to nearly all recorded high discharge episodes in low conductivity waters is a decrease in pH, of up to *c*.2 pH units (a 100-fold increase in H^+ concentration). There are several possible sources of this acidity. Acidity deposited in precipitation can be concentrated during episodes, either by rain washing concentrated dry deposits from vegetation (van Breemen *et al.*, 1982; Miller & Miller, 1982) or by processes in the snowpack which result in release of a large proportion of the acidity in the snow in the first stages of melting (Johannessen & Henriksen, 1977; Schofield & Trojnar, 1980).

Exchanges at the leaf surface in forest foliage may contribute acidity to throughfall and stemflow (Miller & Miller, 1982). An important further source of acidity is the soil. Acidification can occur as a result of the build-up and decomposition of plant debris, especially in heathland and coniferous forest (Rueslåtten & Jørgensen, 1978; Jones *et al.*, 1983; Catt, 1985; Hornung, 1985). Following dry weather, soil water may have high H^+ concentrations as a result of oxidation of organic sulphur and nitrogen compounds (Rosenqvist, 1978; Ulrich, 1983; UKAWRG, 1986), and this water could contribute to episodic pH decline in runoff. In areas influenced by maritime climate, sea salt deposited in precipitation can displace acidity from already saturated peaty soils, by exchange of Na^+ for H^+ (Rosenqvist, 1978; Langan, 1987).

If soil minerals and bedrock are resistant to weathering, acid runoff may require prolonged contact with them for neutralisation to occur: if runoff is rapid or soils are frozen or waterlogged such contact will be limited and runoff entering watercourses is likely to cause a decrease in pH (Rosenqvist, 1978; Miller, 1985).

Accompanying this fall in pH in the watercourses may be a fall in concentration of major ions (e.g. Ca^{2+}, K^+, Mg^{2+}, Na^+) if the majority of the runoff has not passed through the soil (Schofield *et al.*, 1985), or a rise in concentration if interaction between runoff and minerals has led to exchange of hydrogen ions for various other cations (Rueslåtten & Jørgensen, 1978). This rise, if present, may be observed in advance of the peak hydrogen ion concentration, and may be followed by a marked fall in concentration as peak hydrogen ion concentration is reached (Bjärnborg, 1983), peak hydrogen ion concentration occurring only after the soil has become waterlogged or the cation exchange process has become saturated. The fall in pH is also usually accompanied by increases in various trace metal concentrations (Tables 1, 2), originating either from metals present in precipitation and dry deposition or by cation exchange at soil minerals and bedrock.

Tables 1 and 2 list some reported episodes, those involving fishkills being shown in Table 2. These lists are not intended to comprise an exhaustive review: many reported episodes are poorly documented, with brief anecdotal observations only, and the literature is rather fragmented, reports often being confined to the internal records of the various environmental monitoring organisations. An attempt has been made to list a selection of the more detailed and more readily available reports, and to include examples from a variety of areas. In Table 1, when an author has described more than one episode of a similar nature, in which fishkills have not been observed, only the most severe episode is listed.

It is evident from the very few reports where continuous or frequent (<3 h) sampling has been carried out that, particularly in the case of rainfall episodes (e.g. Jones *et al.*, 1983; Langan, 1987), changes in pH and other parameters can be very rapid, and that peak values may be sustained only for periods of hours or even minutes. Daily or less frequent sampling, particularly if carried out at the same time of day, can be misleading, e.g. in the case of snowmelt episodes there may be thawing during the day and refreezing at night, giving a series of episodes of short duration. Some of the episodes listed in Tables 1 and 2 may therefore conceal multiple events, and greater peak values than those recorded.

Because of the increasing awareness of the toxicity of aluminium and the relevance of chemical speciation to its toxic action (see Brown & Sadler, McDonald *et al.*, Potts & McWilliams, this volume), the most recent reports of episodes usually provide some information about aluminium levels and speciation: where available, peak values are shown in Tables 1 and 2. There are, however, very few observations of the levels and speciation of other potentially toxic metals in episodes. Where observations have been made, it is usually iron and manganese which have received most attention. Other potentially toxic metals (e.g. cadmium, copper, lead, nickel, zinc) have either been ignored, or measurements have been too close to the detection limits of the techniques employed for any useful conclusions to be drawn. Nevertheless, some or all of these metals are likely to play a part in fish population decline in acid waters (see McDonald *et al.*, this volume).

Two extreme examples of the water quality changes occurring during high discharge episodes in low conductivity waters are summarised in Figures 1 and 2. Figure 1 represents some of the changes occurring in a high altitude stream during a rainfall episode in a coniferous forest catchment in North Carolina, USA, in an area from which fishkills had been reported (Jones *et al.*, 1983). Figure 2 shows changes in a high altitude stream in central Sweden during spring snowmelt (Andersson & Nyberg, 1984). Experimental exposure of brown trout (*Salmo trutta*) to this episode resulted in 100% mortality.

In both episodes, increased discharge was accompanied by decreasing pH and by increasing concentrations of aluminium, iron and manganese. In the rainfall episode (duration *c.* 24 h) calcium concentration increased slightly (as did concentrations of magnesium and potassium: not shown in Figure 1), and the peak concentrations of all these metals occurred in advance of the peak H^+ concentration, implying initial release of metal cations from minerals in exchange for H^+. Peak concentrations of the strong acid anions NO_3^- and SO_4^{2-} (not shown) coincided with peak H^+ concentration.

The snowmelt episode was of much greater duration (*c.* 1.5 months), although it is possible that daily sampling may have concealed a more complex pattern of repeated episodes. A decline in calcium concentration probably reflected the dilution effect of meltwater passing over frozen soil and entering the stream directly. The greater duration and the dilution effect are typical of snowmelt episodes. Aluminium, iron and manganese concentrations appeared to have reached a peak slightly in advance of the peak H^+ concentration, and to have declined before H^+ concentration declined. However, sampling intervals for the metals were too long for reliable conclusions, and the metal analyses did not distinguish particulate and organically complexed material from dissolved inorganic forms. That undissolved and organically complexed material made a significant contribution to the samples is suggested by the extremely high levels of iron and manganese recorded in comparison with typical concentrations in acid waters (cf. McDonald *et al.*, this volume). In both this snowmelt episode and the rainfall episode of Jones *et al.* (1983) the colour of the water suggested high levels of organic material during the periods of rapid runoff. The very high levels of aluminium recorded in the rainfall episode (Figure 1)

Figure 1. Water quality in Raven Fork stream, N. Carolina, during a rainfall episode, 21–22 January 1982. Total concentrations of Mn, Fe, Al, Ca and H^+: μmol l^{-1}; flow rate: $m^3 s^{-1}$; t: time. From data of Jones *et al.* (1983).

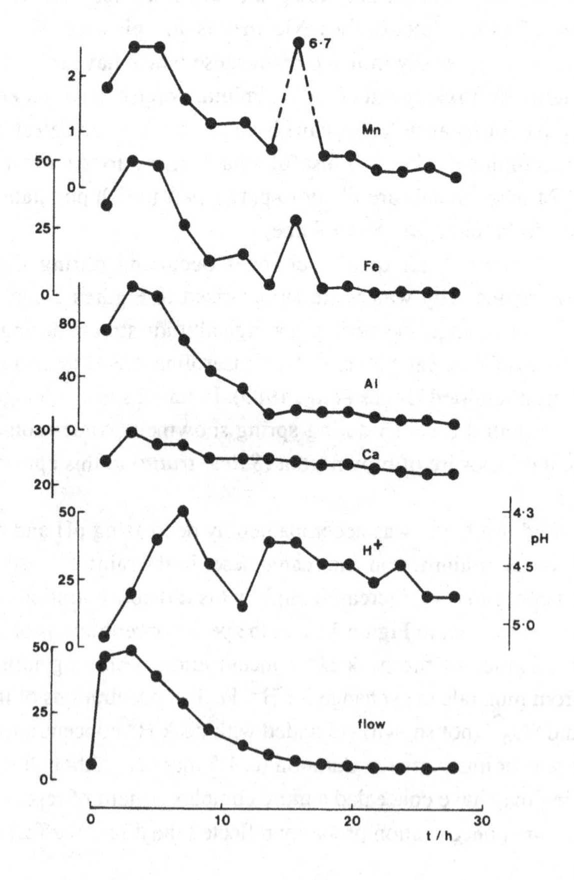

probably reflect a high undissolved fraction. Although not measured in this particular episode, dissolved aluminium in other episodes described by the same authors represented only a small fraction of total aluminium, smaller at greater total aluminium levels, e.g. in one of these episodes, a peak total aluminium concentration of 74 μmol l^{-1} (2000 μg l^{-1}) was accompanied by a dissolved aluminium concentration of 10 μmol l^{-1} (260 μg l^{-1}). Maximum total aluminium concentrations are usually much lower than those shown in Figure 1 (cf. Tables 1, 2).

Effects on fish

There are sporadic reports of fishkills during high discharge episodes from all the main areas in which fish populations in low conductivity waters have declined

Figure 2. Water quality in Bjursvasslan Brook, Central Sweden, during snowmelt in spring 1983, with cumulative mortality of experimentally exposed brown trout. Onset of thaw was in mid-April and maximum runoff occurred in mid-May. Total concentrations of Mn, Fe, Al, Ca and H^+: μmol l^{-1}; a: maximum recorded labile inorganic aluminium concentration, 2.04 μmol l^{-1}. Redrawn from Andersson & Nyberg (1984).

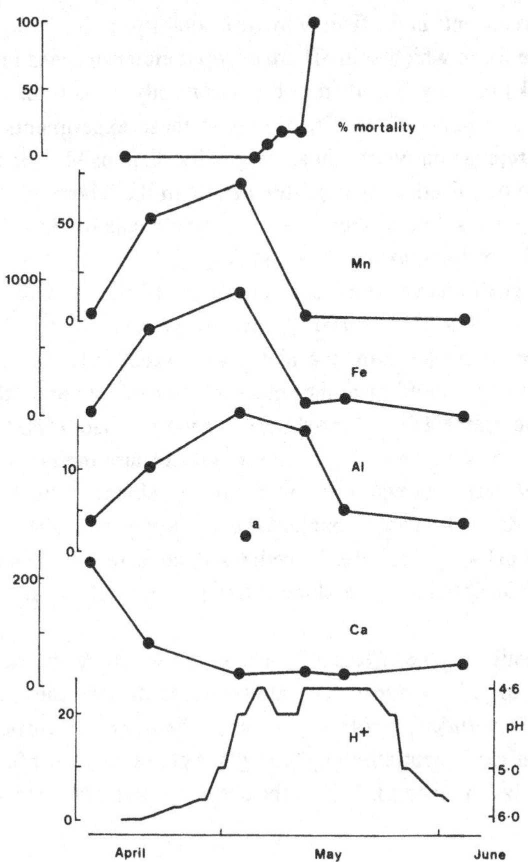

(Table 2). It is, however, difficult to be sure about how frequent and extensive such kills are, or about the mechanisms by which they occur. The areas in question have sparse human populations and relatively poor accessibility, and reports usually refer to deaths of readily visible large juveniles or adults of the economically important species (particularly Salmonidae). It could be that fishkills are more frequent, and affect a larger number of species and a wider spread of their life histories, than is suggested by the literature. Typically, experienced observers with access to suitable measuring equipment are alerted only after a fishkill has passed its period of maximum severity, and frequent or continuous monitoring of the fish populations and water quality variables throughout the episode is not possible.

The episodic fishkill is not apparently a frequent enough occurrence to enable one to be 'captured' by a comprehensive field monitoring exercise already on site: in any case, most of the detailed investigations of water quality in low conductivity acid waters of recent years have been sited on waters which have already lost their fish populations. Explanation of the cause of episodic fishkills tends therefore to be based on inference from very small numbers of observations and measurements, aided by conclusions drawn from experiments in the field or in the laboratory.

Field experiments (Table 2), in which animals are exposed either in cages in the water body itself or in tanks of water piped from the water body, usually involve waters with no surviving fish populations. Conditions in these experiments are therefore not necessarily representative of those originally responsible for fish population loss. Simulation of episodes for exposure of fish in the laboratory (see below) is difficult because of the shortage of comprehensive field data from fishkills upon which to base the choice of conditions for the simulation.

In most of the reported high discharge episodic fishkills in the field, and in all field experiments involving episodes of fish mortality, there is evidence that pH has declined (Table 2). However, the pH minima reached (>4.4, except in lake surface layers in immediate contact with runoff, or in the smallest streams) are not on their own sufficient to explain the deaths. Laboratory-simulated short episodes (≦30 h) at these pH levels, in the absence of high aluminium levels, do not cause mortalities in rainbow trout (*Salmo gairdneri*) or brown trout (*Salmo trutta*) (Jones *et al.*, 1983; Dempsey, 1986; R. Morris & J.P. Reader, unpublished data). Long term laboratory exposure of brown trout to pH ≧4.4, in artificial media with control of trace metal concentrations, do not result in significant mortalities (Dalziel *et al.*, 1986b; Sadler & Lynam, 1986a).

In a field experiment investigating the effects of liming to prevent pH decline, a fall in the pH of stream water to pH 4.4 during a rainfall episode did not cause any mortality in brook trout (*Salvelinus fontinalis*), whereas there were significant mortalities in the same water partly neutralised by liming: the pH of this water fell to 4.9 during the episode (Muniz & Leivestad, 1979). The absolute value of pH reached was clearly not the deciding factor in the mortalities. Other changes associated with the pH decline must have been important.

Several other fishkill episodes have involved decline of pH to apparently innocuous values of 4.8 or above (Table 2). Changes in concentration and toxicity of trace metals probably provide the explanation. Aluminium seems to be at its most toxic to fish in the pH range 5.0–5.5, as a result of speciation changes (Brown & Sadler, this volume). In the experiment described by Muniz & Leivestad (1979) (above), aluminium-rich stream water (total aluminium concentration 18.9 $\mu mol\ l^{-1}$, 508 $\mu g\ l^{-1}$) at very low pH (4.4) became acutely toxic only when the pH was raised by liming to 4.9 or above.

Aluminium toxicity may well have a major part to play in episodic fish kills, but there is a possibility that other metals may be involved. Andersson & Nyberg (1984) reported mortalities of caged brown trout exposed to snowmelt episodes in streams in Sweden even when concentrations of labile inorganic aluminium (generally regarded as the most toxic fraction: see Brown & Sadler, this volume) were below 1.1 $\mu mol\ l^{-1}$ (30 $\mu g\ l^{-1}$), a level which appears to be a threshold for brown trout mortality (Sadler & Lynam, 1986b). The majority of aluminium was organically complexed, and therefore probably not toxic.

High concentrations of iron and manganese were observed in these episodes (Figure 2), and iron, manganese and aluminium accumulated in the gills, iron in particularly large concentrations. The authors suggest that iron may have been the main cause of mortalities. It is, however, also possible that insoluble or colloidal inorganic aluminium was the cause: aluminium-induced mortalities of fish are sometimes blamed on accumulation of precipitated aluminium on the gill surfaces (see below).

The uncertainty about the real cause of mortalities in these experiments, and in those of Muniz & Leivestad (1979), both relatively well documented studies, only serves to emphasise the need for much more information about concentrations and speciation of trace metals during episodic fishkills.

If there is doubt about the exact water quality changes responsible for episodic fishkills, there must also be doubt about the mechanism of death. Evidence of ionoregulatory disturbance, in the form of depleted plasma concentrations of sodium and chloride, has been observed in moribund fish during fishkills (Leivestad & Muniz, 1976; Skogheim *et al.*, 1984b). Experimentally induced severe ionoregulatory disturbance of this kind can lead to death, probably by causing circulatory failure (Wood, this volume).

Aluminium causes disturbance of sodium and chloride balance in laboratory exposures (Neville, 1985; Dalziel *et al.*, 1986a,b). Low pH can cause similar disturbance (see Potts & McWilliams, Wood, this volume), but, at realistic pH levels (>4.4), in trace metal-free media, the effects of low pH alone are very much less severe than when aluminium is present (Dalziel *et al.*, 1986a,b). Some other trace metals elevated in acid waters can cause similar effects (McDonald *et al.*, this volume).

Respiratory problems (mucous clogging, increased ventilation, coughing or hypoxia) have been observed in laboratory exposures of fish to a wide pH range (4.0–6.1) (Muniz & Leivestad, 1980; Rosseland, 1980; Schofield & Trojnar, 1980; Rosseland & Skogheim, 1984; Neville, 1985; Dalziel *et al.*, 1986b). These effects appear to be associated with supersaturation of aluminium (Rosseland & Skogheim, 1984) and may therefore be caused by precipitation on the gill surfaces.

Mucous clogging of the gills has been reported in experimental exposure to an episode in the field (Schofield & Trojnar, 1980), but in most fishkills and field experiments respiratory difficulties have not been observed, even when water quality was such that saturation with aluminium and precipitation might have been expected (Skogheim *et al.*, 1984b).

Apart from peak pH levels and trace metal concentrations, several other variables are likely to influence the effects of an episode on fish. Calcium concentration plays an important role in fish population status (Brown & Sadler, this volume), and high concentrations generally reduce mortality and sublethal physiological disturbances associated with low pH or with various trace metals (see Brown & Sadler, McDonald *et al.*, Potts & McWilliams, Wood, this volume).

The waters susceptible to low pH episodes usually have low calcium concentrations, <100 μmol l^{-1}, and in some episodes, particularly those involving snowmelt, calcium concentration can decline markedly (see above), presumably exacerbating the effects of other water quality changes.

The duration of deleterious conditions is probably important: in laboratory-simulated episodes of low pH (4.5 or 5.6) and high aluminium concentration (nominally 10 μmol l^{-1}, 270 μg l^{-1}) no mortalities of brown trout were observed when peak values were held for 6 h, but mortality was high when this period was extended to 12 h. Aluminium-induced inhibition of sodium uptake was also more severe in the longer episodes (R. Morris & J.P. Reader, unpublished data). In much longer laboratory exposures to constant conditions, aluminium-induced mortalities of rainbow trout and brown trout at low pH (≧4.4) have been observed at concentrations as low as 2.8 μmol l^{-1} (75 μg l^{-1}) and 1.1 μmol l^{-1} (30 μg l^{-1}) respectively (Neville, 1985; Sadler & Lynam, 1986b).

It is possible that the rate of change of water quality parameters in an episode is a factor in the responses of fish. Low pH-induced disturbance of ionoregulation was more severe in brown trout transferred abruptly to low pH (7.0–4.0) than when pH was lowered in a series of smaller steps (Stuart & Morris, 1985). Nearly all laboratory exposures to constant conditions really take the form of abrupt episodes, animals being transferred suddenly to the experimental medium, but direct comparison between laboratory exposures and episodic fishkills is difficult because water quality data in the latter are incomplete.

Until more is known about the nature of the changes occurring during episodic fishkills in the field, and until there has been more study of the effects of laboratory-simulated episodes, it will not be possible to decide whether the rapidity of onset of episodic changes is important, or whether the deaths are simply a function of levels and duration of exposure.

Previous history of the fish appears to affect responses to episodic changes. Atlantic salmon (*Salmo salar*) or sea trout (*Salmo trutta*) undergoing adaptation of the osmoregulatory process during migration from rivers to sea (smolt), or vice versa (adults returning to spawn), are particularly vulnerable (Henriksen *et al.*, 1984; Hesthagen & Skogheim, 1984; Skogheim *et al.*, 1984b).

There is some evidence for differences in susceptibility between populations, and between species. Hatchery-reared brown trout suffered more rapid onset of mortalities than did wild trout in the field experiments of Andersson & Nyberg (1984), but fish from the two populations were exposed to episodes in two different years, and water quality changes were not identical. In a fishkill in a river supporting populations of Atlantic salmon, brown and sea trout and European eel (*Anguilla anguilla*), only salmon migrating upstream to spawn were found dead (Skogheim *et al.*, 1984b). However, there was no conclusive evidence that the other species were present at the time in the reaches of the river affected by the episode. In the River Vikedal in SW Norway in recent years, episodic mortalities of juvenile Atlantic salmon have been significantly higher than those of brown trout, and there has been a decline in the adult salmon population at the same time, suggesting that episodic fishkills are having a distinct effect on population status. There is, however, some evidence of increased commercial exploitation of salmon stocks (Hesthagen, 1986). In field experiments, juvenile brook trout and brown trout suffered little or no mortality during an episode which caused 100% mortality in juvenile Atlantic salmon (Skogheim *et al.*, 1984a). Muniz *et al.* (1979) reported deaths of brown trout in a river during a snowmelt episode, whereas perch (*Perca fluviatilis*) did not appear to be affected.

Behaviour patterns probably influence responses to episodes. There is evidence that fish show avoidance reactions to low pH conditions, and it is possible that during episodes, particularly in large water bodies, they could locate areas of more favourable conditions, e.g. deeper water or upwelling groundwater (Gunn, 1986). In the fishkill reported by Skogheim *et al.* (1984b), salmon spawners were seen moving downstream, the reverse of their expected behaviour pattern, away from the source of the acid aluminium-rich water.

Gunn (1986) describes the differences between spawning site selection behaviour of various salmonid species and points out that these differences may lead to differences in susceptibility among the early life history stages of different species.

For example, lake trout (*Salvelinus namaycush*) usually spawn on gravel or rubble on steep lake shores, and the embryos may therefore be particularly vulnerable to episodes of direct runoff from the banks during spring snowmelt. Such episodes can produce considerable changes at lake margins. Gunn & Keller (1984) reported some mortality of experimentally exposed lake trout fry on the rubble surface during episodes of low pH and elevated trace metals concentrations. Water quality measurements indicated that conditions were probably even more deleterious in the interstices of the rubble, where embryos would normally be lying.

In some very small low order streams in the UK, pH minima as low as 3.5–3.6 have been recorded (S.J. Langan, personal communication; A.W.H. Turnpenny, personal communication). Most monitoring of episodes has ignored the smallest streams, where changes caused by acid runoff are likely to be greatest, but such streams may be vitally important to species (e.g. brown trout) for which they provide spawning sites or refuges from rivers in spate.

Summary

Episodes of high discharge in watercourses in areas with low conductivity water and weathering-resistant bedrock are accompanied by a fall in pH and a rise in concentrations of trace metals. Other changes occurring at the same time may include a dilution of the major inorganic ions (e.g. Ca^{2+}). Of the trace metals, aluminium has been most studied. Other trace metals have received little attention, but several (iron, lead, manganese, zinc) have been found to increase in concentration during episodes, and it is possible that others may behave in the same way.

The wide variety of conditions which determine the nature of an episode, and a paucity of studies providing comprehensive and continuous water analysis, preclude the description of a 'typical' episode. However, episodes associated with snowmelt are usually much longer than those caused by heavy rain, and have a more marked dilution of major ions. In short, rain-associated, episodes this dilution effect may be absent, and increases in concentration of major ions may occur.

Fishkills are sometimes observed during high discharge low pH episodes. However, detailed study of these fishkills and their causes is difficult, because of their sporadic and unpredictable occurrence. It is likely that they play an important part in fish population decline in acid waters. The deaths appear to be the consequence of trace metal toxicity (particularly aluminium), accompanying the concentration increases and speciation changes associated with a fall in pH, and failure of ionoregulation is a likely cause of death. However, unequivocal evidence is lacking, both for the mechanisms of fishkills and for their role in population decline.

References

Andersson, P. & Nyberg, P. (1984). Experiments with brown trout (*Salmo trutta* L.) during spring in mountain streams at low pH and elevated levels of iron, manganese and aluminium. *Rep. Inst. Freshwater Res. Drottningholm*, **61**, 34–47.

Bjärnborg, B. (1983). Dilution and acidification effects during the spring flood of four Swedish mountain brooks. *Hydrobiologia*, **101**, 19–26.

Breemen, N. van, Burrough, P.A., Velthorst, E.J., Dobben, H.F. van, Wit, T. de, Ridder, T.B. & Reijnders, H.F.R. (1982). Soil acidification from atmospheric ammonium sulphate in forest canopy throughfall. *Nature (Lond.)*, **299**, 548–50.

Catt, J.A. (1985). Natural soil acidity. *Soil Use Mgmt.*, **1**, 8–10.

Christopherson, N., Seip, H.M. & Wright, R.F. (1982). A model for streamwater chemistry at Birkenes, Norway. *Water Resources Res.*, **18**, 977–96.

Dale, T., Henriksen, A., Joranger, E. & Krog, S. (1974). Vann- og nedbörkjemiske studier i Birkenesfeltet for perioden 20. juli til 30. april 1973. TN 1/74 Oslo–Ås: SNSF.

Dalziel, T.R.K., Morris, R. & Brown, D.J.A. (1986a). The effects of low pH, low calcium concentrations and elevated aluminium concentrations on sodium fluxes in brown trout, *Salmo trutta* L. *Water Air Soil Pollut.*, **30**, 569–77.

Dalziel, T.R.K., Reader, J.P. & Morris, R. (1986b). Growth, net ion uptake and skeletal calcium deposition in brown trout (*Salmo trutta*) exposed to aluminium and manganese in soft acid water. CERL Report TPRD/L/3041/R86. Leatherhead, Surrey, UK: Central Electricity Research Laboratories.

Dempsey, C.H. (1986). Investigations into the causes of acute toxicity to brown trout of episodes of reduced pH. CERL Report TPRD/L/3077/R86. Leatherhead, Surrey, UK: Central Electricity Research Laboratories.

Driscoll, C.T., Jr, Baker, J.P., Bisogni, J.J., Jr & Schofield, C.L. (1980). Effect of aluminium speciation on fish in dilute acidified waters. *Nature (Lond.)*, **284**, 161–4.

Gunn, J.M. (1986). Behaviour and ecology of salmonid fishes exposed to episodic pH depressions. *Envir. Biol. Fish.*, **17**, 241–52.

Gunn, J.M. & Keller, W. (1984). Spawning site water chemistry and lake trout (*Salvelinus namaycush*) sac fry survival during spring snowmelt. *Can. J. Fish. Aquat. Sci.*, **41**, 319–29.

Hagan, A. & Langeland, A. (1973). Polluted snow in southern Norway and the effect of meltwater on freshwater and aquatic organisms. *Envir. Pollut.*, **5**, 45–57.

Haines, T.A. (1981). Acidic precipitation and its consequences for aquatic ecosystems: a review. *Trans. Am. Fish. Soc.*, **110**, 669–707.

Harvey, H.H. (1982). Population responses of fish in acidified waters. In *Acid Rain/Fisheries*, ed. R.E. Johnson, pp. 227–42. Bethesda, Md: American Fisheries Society.

Harvey, H.H. & Lee. C. (1982). Historical fisheries changes related to surface water pH changes in Canada. In *Acid Rain/Fisheries*, ed. R.E. Johnson, pp. 45–55. Bethesda, Md: American Fisheries Society.

Henriksen, A., Skogheim, O.K. & Rosseland, B.O. (1984). Episodic changes in pH and aluminium-speciation kill fish in a Norwegian salmon river. *Vatten*, **40**, 255–60.

Henriksen, A. & Wright, R.F. (1977). Effects of acid precipitation on a small acid lake in southern Norway. *Nord. Hydrol.*, **8**, 1–10.

Hesthagen, T. (1986). Fish kills of Atlantic salmon (*Salmo salar*) and brown trout (*Salmo trutta*) in an acidified river of SW Norway. *Water Air Soil Pollut.*, **30**, 619–28.

Hesthagen, T. & Skogheim, O.K. (1984). High mortality of presmolt of Atlantic salmon, *Salmo salar* L. and sea trout, *Salmo trutta* L. during spring snowmelt

1982 in River Vikedalselva, Western Norway. Manuscript cited by Rosseland & Skogheim (1984).

Hornung, M. (1985). Acidification of soils by trees and forests. *Soil Use Mgmt.*, **1**, 24–8.

Howells, G. (1983). Acid waters – the effect of low pH and acid associated factors on fisheries. *Adv. Appl. Biol.*, **9**, 143–255.

Hultberg, H. (1977). Thermally stratified acid water in late Winter – a key factor inducing self-accelerating processes which increase acidification. *Water Acid Soil Pollut.*, **7**, 279–94.

Jeffries, D.S., Cox, C.M. & Dillon, P.J. (1979). Depression of pH in lakes and streams in central Ontario during snowmelt. *J. Fish. Res. Bd. Can.*, **36**, 640–6.

Johannesen, M. & Henriksen, A. (1977). Chemistry of snowmelt water: changes in concentration during melting. FR 11/77. Oslo–Ås: SNSF.

Jones, H.C., Noggle, J.C., Young, R.C., Kelly, J.M., Olem, H., Ruane, R.J., Pasch, R.W., Hyfantis, G.J. & Parkhurst, W.J. (1983). Investigations of the cause of fishkills in fish-rearing facilities in Raven Fork watershed. TVA/ONR/WR–83/9. Tennessee Valley Authority, Division of Air and Water Resources.

Langan, S.J. (1987). Episodic acidification of streams at Loch Dee. *Proc. Roy. Soc. Edinburgh Ser. B.* (In press).

Leibfried, R.T., Sharp, W.E. & DeWalle, D.R. (1984). The effect of acid precipitation runoff on source water quality. *J. Am. Waterworks Assoc.*, **76**, 50–3.

Leivestad, H. & Muniz, I.P. (1976). Fish kill at low pH in a Norwegian river. *Nature (Lond.)*, **259**, 391–2.

Miller, H.G. (1985). The possible role of forests in streamwater acidification. *Soil Use Mgmt.*, **1**, 28–9.

Miller, H.G. & Miller, J.D. (1982). The interaction of acid precipitation and forest vegetation in northern Britain. *Water Qual. Bull.*, **8**, 121–6.

Muniz, I.P. & Leivestad, H. (1979). Langtidseksponering av fisk til surt vann. Forsøk med bekkerøye *Salvelinus fontinalis* Mitchill. IR 44/79. Oslo–Ås: SNSF.

Muniz, I.P. & Leivestad, H. (1980). Toxic effects of aluminium on the brown trout, *Salmo trutta* L. In *Proc. Int. Conf. Ecological Impact of Acid Precipitation*, ed. D. Drabløs & A. Tollan, pp. 320–1. Oslo–Ås: SNSF Project.

Muniz, I.P., Leivestad, H. & Bjerknes, V. (1979). Fiskedød i Nidelva (Arendalvassdraget) våren 1979. TN 48/79. Oslo–Ås: SNSF.

Neville, C.M. (1985). Physiological response of juvenile rainbow trout, *Salmo gairdneri*, to acid and aluminum – prediction of field responses from laboratory data. *Can. J. Fish. Aquat. Sci.*, **42**, 2004–19.

Prigg, R.F. (1983). Juvenile salmonid populations and biological quality of upland streams in Cumbria with particular reference to low pH effects. Report no. BN 77–2–83 (unpublished). Carlisle, UK: North West Water Authority.

Rosenqvist, I.Th. (1978). Alternative sources for acidification of river water in Norway. *Sci. Total Envir.*, **10**, 39–49.

Rosseland, B.O. (1980). Physiological responses to acid water in fish. 2. Effects of acid water on metabolism and gill ventilation in brown trout, *Salmo trutta* L. and brook trout, *Salvelinus fontinalis* Mitchill. In *Proc. Int. Conf. Ecological Impact of Acid Precipitation*, ed. D. Drabløs & A. Tollan, pp. 348–9. Oslo–Ås: SNSF Project.

Rosseland, B.O. & Skogheim, O.K. (1984). A comparative study on salmonid fish species in acid aluminium-rich water. II. Physiological stress and mortality of one- and two-year-old fish. *Rep. Inst. Freshwater Res. Drottningholm*, **61**, 186–94.

Rueslåtten, H.G. & Jørgensen, P. (1978). Interaction between bedrock and precipitation at temperatures close to 0°C. *Nord. Hydrol.*, **9**, 1–6.
Sadler, K. & Lynam, S. (1986a). Some effects of low pH and calcium on the growth and tissue mineral content of yearling brown trout (*Salmo trutta*). *J. Fish Biol.*, **29**, 313–24.
Sadler, K. & Lynam, S. (1986b). Some effects on the growth of brown trout from exposure to aluminium at different pH levels. CERL Report TPRD/L/2982/R86. Leatherhead, Surrey, UK: Central Electricity Research Laboratories.
Schofield, C.L., Galloway, J.N. & Hendry, G.R. (1985). Surface water chemistry in the ILWAS basins. *Water Air Soil Pollut.*, **26**, 403 – 23.
Schofield, C.L. & Trojnar, J.R. (1980). Aluminum toxicity to brook trout (*Salvelinus fontinalis*) in acidified waters. In *Polluted Rain*, ed. T.Y. Toribara, M.W. Miller & P.E. Morrow, pp. 341–66. New York: Plenum Press.
Skogheim, O.K., Rosseland, B.O., Hafsund, F., Kroglund, F. & Hagenlund, G. (1984a). Eksponering av bleke, aure og bekkerøye i surt vann. *Rapport Fra Fiskeforskningen* 1984/2. Oslo–Ås: Direktoratet for Vilt og Ferskvannsfisk.
Skogheim, O.K., Rosseland, B.O. & Sevaldrud, I. (1984b). Deaths of spawners of Atlantic salmon in River Ogna, S.W. Norway, caused by acidified aluminium-rich water. *Rep. Inst. Freshwater Res. Drottningholm*, **61**, 195–202.
Spry, D.J., Wood, C.M. & Hodson, P.V. (1981). The effects of environmental acid on freshwater fish with particular reference to the softwater lakes in Ontario and the modifying effects of heavy metals. A literature review. *Can. Tech. Rep. Fish. Aquat. Sci.*, **999**, 145 pp.
Stuart, S. & Morris, R. (1985). The effects of season and exposure to reduced pH (abrupt and gradual) on some physiological parameters in brown trout (*Salmo trutta*). *Can. J. Zool.*, **63**, 1078–83.
Ulrich, B, (1983). A concept of forest ecosystem stability and of acid deposition as driving force for destabilisation. In *Effects of Accumulation of Air Pollutants in Forest Ecosystems*, ed. B. Ulrich & J. Pankrath, pp. 1–29. Dordrecht: D. Reidel.
United Kingdom Acid Water Review Group (UKAWRG) (1986). *Acidity in United Kingdom Fresh Waters*. London: Dept of the Environment.

NORBERT HEISLER

Acid–base regulation in fishes. 1. Mechanisms

Introduction

Adjustment of pH is one of the central tasks for homeostatic regulation. Deviations from certain set-point values may result in reduced metabolic performance, due to the enzyme activity of metabolic energy-producing processes having pronounced pH optima. Accordingly, any net endogenous production or exogenous induction of acid–base relevant ions has to be counteracted by equivalent removal from the body fluids in order to maintain steady state conditions.

The regulatory mechanisms available for this purpose are in principle the same in all classes of animals, but are (or can be) utilized to variable extents. The situation of fishes is characterized by their intimate contact with the aqueous environment, including utilization of water as a gas exchange medium. Immersion in water favours ion transfer mechanisms supporting acid–base regulation, but also entails severe restrictions for the regulation of P_{CO_2} in the body fluids. This chapter will briefly delineate and discuss the basic principles of acid–base regulation in fishes with respect to their theoretical limitations, and to their relative importance for the regulation of acid–base homoiostasis in fishes. The selection of references was limited by space, and review articles have been cited wherever possible to provide greater access to the subject. This chapter leads directly into Chapter 6 by J.N. Cameron, which explores the exogenous and endogenous variables affecting acid–base regulation in fishes.

Buffering

Buffering is a mechanism for transient acid–base regulation. Surplus H^+ are transferred into the non-dissociated state and masked by association with buffer bases. Accordingly, this mechanism is capable of reducing pH changes as compared to a non-buffered fluid system, but cannot restore the original fluid pH. Buffering serves valuable functions especially during the time period between introduction of surplus H^+ and OH^- into the body fluids, and their final elimination. It also supports the limitation of pH changes to values still compatible with life functions until the cause of disturbance is eliminated or pH is compensated by other means (for more detailed treatment of the subject see Siggaard-Andersen, 1974; Heisler, 1986a).

Non-bicarbonate buffering

Biologically relevant non-bicarbonate buffers are mainly protein residues (histidine, cysteine and terminal NH_2-groups), characterized by pH values close to physiological pK values. Buffering usually takes place in a system closed for acid and base forms of the buffer (for definitions of acids and bases see Brønsted, 1923), such that the total buffer concentration is constant. The buffer value (β) as a measure for the buffering capability is defined as the change in base (B^-) or acid form (HB) of the buffer system per change in pH (Van Slyke, 1922; Figure 1). On the basis of the sigmoidally shaped relationship between the concentration of buffer bases (buffer

Figure 1. Characteristics of a closed buffer system. Buffer reaction (B^- = buffer base, HB = acid buffer form), common buffer equation (Henderson–Hasselbalch equation), definition of buffer value (β = buffer value, C = total buffer concentration). (*a*) Sigmoidal buffer curve. (*b*) Distribution of the buffering capability around the pK' value.

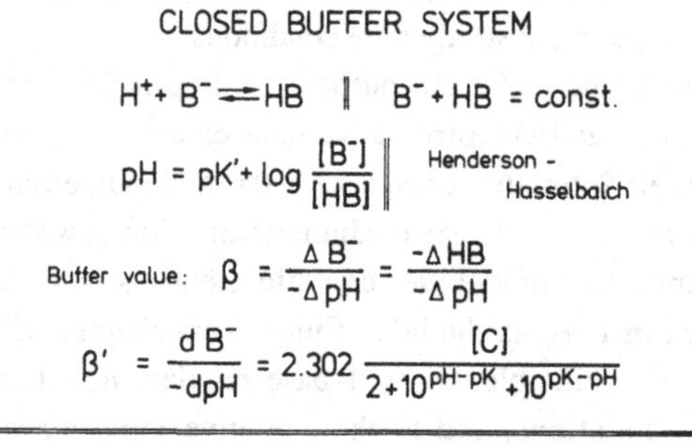

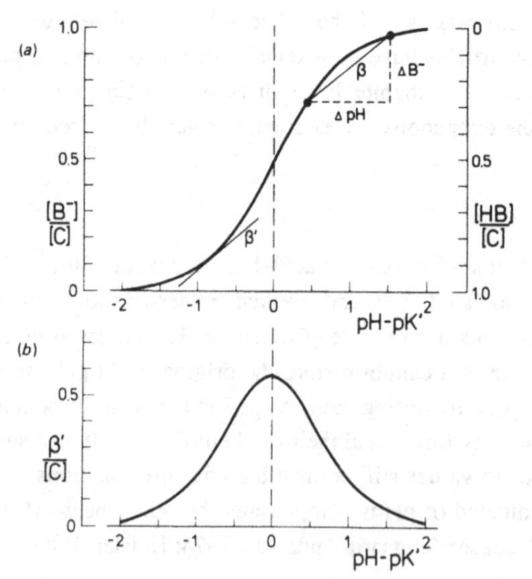

anions, B^-) or the acid form of the buffer system (HB), and the fluid pH (Figure 1), the buffer value is highest at the centre of the buffer curve (pK' value), falling in a bell-shaped manner towards higher and lower pH values (for details see Heisler, 1986a).

The non-bicarbonate buffer values of blood and intracellular tissue compartments are generally lower in fishes than in higher vertebrates by a factor of 1.5–4, (Heisler, 1986a). The intracellular tissue buffer values are much higher than those for blood and extracellular space because the protein residues are predominantly intracellular. Accordingly by far the largest proportion of the buffer capacity, which is defined as the product of fluid compartment volume and buffer value ($\kappa = \beta.V$), is located in the intracellular body compartments.

Bicarbonate buffering

The CO_2–bicarbonate buffer system may be described in a simplified manner as dissociation of CO_2 into H^+ and HCO_3^- or vice versa (for details see Heisler, 1986a). The peculiarity of this system resides in the volatile nature of the acid form of the buffer system (CO_2), which is regulated by the gas exchange apparatus. Since the total buffer concentration is variable, this type of buffer system is designated as open.

The possibility of selectively modulating the concentration of one of the buffer components (acid or base form) improves the effectiveness of a buffer considerably. The amount of H^+ bound for a given change in pH is much larger in an open buffer system, where a large proportion (Figure 2(*a*), (1)+(2)) of the buffer bases (B^-) can be used up without changing the concentration of the acid form (HB) of the buffer, than in a closed buffer system (Figure 2(*a*),(1)), where H^+-equimolar amounts of base are transferred into the acid form of the buffer.

These characteristics apply to the CO_2–bicarbonate buffer system, as long as CO_2 produced during buffering of H^+ can be eliminated, e.g. from blood passing the gas exchange structures. Removal of CO_2 over the gills renders the system highly effective in reducing the pH changes during buffering of H^+, compared with the large pH shifts expected in a comparable closed buffer system (Figure 3). This remarkable buffer capability is further improved by hyperventilation and concomitant reduction of the CO_2 concentration (respiratory pH compensation). Blood passing through tissues, however, has to be considered as a closed buffer system, since CO_2 cannot be removed, but is actually added from the tissue metabolic output. Then the increase in P_{CO_2} from bicarbonate buffering is considerable (Figure 3), and the fall in perfusate pH induced by addition of H^+ during tissue passage may become the limiting factor for fast washout of any additional H^+ accumulated in the intracellular compartments (Heisler, 1986c).

The bicarbonate buffer value is a direct function of the bicarbonate concentration of the fluid compartment (Figure 3). Since bicarbonate is generally lower in fishes than in higher vertebrates, and lower in the relatively more important intracellular than in the extracellular space (Heisler, 1984, 1986b,c), the contribution of bicarbonate buffering to fish acid–base regulation is generally limited as expressed by the bicarbonate buffer capacity in comparison to the non-bicarbonate buffer capacity (for details see Heisler, 1986a).

Figure 2. Characteristics of an open buffer system (one component, the acid form of the buffer system constant). (*a*) The sigmoidal buffer curve is flattened by removal of the acid buffer form produced during buffering. The constancy of the acid form is indicated by the stippled area. The amount of H^+ buffered for a given change in pH is much larger for the open system (1+2: $\Delta[H^+]_{op}$, [HB] = const.) than for the closed buffer system (1:$\Delta[H^+]_{cl}$, [HB] + [B^-] = const.). (*b*) Comparison of the specific buffer values (β'/[C]) for closed and open buffer systems.

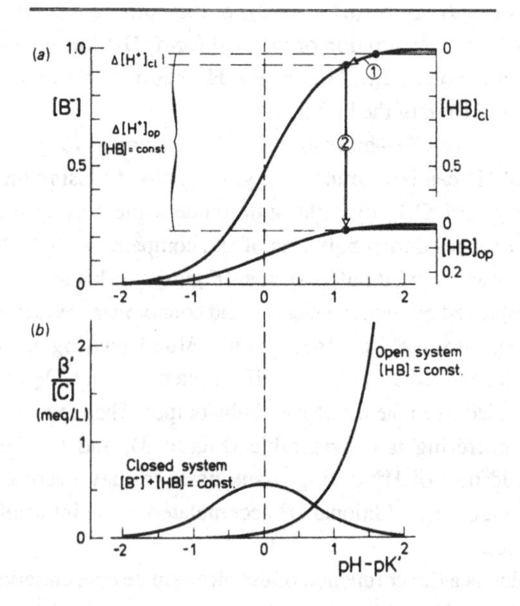

Changes in P_{CO_2}

Terrestrial animals often utilize changes in P_{CO_2} induced by changes in pulmonary ventilation to compensate non-respiratory acid–base disturbances (Woodbury, 1965) on the basis of open system bicarbonate buffering (cf. Figure 3). Water-breathing animals, however, are handicapped in applying this mechanism of P_{CO_2} adjustment by the physical properties of water as gas exchange medium (Rahn, 1966; Dejours, 1981; Piiper, 1986; Heisler, 1986b).

The main factor involved is the much lower capacitance coefficient ratio of O_2/CO_2 in water (*c.* 0.03) than in air (*c.* 1) (Figure 4). On the basis of the relationships between O_2 consumption and CO_2 production, and the capitance coefficients of the

Figure 3. CO_2–bicarbonate buffer system. (*a*) Comparison of the specific buffer values for open and closed systems. (*b*) Changes in pH upon binding of H^+ in open and closed systems. Also indicated the rise in P_{CO_2} by buffering in the closed system by transfer of base into acid form of the buffer (e.g. during tissue passage).

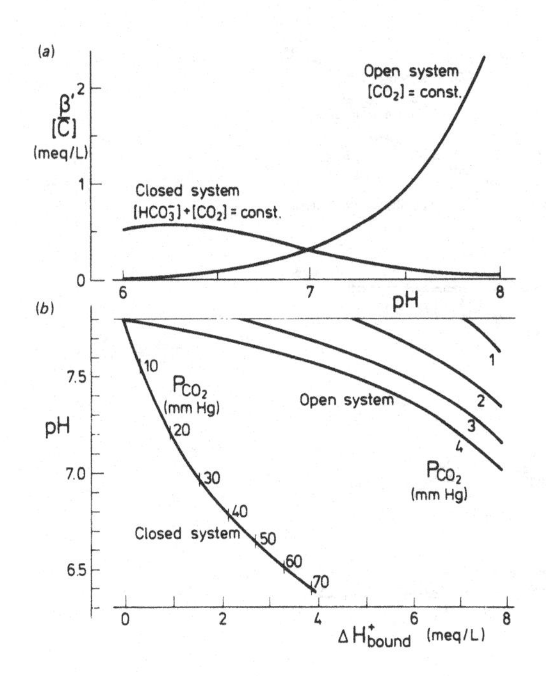

gas exchange medium (Figure 4(*a*)), the net ventilation of the gas exchange structures has to be much higher in water than in air-breathers in order to suffice the animal's oxygen demand. Since the CO_2 capacitance coefficients for water and air are similar, the comparatively high ventilation rate results in a considerable reduction of the inspired/expired P_{CO_2} difference, a factor largely limiting further adjustment of expired P_{CO_2} (Figure 4(*b*)).

Figure 4. Relationships between oxygen consumption ($\dot{M}_{O_2}$), CO_2 production ($\dot{M}_{CO_2}$), ventilation rate of the gas exchange medium ($\dot{V}_m$), capacitance coefficients (b), inspired and expired partial pressures (P_I, P_E), and respiratory gas exchange ratio (RQ). (*a*) O_2 and CO_2 capacitance curves for air, pure water and seawater. (*b*). P_{O_2}/P_{CO_2} diagram for air breathers and water breathers (RQ = 1). Areas labelled 'E' indicate the range of reported expired partial pressure values. Insert: Relationship between P_{CO_2} and water ventilation rate ($\dot{V}_m$) for pure water, and for seawater (SW) or carbonated water (CbW).

$$\dot{M}_{CO_2} = \dot{V}_m \cdot b_{mCO_2}(P_E - P_I)_{CO_2}$$

$$\dot{M}_{O_2} = \dot{V}_m \cdot b_{mO_2}(P_I - P_E)_{O_2}$$

$$\dot{V}_m = \frac{\dot{M}_{O_2}}{b_{mO_2}(P_E - P_I)_{O_2}}$$

$$RQ = \frac{\dot{M}_{CO_2}}{\dot{M}_{O_2}} = \frac{b_{CO_2}}{b_{O_2}} \cdot \frac{(P_E - P_I)_{CO_2}}{(P_I - P_E)_{O_2}}$$

$$\text{AIR:} \quad \frac{b_{O_2}}{b_{CO_2}} \approx 1$$

$$(P_E - P_I)_{CO_2} = RQ \cdot \frac{b_{O_2}}{b_{CO_2}} \cdot (P_I - P_E)_{O_2}$$

$$\text{WATER:} \quad \frac{b_{O_2}}{b_{CO_2}} \approx \frac{1}{30}$$

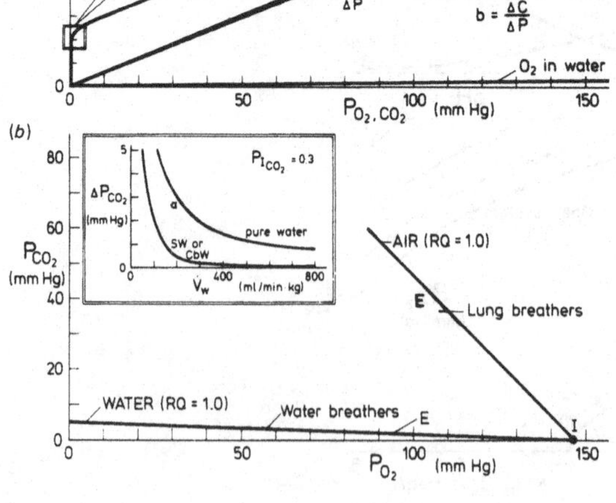

In seawater and carbonated freshwater (the habitats of the majority of fish species) the difference between the effective O_2 and CO_2 capacitance coefficients is enhanced by carbonate- and other non-bicarbonate buffering of CO_2 (cf. Figure 6) in the very low range of P_{CO_2} values (0–1 mm Hg; Figure 4(*a*), inset). Due to this mechanism, the relationship between ventilation and expired P_{CO_2} is shifted to even lower expired P_{CO_2} values (Figure 4(*b*), insert), further reducing the range of P_{CO_2} available for acid–base adjustments via changes in ventilation.

On the basis of the counter-current gas exchange system arterial P_{CO_2} in fishes would be expected to be much lower than expired, and close to inspired P_{CO_2} (Figure 5(*a*); see also Piiper, 1986). However, P_{CO_2} values in arterial fish blood are usually about 2–3 mm Hg higher than those of inspired water (Figure 4(*b*), insert, shaded area 'a'; for references see Heisler, 1984, 1986b). This discrepancy, which is of

Figure 5. Gas exchange in the gills of teleost fish. (*a*) Perfect counter-current model of blood and water flow. With lack of diffusion limitation there is close equilibrium between blood and water P_{CO_2} with almost complete overlap of inspired/expired and arterial/venous partial pressure ranges. (*b*) Partial bypass of both water and blood of the respiratory gas exchange surface. Admixture of inspired water and of venous blood (pointed lines) lead to larger deviation between blood and water P_{CO_2} values and less overlap of partial pressures.

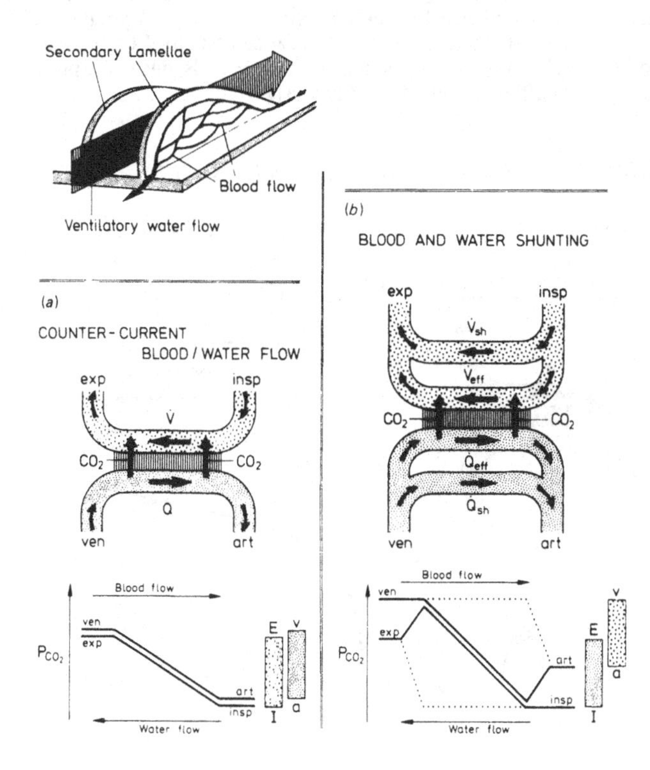

considerable importance for the evaluation of changes in ventilation as a mechanism for acid–base adjustment, may be explained on the basis of a number of factors. Beside the effects of unstirred layers of water close to the gill gas exchange surface, and the resultant diffusion limitation, this phenomenon could be produced by water bypassing the gas exchange surface. Admixture of inspired water would lower mixed expired P_{CO_2}, and admixture of venous blood increase the mixed arterialized blood P_{CO_2}, thus resulting in enhanced blood/water P_{CO_2} differences (Figure 5(*b*). Another possibility would be reaction limitation of CO_2 in the water due to lack of carbonic anhydrase. CO_2 eliminated would be taken up by the water exclusively on the basis of physical dissolution along the slope of the capacitance line for pure water (Figure 6, E_{dis}), rather than being buffered according to the steep steady state water CO_2 dissociation curve. After exiting the gas exchange area P_{CO_2} in the expired water would then slowly be reduced due to uncatalysed hydration to the expired equilibrium point (Figure 6, E_{eq}). This effect would be most pronounced in the range of low P_{CO_2} values (and accordingly high effective CO_2 capacitance coefficients).

Figure 6. Effect of reaction limitation of CO_2 hydration in the water. (*a*) The lack of carbonic anhydrase allows equilibrium between water and blood during gill passage for only P_{CO_2}. After leaving the gills water P_{CO_2} falls as a result of slow uncatalysed hydration of CO_2. (*b*) During gill passage the water is loaded with CO_2 only according to the physical water solubility (I → E_{dis}). Hydration of CO_2 (E_{dis} → E_{eq}) after gill passage reduces P_{CO_2} at constant total CO_2 content. The effect of this disequilibrium is the larger the lower P_{CO_2} is, due to the pronounced non-bicarbonate buffering at low P_{CO_2} (high pH).

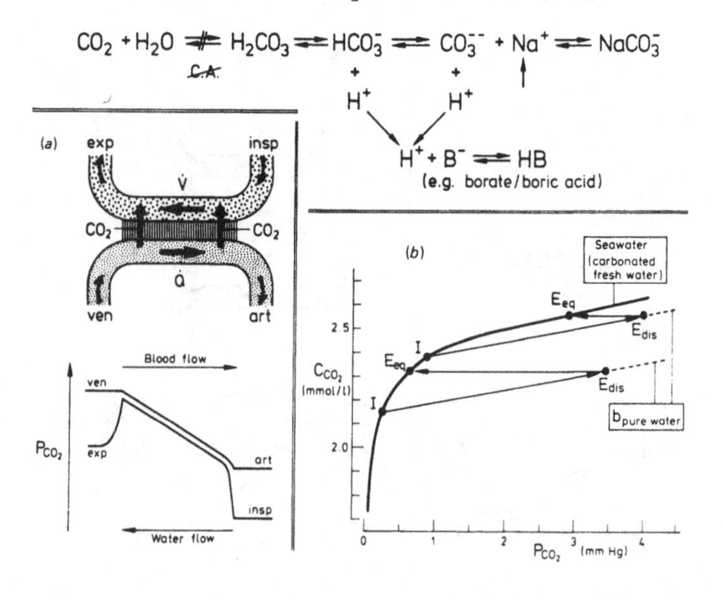

Regardless of the involved mechanisms, the relatively high water/blood P_{CO_2} differences reduce the effect of any changes in expired P_{CO_2}. Although ventilation-induced changes in expired water P_{CO_2} can be large in relation to a low inspired water level, the magnitude of possible changes in arterial P_{CO_2} based on directly transmitted changes in expired P_{CO_2} is small in relative terms on the background of the arterial P_{CO_2} values. Changes in ventilation are without any significance for arterial P_{CO_2} when the inspired water is not in equilibrium with air, and water P_{CO_2} is elevated, conditions often met in natural habitats (e.g. Heisler *et al.*, 1982). The role of ventilation for fish acid–base regulation has accordingly to be viewed as relatively unimportant, and energetically inefficient in face of *a priori* high ventilation rates of the high-viscosity gas exchange medium water. However, in some rare cases, such as the reduction of blood P_{CO_2} during environmental hypoxia this mechanism is important in raising blood pH and thus aiding oxygen transport (Eddy, 1974).

Acid–base relevant ion transfer processes

Transmembrane and transepithelial ion transfer processes are the only mechanisms capable of restoring the original conditions after a non-respiratory acid–base disturbance. The steady state metabolic production of acid–base relevant ions (cf. Heisler, 1984), 1986b) is eliminated via this route, but the transfer mechanisms are taxed to their limits only during extreme stress conditions (Heisler, 1986b,c). The transfer rates of H^+-equivalent ions observed between intracellular and extracellular body compartments during acute and severe disturbances are comparable to the rate of diffusional CO_2 elimination (e.g. Benadé & Heisler, 1978; Holeton & Heisler, 1983; Holeton *et al.*, 1983; Heisler, 1985). According to these high membrane transfer rates, the H^+ elimination from the tissue may become perfusion and equilibrium limited, when the capacity of the perfusate to buffer H^+ ions is too small for the amount transferred across the cell membrane (Heisler, 1986c; cf. A 2).

The capacity of transepithelial ion transfer mechanisms is much smaller than that of transmembrane transfer, which is attributable primarily to the much larger interface area between cells and interstitial fluid, rather than to a lower specific transfer rate of the epithelia involved. The metabolic rate-specific transepithelial transfer capacity for acid–base relevant ions is larger by orders of magnitude in fishes than that in higher vertebrates (for review, see Heisler, 1986d). The ionic transfer is mainly performed at the branchial epithelium, with most of the studies (>80%) indicating a less than 7% contribution from renal and other excretory sites (e.g. skin, rectal glands, abdominal pores) (for review, see Heisler, 1984, 1986b).

Three major branchial ion transfer mechanisms are involved in this very effective acid–base regulation process. Transfer of HCO^-_3, or H^+ ions in opposite directions have to be exchanged against counterions of the same charge in order to maintain electroneutrality. Co-transfer together with oppositely charged ions is in conflict with the requirements of osmoregulation. Accordingly acid–base relevant ion transfer is most likely performed as net 1:1 HCO_3^-/Cl^-, and/or H^+/NH_4^+ ion exchange (for review, see Maetz, 1974; Evans, 1979, 1980, 1984, 1986). Not all of these mechanisms, however, seem to be utilized during physiological acid–base regulation.

The NH_4^+/Na^+ exchange mechanism appears to be hardly ever exploited during acid–base stress conditions, because the rate of ammonia release usually remains constant even during severe acid–base perturbations (cf. Heisler, 1984, 1986b). In only a few cases, characterized by special conditions like induction of acid–base disturbances by ammonium chloride infusion, or by low environmental pH, the ammonia elimination was actually enhanced as compared to control conditions. These data, however, are likely to be explained by non-ionic diffusion of NH_3 across the epithelium, similar to the elimination of CO_2 (Claiborne & Heisler, 1986; Heisler, 1986b). It has recently been demonstrated by determination of the gill NH_3 diffusion

Figure 7. Ion-exchange processes and osmolarity. (*a*) Utilization of the H^+/Na^+ ion exchange mechanisms leads to further accumulation of osmotically active substances in the organisms, whereas (*b*) HCO_3^-/Cl^- exchange is osmotically neutral. Small open arrows indicate rise, constancy or fall of the respective ion concentration.

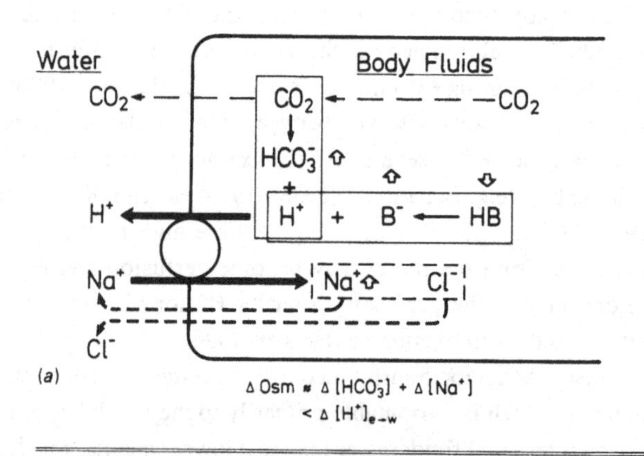

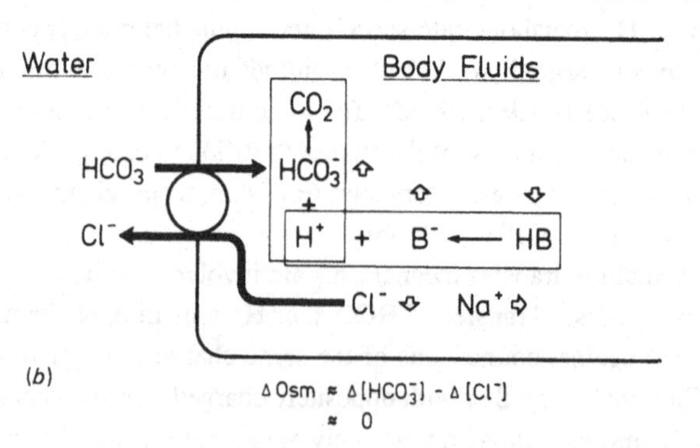

coefficient that with normal environmental conditions (i.e. low NH_3 partial pressure) ammonia is close to completely eliminated by non-ionic diffusion (Cameron & Heisler, 1983, 1985). Based on these data, sparsely observed elevations in ammonia release are probably attributable to an increase in diffusional gradient either by elevation of plasma P_{NH_3}, or by a fall in water P_{NH_3} induced by lowered environmental pH (Claiborne & Heisler, 1986; Heisler, 1984, 1986b). However, Wood (this volume) describes the apparent involvement of NH_4^+ fluxes in acid–base regulation of fishes, during exposure to acid environmental water.

Accordingly, acid–base relevant transepithelial ion transfer seems to be mainly performed by HCO_3^-/Cl^- and H^+/Na^+ ion exchange processes. Experiments in carp appear to indicate that net HCO_3^-/Cl^- ion exchange is preferentially utilized under physiological conditions (Claiborne & Heisler, 1984, 1986; Andersen, Claiborne & Heisler, unpublished data). Such a strategy may be related to the fact that the HCO_3^-/Cl^- ion exchange is osmotically neutral, whereas the H^+/Na^+ exchange leads to accumulation of additional osmotically active molecules (Na^+ + HCO_3^-) in the body fluids of the animal (Figure 7).

Although ion transfer mechanisms are the most efficient tools for fish acid–base regulation, their capacity is not unlimited. Saturation of the carrier-mediated transport systems occurs in intact fishes between 5 and 25 μmol/(min kg^{-1} body water) (for review, see Heisler, 1986d). These maximal transfer rates can be achieved only with optimal conditions. Ion exchange processes are by definition sensitive to the availability of the appropriate counter ions in the environmental water, the lack of which is certainly one of many factors potentially limiting the transfer of ions relevant to acid–base regulation (cf. Heisler, 1986b,c,d). Indeed, it is possible that the osmotic disruption arising from the problem of H^+ excretion in acid environments is a major causal factor in fish death, particularly in naturally soft waters where ion levels are low and acid toxicity is most often encountered. This problem is discussed by Wood (this volume).

Conclusion

Three general mechanisms are available for fish acid–base regulation. Buffering is a very effective process for transient removal of surplus H^+ from the body fluids and during transport from the site of production to the site of elimination, but is very much limited by the available buffer capacity (β.V) of the organism. Buffering of surplus H^+ is always accompanied by changes in the fluid pH, if P_{CO_2} cannot be adjusted to compensate pH changes. Adjustment of P_{CO_2} by changes in ventilation in water-breathing fishes, however, is extremely limited by the physical properties of the gas exchange medium (i.e. water), and correlated factors. Final and also transient adjustment of the acid–base status is performed by comparatively potent branchial ion transfer processes. The contribution of net HCO_3^-/Cl^- ion exchange appears to be predominant, whereas H^+/Na^+ and NH_4^+/Na^+ are of lesser significance and entail an osmotic problem. The limitations of branchial ion transfer mechanisms mainly reside in the availability of external counter ions, which may be low in naturally soft waters prone to acid pollution.

References

Benadé, A.J.S. & Heisler, N. (1978). Comparison of efflux rates of hydrogen and lactate ions from isolated muscles *in vitro*. *Respir. Physiol.*, **32**, 369–80.

Brønsted, J.N. (1923). Einige Bemerkungen über den Begriff der Säuren und Basen. *Rec. Trav. Chim. Pays-Bas*, **42**, 718–28.

Cameron, J.N. & Heisler, N. (1983). Studies of ammonia in rainbow trout: Physico-chemical parameters, acid–base behaviour and respiratory clearance. *J. exp. Biol.*, **105**, 107–25.

Cameron, J.N. & Heisler, N. (1985). Ammonia transfer across fish gills: a review. In *Circulation, Respiration and Metabolism*, pp. 91–100, ed. R. Gilles, Heidelberg: Springer.

Claiborne, J.B. & Heisler, N. (1984). Acid–base regulation in the carp (*Cyprinus carpio*) during and after exposure to environmental hypercapnia. *J. exp. Biol.*, **108**, 25–43.

Claiborne, J.B. & Heisler, N. (1986). Acid–base regulation and ion transfers in the carp (*Cyprinus carpio*): pH compensation during graded long- and short-term environmental hypercapnia and the effect of bicarbonate infusion. *J. exp. Biol.* **126**, 41–61.

Dejours, P. (1981). *Principles of Comparative Respiratory Physiology*. Amsterdam: Elsevier/North-Holland.

Eddy, F.B. (1974). Blood gases of the tench (*Tinca tinca*) in well aerated and oxygen deficient waters. *J. exp. Biol.*, **60**, 71–83.

Evans, D.H. (1979). Fish. In *Comparative Physiology of Osmoregulation in Animals, vol. I*, ed. G.M.O. Maloiy, pp. 305–90. New York: Academic Press.

Evans, D.H. (1980). Kinetic studies of ion transport by fish gill epithelium. *Am. J. Physiol.*, **238**, R224–30.

Evans, D.H. (1984). The role of gill permeability and transport mechanisms in euryhalinity. In *Fish Physiology, vol. X, part B*, ed. W.S. Hoar & D.J. Randall, pp. 315–401. Orlando: Academic Press.

Evans, D.H. (1986). The role of branchial and dermal epithelia in acid–base regulation in aquatic animals. In *Acid–base Regulation in Animals*, ed. N. Heisler, pp. 139–72. Amsterdam: Elsevier Science Publishers B.V.

Heisler, N. (1984). Acid–base regulation in fishes. In *Fish Physiology, vol. X, part A*, ed. W.S. Hoar & D.J. Randall, pp. 315–401, Orlando: Academic Press.

Heisler, N. (1985). Branchial ion transfer processes as mechanisms for fish acid–base regulation. In *Transport Processes, Iono- and Osmoregulation*, ed. R. Gilles & M. Gilles–Baillien, pp. 177–93. Berlin: Springer.

Heisler, N. (1986a). Buffering and transmembrane ion transfer processes. In *Acid–base Regulation in Animals*, ed. N. Heisler, pp. 3–47. Amsterdam: Elsevier Elsevier Science Publishers B.V.

Heisler, N. (1986b). Acid–base regulation in fishes. In *Acid–base Regulation in Animals*, ed. N. Heisler, pp. 309–56. Elsevier Amsterdam: Elsevier Science Publishers B.V.

Heisler, N. (1986c). Comparative aspects of acid–base regulation. In *Acid–base Regulation in Animals*, ed. N. Neisler, pp. 397–450. Elsevier Amsterdam: Elsevier Science Publishers B.V.

Heisler, N. (1986d). Mechanisms and limitations of fish acid–base regulation. In *Fish Physiology: Recent Advances*, ed. S. Nilsson & S. Holmgren, London: Croom Helm.

Heisler, N., Forcht, G., Ultsch, G.F. & Anderson, J.F. (1982). Acid–base regulation in response to environmental hypercapnia in two aquatic salamanders, *Siren lacertina* and *Amphiuma means*. *Respir. Physiol.*, **49**, 141–58.

Holeton, G.F. & Heisler, N. (1983). Contribution of net ion transfer mechanisms to the acid–base regulation after exhausting activity in the larger spotted dogfish (*Scyliorhinus stellaris*). *J. exp. Biol.*, **103**, 31–46.

Holeton, G.F., Neumann, P. & Heisler, N. (1983). Branchial ion exchange and acid–base regulation after strenuous exercise in rainbow trout (*Salmo gairdneri*). *Respir. Physiol.*, **51**, 303–18.

Maetz, J. (1974). Adaptation to hyper-osmotic environments. *Biochem. Biophys. Perspect. Mar. Biol.*, **1**, 91–149.

Neumann, P., Holeton, G.F. & Heisler, N. (1983). Cardiac output and regional blood flow in gills and muscles after exhaustive exercise in rainbow trout (*Salmo gairdneri*). *J. exp. Biol.*, **105**, 1–14.

Piiper, J. (1986). Gas exchange and acid–base status. In *Acid–base Regulation in Animals*, ed. N. Heisler, pp. 49–81, Amsterdam: Elsevier Elsevier Science Publishers B.V.

Rahn, H. (1966b). Aquatic gas exchange: theory. *Respir. Physiol.*, **1**, 1–12.

Siggaard–Andersen, O. (1974). *The Acid–base Status of the Blood*. Copenhagen: Munksgaard.

Van Slyke, D.D. (1922). On the measurement of buffer values and on the relationship of buffer value to the reaction constant of the buffer and the concentration and the reaction of the buffer system. *J. Biol. Chem.*, **73**, 127–47.

Woodbury, J.W. (1965). Regulation of pH. In *Physiology and Biophysics*, ed. T.C. Ruch & H.D. Patton, pp. 899–934. Philadelphia: Saunders.

JAMES N. CAMERON

Acid–base regulation in fishes. 2. Biological responses and limitations

Introduction

The previous chapter (Heisler, this volume) describes the physiological mechanisms responsible for acid–base regulation in fishes. This chapter follows directly from Heisler's account and describes the physiological effects of and responses to exogenous variables (temperature, oxygen and carbon dioxide levels and pH) and endogenous variables (exercise, feeding and anaemia) which affect acid–base regulation and may limit the survival of fish.

Exogenous perturbations

Temperature

(i) *Physico-chemical basis* Temperature affects the acid–base status of fish because it affects the chemical equilibria both of water and of the principal buffer systems of the blood and intracellular fluids. A small part of pure water is normally dissociated into H^+ and OH^-; at 24 °C the quantity of each is 10^{-7}M, and the pH is 7. At 0 °C, however, dissociation is reduced so that the pH of pure water is 7.47, and at 37 °C it is 6.81. Temperature also influences the important physiological buffers to an extent dependent upon the value of ΔH^o (the latent heat of ionization) for each dissociation reaction. This value for water is about 7000 cal mol^{-1}, but varies from only about 2200 for the bicarbonate system (which has a very flat temperature response) to more than 8000 for certain protein groups (Cameron, 1984; Reeves, 1976; Reeves & Malan, 1976). An important aspect of these temperature-induced changes in acid–base status is that they determine the net charge of proteins; the net charge in turn affects many biochemical properties, especially of enzymes (Somero, 1981).

(ii) *Physiological responses* The important physiological buffers are proteins, phosphate compounds, and the CO_2–bicarbonate system. With just the right mix of these different buffers and the right ΔH^o values, net protein charge could be maintained as temperature changes with no acid–base relevant transport from the various physiological fluid compartments (Reeves, 1972, 1976; Reeves & Malan, 1976; Cameron, 1984). Summaries of data from actual studies of both extracellular and intracellular pH as a function of temperature in a variety of fishes (Figures 1, 2) show that pH and temperature vary inversely. The slope in most studies is close to that for the neutral pH of pure water, and close to that of many (histidine-containing) proteins, but there is no general agreement that this can be explained by passive

Figure 1. Summary of arterial pH, arterial CO_2 partial pressure, and $[HCO_3^-]$ for four water-breathing fish: rainbow trout *Salmo gairdneri* (Randall & Cameron, 1973), channel catfish *Ictalurus punctatus* (Cameron & Kormanik, 1982), silver seatrout *Cynoscion arenarius* (Cameron, 1978), and spotted dogfish *Scyliorhinus stellaris* (Heisler *et al.*, 1976b). The pH–temperature slopes are given next to the regression lines in the top panel.

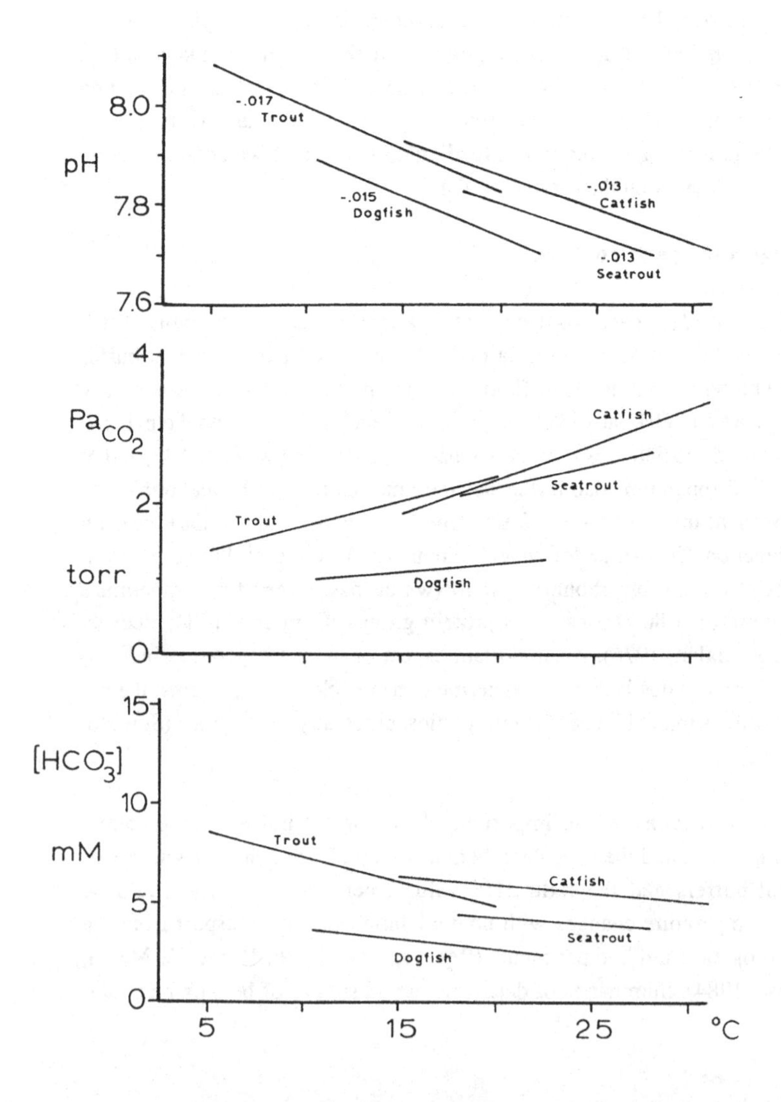

buffering changes (cf. Reeves, 1972; Heisler, Weitz & Weitz, 1976b; Heisler, 1980, 1982; Cameron & Kormanik, 1982; Cameron, 1984). In fact significant transfer of acid–base relevant ions following temperature change, both between the external medium and the blood, and between the blood and intracellular fluids, have been measured for quite a few fish species (Cameron & Kormanik, 1982; Heisler *et al.*, 1976b; Heisler & Neumann, 1980; Heisler, 1980, 1982), so it is now clear that 'closed-system' models such as that of Reeves (1972) are inadequate to explain the acid–base adjustments of fish to temperature change. The appropriate changes are brought about by a combination of passive buffering adjustments and active inter-compartmental ion transfers.

Figure 2. Observed $\Delta pH/\Delta T^{o}$ for red muscle, white muscle and heart muscle from dogfish (Heisler *et al.*, 1976b), carp (Heisler, 1980) catfish (Cameron & Kormanik, 1982), and eel (Walsh & Moon, 1982). The calculated enthalphy (ΔH^{o}) values corresponding to the temperature slopes are indicated at the far left.

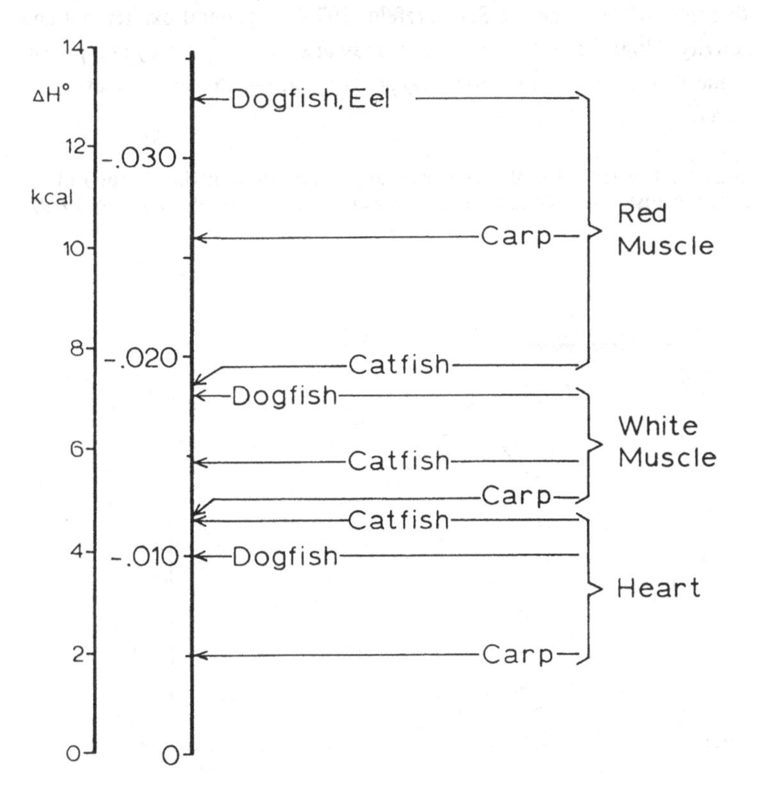

There are also some interesting differences among different tissue types (Figure 2); red muscle has a consistently higher pH/temperature slope than white muscle, for example, suggesting biochemical differences that should be worth investigating.

(iii) *Limitations* The viable temperature range is probably not affected by acid–base considerations. Acid–base status has now been studied from animals over a broad temperature range (up to 40 °C, including birds) without any hints of an upper or lower limit. The transport requirements impose a certain energy requirement, but this is not likely to be a significant factor.

Hypoxia

(i) *Primary physiological effects* Fishes vary in their ability to extract oxygen from progressively hypoxic waters, but for all of them a point is reached when the oxygen supply falls below their metabolic requirements. At this point metabolism shifts away from oxidative pathways, significant amounts of pyruvate are shunted to lactate, and H^+ are produced. In some circumstances the total metabolic rate may also increase, due either to the work of increased ventilation (Cameron & Cech, 1970; Jones, 1971; Jones & Schwarzfeld, 1974) or general excitement and increased activity (Hoglund, 1961). If so, there may also be a respiratory component of acidosis due to increased metabolic CO_2 production, but the primary effect in hypoxia is lactacidosis.

Figure 3. The response of blood pH and [lactate] to hypoxia (79 torr) in the channel catfish. Lines fitted by eye. (Redrawn from Burggren & Cameron, 1979).

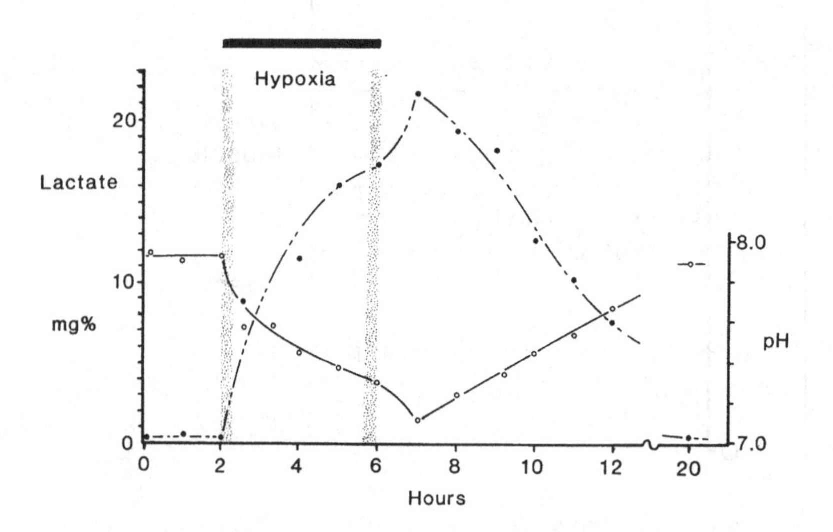

The primary acid–base responses of channel catfish (Figure 3) to hypoxia are typical: there is a progressive decrease in both blood pH and [HCO_3^-], and an increase in blood [lactate]. Often the maximum changes are not observed until just after the return to normal oxygen concentrations (Figure 3; Black *et al.*, 1959; Burggren & Cameron, 1979); this is probably due to limitations on the blood perfusion of white muscle during hypoxia. After the return to normal oxygen, the lactate is re-metabolized and very little is excreted. Hydrogen ions are probably transported to the environment at a fairly rapid rate during hypoxia as they are during exercise (see below), limiting the severity of the acidosis.

(ii)*Limitations* Few fish can survive severe hypoxia or anoxia for any extended period. There are some old reports of high anaerobic capabilities in certain carp (Blazka, 1958), but these have not been recently confirmed. There are also anecdotal reports of overwintering by various fish in totally anoxic ponds and lakes. In these fish, there must be some means of limiting the extent of the lactacidosis, but nothing is presently known.

Acid–base disturbances are probably a primary cause of hypoxia-related death. In catfish studies the pH in some fish fell from 7.8 to about 6.9 during severe hypoxia, sufficient to cause death (Burggren & Cameron, 1979). In some of the fish, the maximum acidosis and death occurred **after** hypoxia, when there appeared to be a rapid release of acidic metabolites from the tissues.

Hypercapnia

(i) *Primary effects* An increase in the ambient CO_2 (hypercapnia) causes a shift in the equilibrium reactions: $CO_2 + H_2O \rightarrow H_2CO_3 \rightarrow HCO_3^- + H^+$. Most of the H^+ produced are buffered, but there is an increase in the steady state [HCO_3^-] and a decrease in the pH of the blood and intracellular fluids. A secondary effect of hypercapnia is to reduce oxygen availability in those species with a significant haemoglobin Bohr shift. In air-breathing animals, ventilation increases in order to reduce the CO_2 gradient between blood and air, ameliorating the effects of all but the most severe hypercapnia. In water-breathing animals, however, normal partial pressure of CO_2 is only a few torr (Randall, Holeton & Stevens, 1967; Randall & Cameron, 1973), so for any external hypercapnia greater than about 0.5% (3.8 torr), changes in ventilation cannot eliminate the hypercapnia. Generally ventilation must be maintained at a high level in order to obtain oxygen, and hypercapnia does not directly lead to an increase in ventilation (Randall & Cameron, 1973; Randall, Heisler & Drees, 1976).

(iii) *Physiological responses of fish to hypercapnia* The changes in blood acid–base status in response to hypercapnia in carp (*Cyprinus carpio*) and the spotted dogfish (*Scyliorhinus stellaris*) are similar to those for the rainbow trout (*Salmo gairdneri*)

(Figure 4). In all three, there is an initial respiratory acidosis along the passive buffer line, followed by an increase in [HCO_3^-] and pH at constant P_{CO_2}. The compensatory shift (A–B) occurs fastest in the dogfish, and slowest in the rainbow trout (Cameron & Randall, 1972; Heisler, Weitz & Weitz, 1976a; Claiborne & Heisler, 1984), an order similar to that of their gill permeabilities and to normal ion fluxes. In none of the species is there 100% compensation of the acidosis; i.e. the steady state pH reached is below the control value, no matter how long the animals are held in hypercapnia.

Intracellular acid–base status shows a similar pattern of change (Figure 5) except that the passive buffer value is much higher, and the compensation nearly 100% in most tissues (Heisler, 1980, 1982; Cameron & Kormanik, 1982; Walsh & Moon, 1982).

The compensation of hypercapnia involves ion transfers between intra- and extracellular fluid pools, and between the extracellular fluids and the environment (De Renzis & Maetz, 1973; Cameron, 1976, 1980; Heisler, 1980, 1982; Evans, 1982; Claiborne & Heisler, 1984). These are thought to occur principally via Na^+/H^+ and Cl^-/HCO_3^- exchanges, which seem to be nearly universal mechanisms of pH

Figure 4. A pH–bicarbonate diagram showing the response to ambient hypercapnia (1% CO_2 in air) in the rainbow trout. The mean resting value ± SE is shown as A, and the 24 h compensated mean at B. The solid line shows the passive buffer line for the blood (slope −8.6), and the dotted line the effective (or achieved) buffer value after compensation. (Redrawn from Cameron & Randall, 1972.)

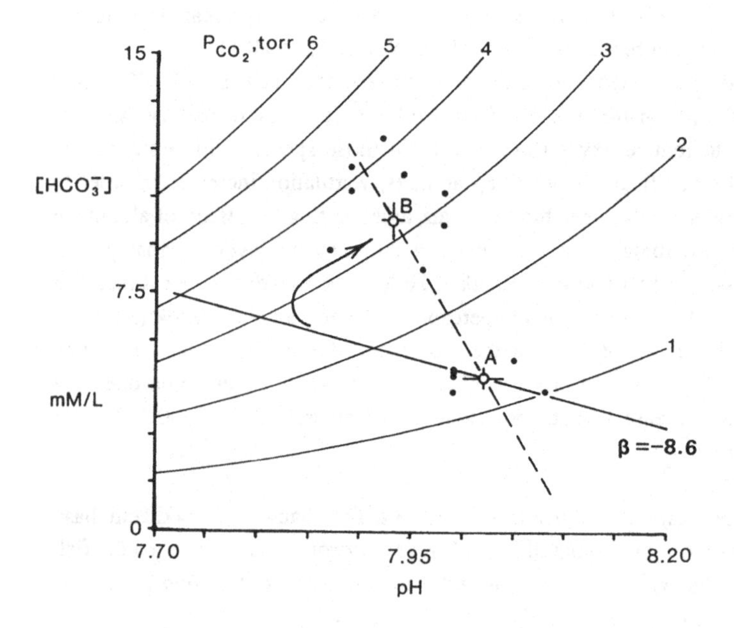

compensation in invertebrates and vertebrates (Cameron, 1976, 1986; Hinke & Menard, 1976; Maetz, Payan & De Renzis, 1976; Heisler, 1980, 1982, 1984; Roos & Boron, 1981; Boron, 1985; Schlue & Thomas, 1985).

(iii) *Limitations* The incomplete pH compensation observed in fishes during long term hypercapnia seems to suggest that the compensatory process is in some way limited. That is, the fish may not be able to achieve a higher HCO_3^- concentration and pH in the blood, either due to limitations on the ionic supplies in the environment or internal fluids, or because of problems maintaining larger gradients across either cell membranes, gills, or kidneys. Two or three pieces of evidence suggest, however, that under any particular hypercapnic regime a new pH 'set-point' is established as a steady state. For example, 24 h and 48 h values for blood pH and [HCO_3^-] are not significantly different (Claiborne & Heisler, 1984; Cameron, unpublished data), and do not show 100% pH compensation to 1% external CO_2. Increasing the external CO_2 to 2%, however, brings about a second compensatory shift, with a higher pH and [HCO_3^-] being reached after another 24 h or so (Figure 6). A second piece of evidence is that infusion of HCO_3^- salts after some period of compensation does

Figure 5. A pH–bicarbonate diagram showing intracellular pH compensation of hypercapnia in various tissues from the channel catfish. The lines connecting the control values (lower points) with values determined after 24 h hypercapnia (upper) indicate the achieved buffering. The slope shown for plasma (right) was determined simultaneously, but the hypercapnia point is off scale. Redrawn from Cameron, 1985.

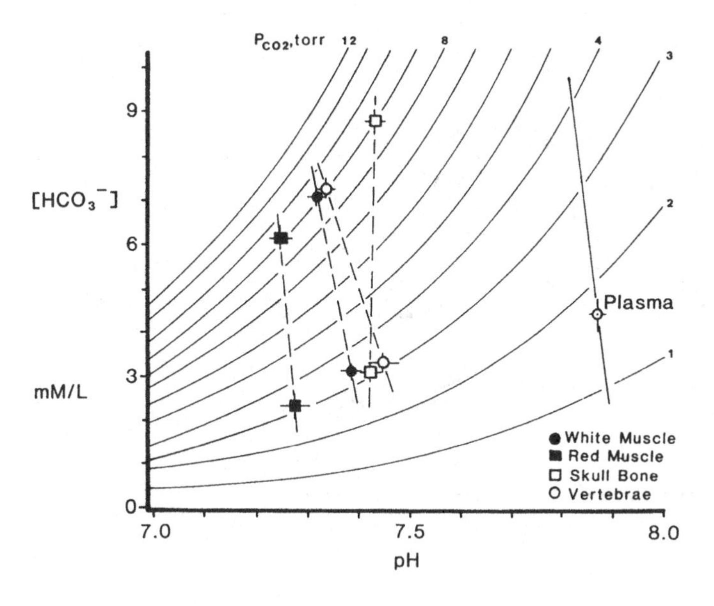

not lead to any lasting increase in plasma [HCO_3^-]; the fish appear to be actively regulating at the new set-point. With progressively higher ambient CO_2, the channel catfish can in fact reach quite high [HCO_3^-] values, but does not achieve a pH compensation of more than about 60% at any level of hypercapnia (Figure 6; Cameron, unpublished data).

In certain circumstances the compensation of hypercapnic acidosis appears to be limited by the availability of ions in the environment. In his study of the Amazonian *Synbranchus marmoratus*, there was virtually no compensation of hypercapnia when external ions were present in very low concentrations (Heisler, 1980).

External pH

The influence of external pH has been examined closely as part of various recent studies on the effects of acid rain (see Wood, Vangenechten *et al.*, this volume). Acid–base effects on fish can be expected to be both direct and indirect. Direct effects would result from the actual H^+ permeability of the external surfaces,

Figure 6. A pH–bicarbonate diagram showing the response to progressive hypercapnia in the channel catfish. After 24–48 h normocapnic control values were determined (lower right point, ± 1 SE), the ambient CO_2 was increased to 1%, 2%, 4% and 6% with 24 h at each concentration. At each level there was an initial response parallel to the passive buffer line (β), then a compensatory shift upwards and to the right. Although complete pH compensation was not achieved at any CO_2 concentration, the fish is capable of further increase, as shown by the final plasma [HCO_3^-] values of nearly 40 mM. (Cameron, unpublished data.)

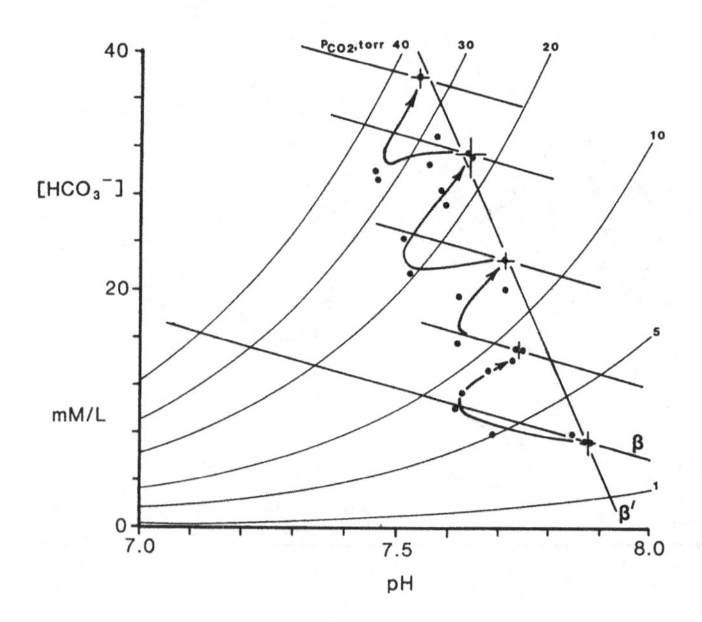

especially the gills, and the relationship between internal and external [HCO_3^-]. at pH values much below 6.0, there is virtually no HCO_3^- present, and at pH 4.0, the H^+ gradient between blood and water is more than 1000:1. Indirect effects may result in two ways: the permeability to various other 'strong' ions (Na^+, Cl^-, etc.) are changed, and internal imbalances of these ions will often result in changes in the acid–base status; and many of the transport proteins in external membranes have hydrophilic tails whose configuration and affinity are strongly affected by pH (W. Philpott, personal communication). It is somewhat surprising, then, that the effects of external pH are no larger than they are (Eddy, 1976; Packer, 1979; Holeton, Booth & Jansz, 1983), and that many workers now regard the acid–base disturbance as secondary in importance to ion and water imbalances (Wood, Vangenechten *et al.*, this volume).

Endogenous perturbations

Exercise

(i) *General respiratory and metabolic consequences* The increased energy expenditure during vigorous 'burst' swimming in fishes is largely met by anaerobic metabolism, i.e. glycolysis leading to lactic acid. Since the pK of lactic acid is around 4, Lac^- and H^+ are produced equally. There is also an increase in the metabolic production of CO_2, which equilibrates, forming HCO_3^- and additonal H^+. The build up of these products in white muscle and some other tissues is exacerbated by low rates of blood flow (Cameron, 1975; Stevens, 1968).

The lactate buildup represents an 'oxygen debt'; i.e. after exercise extra oxygen is required to re-metabolize the lactate to glucose and glycogen. Measurements of oxygen consumption after exercise typically show a prolonged increase which can be attributed to the oxygen debt.

(ii) *Experimental observations* Actual data for changes in blood parameters during and after exercise in various water-breathers (Figure 7) show a common pattern: P_{CO_2} and [Lac^-] increase, and pH decreases at rates depending on temperature, degree and duration of the exercise (Black, 1955; Black *et al.*, 1959; Black, Manning & Hayashi, 1966; Randall, Holeton & Stevens, 1967; Piiper, Meyer & Drees, 1972; Wood, McMahon & McDonald, 1977; Holeton & Heisler, 1983; Holeton, Neumann & Heisler, 1983; Jensen, Nikinmaa & Weber, 1983). Peak values for the changes often occur after the exercise has ended, but the apparent H^+ load appearing in the blood is considerably smaller than the Lac^- load in all studies except one (Wood *et al.*, 1977). The lactate appearing in blood usually peaks after exercise, which probably reflects both a low permeability of lactate and increases in blood flow to white muscle after exercise (G.F. Holeton & N. Heisler, unpublished data). Judging from early reports that exercise could cause death in some salmonids (Black, 1958), the lactate 'flush' may sometimes be quite pronounced, as was observed following hypoxia (above).

The mismatch between the apparent total H^+ and lactate loads is evidently due to rapid exchange of H^+ across the gills, in effect a temporary shuttle to the environment (Holeton *et al.*, 1983; Heisler, 1984). As the lactate is re-metabolized later, the H^+ would return across the gills to maintain acid–base balance, but this apparently occurs at a lower rate and has not been successfully measured.

(iii) *Limitations* Besides the obvious metabolic limits on energy production by anaerobic pathways, the acid–base changes intracellularly may strongly limit the work output of muscle tissues. Quite severe acidosis is often observed after bouts of strenuous exercise, and acidosis is probably an important facet of muscle fatigue. In extreme cases, the massive release of lactate after exercise appears capable of causing death from acid–base imbalance.

Other endogenous perturbations

(i)*Feeding* There do not seem to be any published studies of the direct acid–base consequences of feeding. Some sort of exercise is usually involved, and the

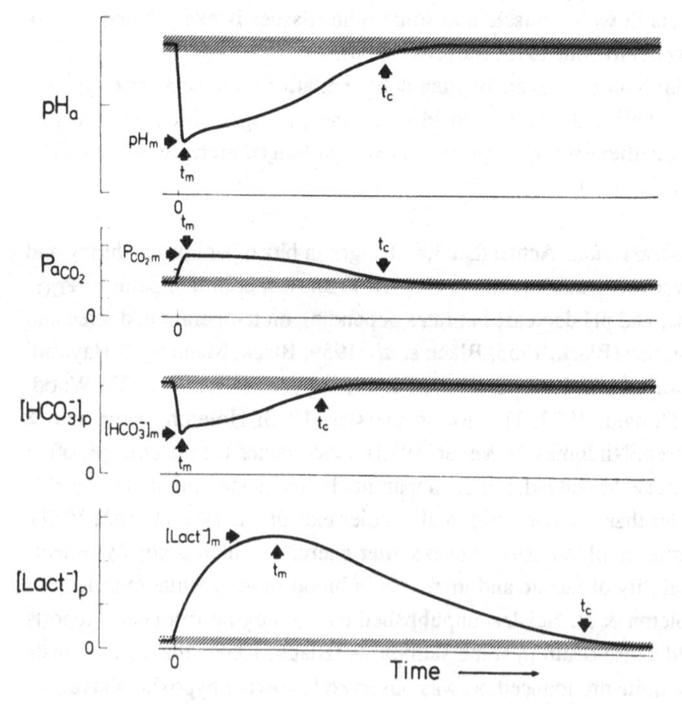

Figure 7. Generalized acid–base response of fish to strenuous exercise. The subscript *m* indicates the point at which maximum deviation of each variable is reached, and *c* indicates the time of return to control values. Note the different time courses for pHa and [Lact$^-$]. (From Heisler, 1984, with permission.)

food ingested often imposes an acid–base load upon metabolism, especially if the meal is high in protein. It would be of interest to study the effects of feeding in carnivores like the pike (*Esox lucius*), which may take an hour or more to swallow a fish up to half its own size. During this swallowing period, ventilation is suspended, and there must be a considerable hypoxic and hypercapnic acidosis (personal observation).

(ii) *Anaemia* Although anaemia has mainly been employed in experimental work as a tool to study the blood gas transport system (Cameron & Wohlschlag, 1969; Cameron & Davis, 1970; Wood, McDonald & McMahon, 1982), natural populations of fishes show a wide range of haematocrits, including values which are quite anaemic (Cameron & Wohlschlag, 1969; Wood *et al.*, 1982). At very low haematocrits, cardiac output is greatly increased, the PCO_2 is decreased, and a chronic alkalosis develops. The lack of circulating erythrocytes may also cause kinetic disturbances in normal CO_2 excretion due to the lack of carbonic anhydrase, and the buffer capacity of the blood will also be much reduced.

Conclusions

The acid–base status of internal fluids of fishes is affected by many different factors, both internal and external. These include temperature, oxygen concentration, carbon dioxide concentration, metabolic substrate, exercise, feeding, anaemia, and external pH. Fish have a variety of compensatory mechanisms which act to maintain the acid–base status in a steady state, the set-point of which is a complex function of environmental and internal factors. In most cases, the acid–base regulating system itself is not the primary factor in the ability of fish to survive adverse conditions, but extreme imbalances can lead to death, especially after exercise or hypoxia.

References

Black, E.C. (1955). Blood levels of hemoglobin and lactic acid in some freshwater fishes following exercise. *J. Fish. Res. Bd. Can.*, **12**, 917–29.

Black, E.C. (1958). Hyperactivity as lethal factor in fish. *J. Fish. Res. Bd Can.*, **15**, 573–86.

Black, E.C., Chiu, W.G., Forbes, F.D. & Hanslip, W.G. (1959). Changes in pH, carbonate, and lactate of the blood of yearling Kamloops trout, *Salmo gairdneri*, during and following severe muscular activity. *J. Fish. Res. Bd. Can.*, **16**, 391–402.

Black, E.C., Manning, G.T. & Hayashi, K. (1966). Changes in levels of hemoglobin oxygen, carbon dioxide, pyruvate, and lactate in venous blood of rainbow trout (*Salmo gairdneri*) during and following severe muscular activity. *J. Fish. Res. Bd. Can.*, **23**, 783–95.

Blazka, P. (1958). The anaerobic metabolism of fish. *Physiol. Zool.*, **31**, 117–28.

Boron, W F. (1985). Intracellular pH regulating mechanism of the squid axon – relation between the external Na^+ and HCO_3^- dependencies. *J. gen. Physiol.*, **85**, 325–46.

Burggren, W.W. & Cameron, J.N. (1979). Anaerobic metabolism, gas exchange, and acid–base balance during hypoxic exposure in the channel catfish, *Ictalurus punctatus*. *J. exp. Zool.*, **213**, 405–16.

Cameron, J.N. (1975). Blood flow distribution as indicated by tracer microspheres in resting and hypoxic Arctic Grayling. (*Thymallus arcticus*). *Comp. Biochem. Physiol.*, **52A**, 441–4.

Cameron, J.N. (1976). Branchial ion uptake in Arctic grayling: resting values and effects of acid–base disturbance. *J. exp. Biol.*, **64**, 711–25.

Cameron, J.N. (1978). Regulation of blood pH in teleost fish. *Respir. Physiol.*, **33**, 129–44.

Cameron, J.N. (1980). Body fluid pools, kidney function, and acid–base regulation in the freshwater catfish *Ictalurus punctatus*. *J. exp. Biol.*, **86**, 171–85.

Cameron, J.N. (1984). The acid–base status of fish at different temperatures. *Am. J. Physiol.*, **246**, R452–9.

Cameron, J.N. (1986). Acid–base equilibria in invertebrates. In *Acid–base Regulation in Animals*, ed. N. Heisler, pp. 357–94. Amsterdam: Elsevier North-Holland.

Cameron, J.N. & Cech, J.J., Jr (1970). Notes on the energy cost of gill ventilation in teleosts. *Comp. Biochem. Physiol.*, **34**, 447–55.

Cameron, J.N. & Davis, J.C. (1970). Gas exchange in rainbow trout with varying blood oxygen capacity. *J. Fish. Res. Bd. Can.*, **27**, 1069–85.

Cameron, J.N. & Kormanik, G.A. (1982). Intracellular and extracellular acid–base status as a function of temperature in the freshwater channel catfish, *Ictalurus punctatus*. *J. exp. Biol.*, **99**, 127–42.

Cameron, J.N. & Randall, D.J. (1972). The effect of increased ambient CO_2 on arterial CO_2 tension, CO_2 content and pH in rainbow trout. *J. exp. Biol.*, **57**, 673–80.

Cameron, J.N. & Wohlschlag, D.E. (1969). Respiratory response to experimentally induced anemia in the pinfish, *Lagodon rhomboides*. *J. exp. Biol.*, **50**, 307–17.

Claiborne, J.B. & Heisler, N. (1984). Acid–base regulation and ion transfer in the carp (*Cyprinus carpio*) during and after exposure to environmental hypercapnia. *J. exp. Biol.*, **108**, 25–44.

DeRenzis, G. & Maetz, J. (1973). Studies on the mechanism of chloride absorption by the goldfish gill: relation with acid–base regulation. *J. exp. Biol.*, **59**, 339–58.

Eddy, F.B. (1976). Acid–base balance in rainbow trout, *Salmo gairdneri*, subjected to acid stresses. *J. exp. Biol.*, **64**, 159–71.

Evans, D.H. (1982). Mechanisms of acid extrusion by two marine fishes: the teleost, *Opsanus beta*, and the elasmobranch, *Squalus acanthias*. *J. exp. Biol.*, **97**, 289–99.

Heisler, N. (1978). Bicarbonate exchange between body compartments after changes of temperature in the larger spotted dogfish (*Scyliorhinus stellaris*). *Respir. Physiol.*, **33**, 145–60.

Heisler, N. (1980). Regulation of the acid–base status in fishes. In *Environmental Physiology of Fishes*, ed. M.A. Ali, pp. 123–62. NATO Adv. Study Inst., Series A,1 vol. 35. New York: Plenum Press.

Heisler, N. (1982). Transepithelial ion transfer processes as mechanisms for fish acid–base regulation in hypercapnia and lactacidosis. *Can. J. Zool.*, **60**, 1108–22.

Heisler, N. (1984). Acid–base regulation in fishes. In *Fish Physiology, vol. X, part A*, ed. W.S. Hoar & D.J. Randall, pp. 315–401. New York: Academic Press.

Heisler, N. & Neumann, P. (1980). The role of physico-chemical buffering and of bicarbonate transfer processes in intracellular pH regulation in response to changes of temperature in the larger spotted dogfish (*Scyliorhinus stellaris*). *J. exp. Biol.*, **85**, 99–110.

Heisler, N., Weitz, H. & Weitz, A.M. (1976a). Hypercapnia and resultant bicarbonate transfer processes in an elasmobranch fish. *Bull. Eur. Physiopathol. Respir.*, **12**, 77–85.

Heisler, N., Weitz, H. & Weitz, A.M. (1976b). Extracellular and intracellular pH with changes of temperature in the dogfish *Scyliorhinus stellaris*. *Respir. Physiol.*, **26**, 249–63.

Hinke, J.A.M. & Menard, M.R. (1976). Intracellular pH of single crustacean muscle fibers by the DMO and electrode methods during acid and alkaline conditions. *J. Physiol. (Lond.)*, **262**, 533–52.

Hoglund, L.B. (1961). The reactions of fish in concentration gradients. *Rep. Inst. Freshwater Res. Drottningholm*, **42**, 147 pp.

Holeton, G.F., Booth, J.H. & Jansz, G.F. (1983). Acid–base balance and and Na^+ regulation in rainbow trout during exposure to, and recovery from, low environmental pH. *J. exp Zool.*, **228**, 11–20.

Holeton, G.F. & Heisler, N. (1983). Contribution of net ion transfer mechanisms to acid–base regulation after exhausting activity in the larger spotted dogfish (*Scyliorhinus stellaris*). *J. exp. Biol.*, **103**, 31–46.

Holeton, G.F., Neumann, P. & Heisler, N. (1983). Branchial ion exchange and acid–base regulation after strenuous exercise in rainbow trout (*Salmo gairdneri*). *Respir. Physiol.*, **51**, 303–18.

Jensen, F.B., Nikinmaa, M. & Weber, R.E. (1983). Effects of exercise stress on acid–base balance and respiratory function in blood of the teleost *Tinca tinca*. *Respir. Physiol.*, **51**, 291–302.

Jones, D.R. (1971). Theoretical analysis of factors which may limit the maximum oxygen uptake of fish: the oxygen cost of the cardiac and branchial pumps. *J. Theor. Biol.*, **32**, 341–9.

Jones, D.R. & Schwarzfeld, T. (1974). The oxygen cost to the metabolism and efficiency of breathing in the trout. (*Salmo gairdneri.*). *Respir. Physiol.*, **21**, 241–54.

Maetz, J., Payan, P. & de Renzis, G. (1976). Controversial aspects of ionic uptake in freshwater animals. In *Perspectives in Experimental Zoology*, vol. I, ed. P.S. Davies, pp. 77–92.

Packer, R.K. (1979). Acid–base balance and gas exchange in brook trout (*Salvelinus fontinalis*) exposed to acidic environments. *J. exp. Biol.*, **79**, 127–34.

Piiper, J., Meyer, M. & Drees, F. (1972). Hydrogen ion balance in the elasmobranch, *Scyliorhinus stellaris*, after exhausting activity. *Respir. Physiol.*, **16**, 290–303.

Randall, D.J. & Cameron, J.N. (1973). Respiratory control of arterial pH as temperature changes in rainbow trout *Salmo gairdneri*. *Am. J. Physiol.*, **225**, 997–1002.

Randall, D.J., Heisler, N. & Drees, F. (1976). Ventilatory response to hypercapnia in the larger spotted dogfish *Scyliorhinus stellaris*. *Am. J. Physiol.*, **230**, 590–4.

Randall, D.J., Holeton, G.F. & Stevens, E.D. (1967). The exchange of oxygen and carbon dioxide across the gills of rainbow trout *J. exp. Biol.*, **46**, 339–48.

Reeves, R.B. (1972). An imidazole alphastat hypothesis for vertebrate acid–base regulation: tissue carbon dioxide content and body temperature in bull frogs. *Respir. Physiol.*, **14**, 219–36.

Reeves, R.B. (1976). Temperature-induced changes in blood acid–base status: pH and pCO_2 in a binary buffer. *J. Appl. Physiol.*, **40**, 752–61.
Reeves, R.B. & Malan, A. (1976). Model studies of intracellular acid–base temperature responses in ectotherms. *Respir. Physiol.*, **28**, 49–64.
Roos, A. & Boron, W. (1981). Intracellular pH. *Physiol. Rev.*, **61**, 296–435.
Schlue, W.R. & Thomas, R.C. (1985). A dual mechanism for intracellular pH regulation by leech neurones. *J. Physiol. (Lond.)*, **364**, 327–38.
Somero, G.N. (1981). pH–temperature interactions of proteins: principles of optimal pH and buffer system design. *Mar. Biol. Lett.*, **2**, 163–78.
Stevens, E.D. (1968). The effect of exercise on the distribution of blood to various organs in rainbow trout. *Comp. Biochem. Physiol.*, **25**, 615–25.
Walsh, P.J. & Moon, T.W. (1982). The influence of temperature on extracellular and intracellular pH in the American eel, *Anguilla rostrata* (Le Sueur). *Respir. Physiol.*, **50**, 129–40.
Wood, C.M., McMahon, B.R. & McDonald, D.G. (1977). An analysis of changes in blood pH following exhausting activity in the starry flounder, *Platichthys stellatus*. *J. exp. Biol.*, **69**, 173–86.
Wood, C.M., McDonald, D.G. & McMahon, B.R. (1982). The influence of experimental anaemia on blood acid–base regulation in vivo and in vitro in the starry flounder (*Platichthys stellatus*) and the rainbow trout (*Salmo gairdneri*). *J. exp. Biol.*, **96**, 221–37.

R.C. THOMAS

Intracellular pH regulation and the effects of external acidification

Introduction

Most measurements of the effects of acid toxicity on aquatic animals concentrate upon changes in body fluid pH and the flux of ions between water and blood. One problem posed by a low blood pH is the potential acidification of the tissues which could cause undesirable deviation from the optimum pH of intracellular enzymes. This chapter examines the techniques and results of investigations into the regulation of intracellular pH (pH_i).

Relatively little is known about pH_i in most animals, experiments being so far generally confined to those with large neurones. This is because until recently the only way of following pH_i over long periods was with pH-sensitive microelectrodes. This method is still the best, but requires both skill and a large cell. For small cells fluorescent dyes are very promising.

In this chapter I will describe some of the evidence on which the present understanding of pH_i regulatory mechanisms in snail, crayfish and leech neurones is based. I will confine this chapter to these preparations because they are reasonably typical, and I lack the space for a full review. I will then consider the effects of external acidification before concluding that maintenance of a constant pH_i depends very much on a constant external pH. For a detailed review of intracellular pH, see Roos & Boron (1981), but for shorter and more recent accounts of the subject see Thomas (1984, 1986).

Methods

The experiments were done on neurones in isolated ganglia from the common snail, *Helix aspersa*, the crayfish, *Procambarus clarkii* or the medicinal leech, *Hirudo medicinalis* as previously described (Thomas, 1977; Moody, 1981; Schlue & Thomas, 1985). In brief, the ganglia were mounted in an experimental chamber, superfused with the appropriate saline solution, and the connective tissue dissected to expose large neurones. Suitable neurones were penetrated with a number of micropipettes, always including a reference micropipette, filled usually with 3M KCl to measure the membrane potential, and a pH-sensitive microelectrode. In the leech experiments a double-barrelled electrode was used. With the snail and crayfish neurones, some experiments were done with additional electrodes to inject HCl or to measure Na^+ and Cl^- ion levels.

Details about ion-sensitive microelectrodes and the electrical recording methods used are given in Thomas (1978). The basic set-up is illustrated diagrammatically in Figure 1. Signals from the electrodes were displayed on an oscilloscope and recorded on a pen-recorder.

Results

Intracellular pH regulation by snail neurones

The first detailed study of this question was published in 1977. Before reporting the original conclusion I will describe three of the experiments I did at that time to show the kind of evidence available.

When I first started working on snail neurone pH, I superfused the preparation with a bicarbonate-free saline. Bubbling solutions with CO_2 was too much trouble. I found that pH_i recovered only very slowly from any acid load. Then I started using the more physiologically correct bicarbonate:CO_2 buffers, and found pH_i regulation transformed!

A typical experiment showing the effects of reducing bicarbonate is shown in Figure 2. In the part of the record illustrated, the traces begin with all microelectrodes poised above a suitable cell. Then the pH electrode was inserted, causing the

Figure 1. Diagram of experimental arrangement used to study intracellular pH (pH_i) with microelectrodes. The two electrodes on the right measure membrane potential (E_m) and pH_i, the two on the left are for acid injection.

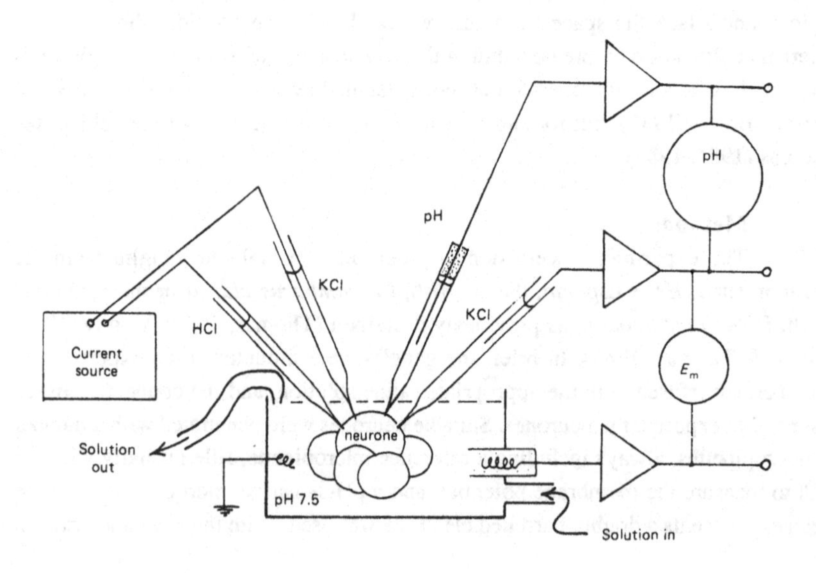

downward deflection seen. One minute later the reference microelectrode was inserted to measure the membrane potential E_m. This potential was immediately subtracted from the pH record, causing the upward jump. The pH_i was between 7.3 and 7.4.

About 15 min later the two current-passing electrodes were inserted, and a backing-off current switched on to stop HCl leaking into the cell. The pH record became noisy during the injection as the KCl current electrode kept blocking.

The HCl injection reduced the pH_i by about 0.3 units, but it recovered within a short time. After a second injection, the superfusate was changed to one with less bicarbonate and CO_2, and there was a transient alkalinization as bicarbonate left the cell. A third HCl injection, half the size of the second, reduced pH_i by a similar amount, showing buffering power had decreased. The recovery was again quite fast. The final HCl injection was made with the cell in bicarbonate-free saline, and this time pH_i recovered much more slowly. In many similar experiments bicarbonate removal always slowed pH_i regulation, usually reducing the rate of acid extrusion to less than 10% of its rate in 20 mM bicarbonate, pH 7.5.

In the experiment shown in Figure 3, the cell was in bicarbonate saline throughout. Removal of all sodium from the saline (replaced with bis (2-hydroxethyl)dimethyl ammonium chloride, BDAC) almost blocked pH_i recovery from the fourth acid injection. The anion-exchange blocker SITS (4-acetamido-4'-isothiocyanato-stilbene-2,2'-disulphonic acid) completely blocked recovery from the last injection.

Figure 2. Snail neurone pH_i regulation. The effect of reducing and then removing bicarbonate from the superfusate on pH_i and its recovery from acid injection. Membrane potential (E_m) shown at the top, injection current in the middle and pH_i at the bottom. Bicarbonate level indicated below the pH_i record. Arrows show where KCl- and HCl-filled micropipettes inserted (R.C. Thomas, unpublished data).

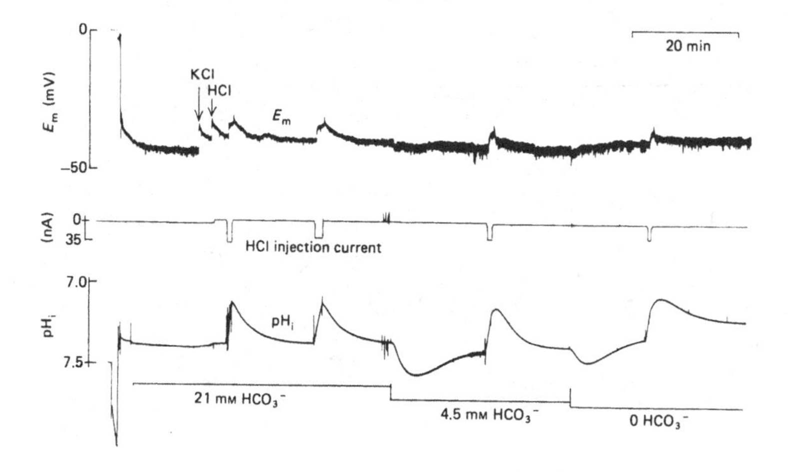

If sodium removal blocks because Na^+ normally enters the cell as acid is extruded, it should be possible to see an increase in internal Na^+ as pH_i recovers from an acid load. In the experiment shown in Figure 4 a cell was acidified by applying CO_2. The pH_i trace shows a transient acidification as CO_2 entered, followed by recovery beyond the pre-CO_2 level. As pH_i increased, internal Cl^- decreased and internal Na^+, transiently, increased. In other experiments I found that such a pH_i recovery could be prevented by depleting the cell of Cl^- (Thomas, 1977).

Thus in snail neurones pH_i recovery from an acid load was inhibited by removing bicarbonate, or external Na^+, or internal Cl^-, and was accompanied by a fall in internal Cl^- and a rise in internal Na. These findings suggested a Na-dependent Cl/HCO_3^- exchange as the prime mechanism for acid extrusion. For symmetry I proposed a four-ion carrier (Thomas, 1977). I could later establish no requirement for ATP, and no sign of any effect of membrane potential on the mechanism (Thomas, 1978). The energy in the Na^+ gradient would be more than enough to drive the other ion movements.

Figure 3. Snail neurone pH_i regulation: the effect of removing external sodium and of the anion exchange inhibitor SITS. Superfusate equilibrated with 2.3% CO_2 throughout. Modified from Thomas (1977).

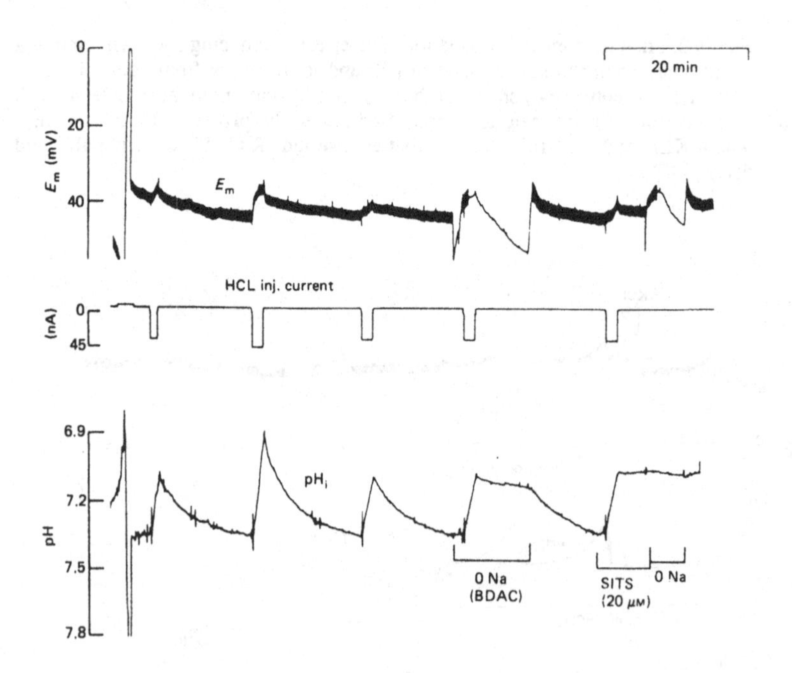

Crayfish neurone pH_i regulation

A careful study by Moody (1981) showed that freshwater crayfish neurones had a dual mechanism of pH_i-regulation. I show two of his results in Figure 5. In Figure 5(*a*) removal of external Na blocked pH_i recovery from acidification.

The acidification was achieved by exposing the preparation to 20 mM NH_4Cl in crayfish saline, the ammonium prepulse method of Boron & de Weer (1976). The cell accumulated ammonium ions. When the external NH_4Cl was removed, the internal ammonium ions shed a H^+ and left as ammonia. When external Na was replaced, pH_i recovered remarkably rapidly.

In Figure 5(*b*), Moody showed that pH_i recovery was blocked by reducing external pH from the normal 7.4 to 6.7. Such blockage of pH_i-recovery is the only

Figure 4. Snail neurone pH_i regulation and its effects on internal Cl^- and Na^+. The neurone was penetrated with a K_2SO_4 microelectrode and pH, Cl^- and Na^+ ion-sensitive microelectrodes. An acid load was imposed by changing the superfusate from CO_2-free to one equilibrated with 2.2% CO_2 in oxygen. (R.C. Thomas, unpublished data).

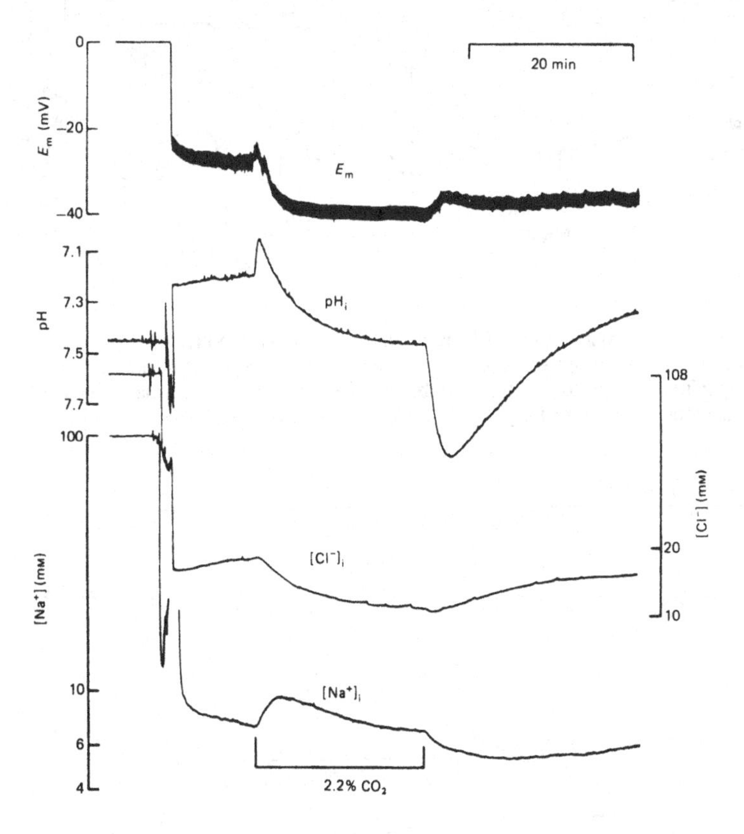

effect of external acidification shown by Moody; he did not explore the effect of acidification without acid loading.

If Figure 5(*a*) showed a complete dependence of pH_i-recovery on external Na, Figure 6 shows that pH_i-recovery is only partially blocked by either bicarbonate removal or a high level of SITS. The recovery from the second bicarbonate (and CO_2) application was much slower than before, but it was not blocked. On the other hand, SITS made little difference to pH_i recovery from acid injection in the absence of bicarbonate. Moody (1981) thus concluded that a fraction of pH_i regulation in

Figure 5. Crayfish neurone pH_i regulation. (*a*) The effects of Na removal on pH_i recovery from an acid load induced by exposure to NH_4Cl. (*b*) The effect of external acidification on pH_i recovery from CO_2-induced pH_i decrease. In both experiments bicarbonate was 5 mM except when NH_4Cl or 5% CO_2 was present. Modified from Moody (1981).

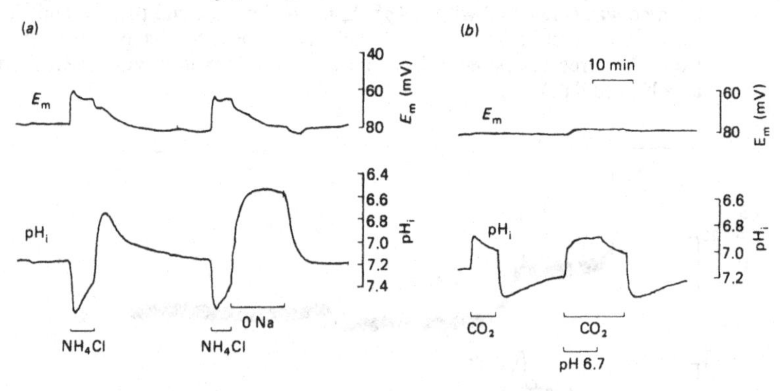

Figure 6. Crayfish neurone pH_i regulation: the effect of SITS with and without bicarbonate. Between and after the two periods in 25 mM bicarbonate the cell was superfused with nominally bicarbonate-free saline. Upper line shows HCl injections, lower record is pH_i. Modified from Moody (1981).

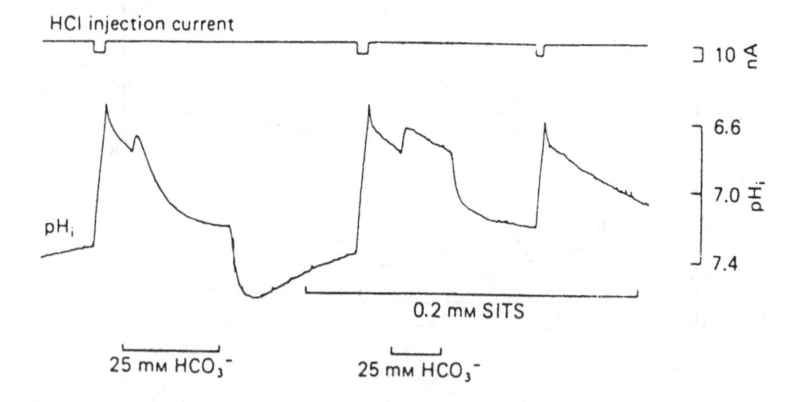

bicarbonate saline was SITS insensitive, probably being simple Na^+/H^+ exchange. The major component, however, was apparently very similar to that proposed for snail neurones.

Leech neurone pH_i regulation

In an extensive series of experiments on leech neurones, of which I show two, Rudiger Schlue, with my intermittent collaboration, concluded that their pH_i regulatory mechanism was very like that in crayfish neurones (Schlue & Thomas, 1985).

The main part of the experiment shown in Figure 7 was done with the preparation in bicarbonate saline. The pH_i was decreased by brief exposures to NH_4Cl. After the first, pH_i fell by about 0.3 units but rapidly recovered. After the second acid loading, pH_i recovery was completely blocked by removing external Na. When normal Na was replaced, pH_i recovered as rapidly as before.

Moody had been unable to obtain clear effects with amiloride, which blocks Na^+/H^+ exchange in mouse skeletal muscle (Aickin & Thomas, 1977) and many other preparations, including leech neurones. This is shown in Figure 8 where

Figure 7. Leech neurone pH_i regulation: the effect of removing external Na on pH_i recorery from NH_4Cl-induced acidification. Top trace membrane potential, bottom trace pH_i, both recorded with a double-barrelled pH microelectrode. Modified from Schlue & Thomas (1985).

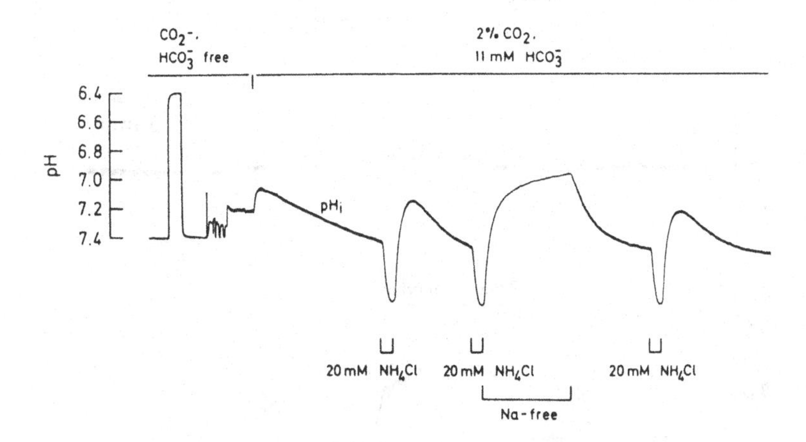

amiloride blocked recovery from an acid load if bicarbonate was absent. Bicarbonate overcame this block, although in amiloride and bicarbonate recovery was slower than in bicarbonate alone. Schlue and I therefore concluded that leech neurones had two mechanisms of pH_i regulation, both dependent on external Na. One, inhibited by amiloride, was presumably simple Na^+/H^+ exchange, while the other was probably the same as in snail neurones. The two mechanisms which explain pH_i regulation in snail, crayfish and leech neurones are shown in Figure 9.

In mammals and other vertebrates so far investigated, pH_i regulation appears to be mainly by an amiloride-sensitive Na^+/H^+ exchange mechanism, but too many experiments on these preparations have been done in bicarbonate-free salines. Quite why the two aquatic animals so far examined both have a dual mechanism is far from clear; perhaps an aquatic life requires a choice of mechanism.

The effect of external acidification

Of the three preparations described here, only snail neurones have been exposed to acid solutions more than once or twice. With snail neurones, not only does external acid tend to block pH_i regulation, it even reverses it if the acidification is large. Small acidifications have proportionally small effects, but do not have no effects at all (M.S. Szatkowski, personal communication).

Figure 8. Leech neurone pH_i regulation: the effect of amiloride in the absence and then presence of bicarbonate on pH_irecovery from an NH_4Cl-induced acidification. Except where shown cell superfused with bicarbonate free saline. Modified from Schlue & Thomas (1985).

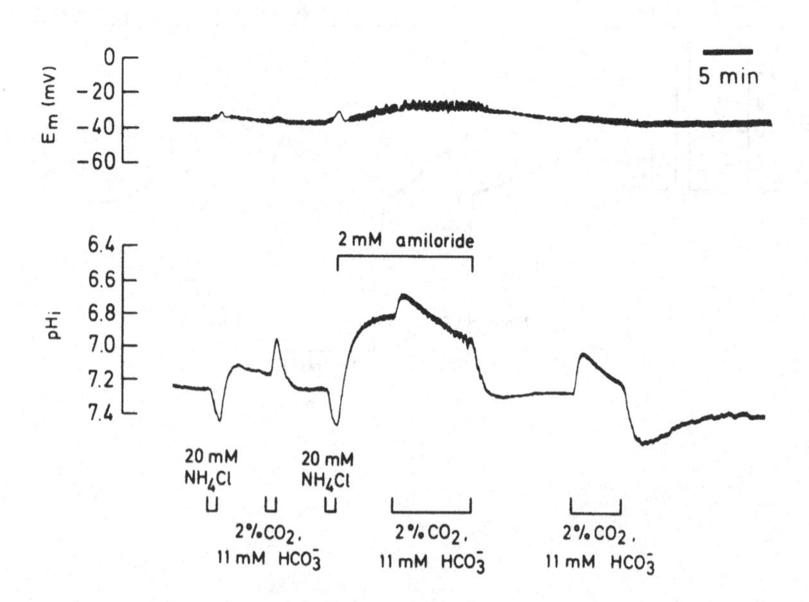

An experiment done over 10 years ago is shown in Figure 10. In those days I believed the normal snail blood pH was about 8, and I knew nothing about bicarbonate. Although this external pH was half a unit more alkaline than I now use, the steady-state pH_i was about 7.3, well in the normal range seen with an external pH of 7.5. Periods of 25 min in either pH 7 or pH 6 (weakly-buffered) solution slowly acidified the pH_i, quite reversibly.

A more recent experiment is shown in Figure 11. In this external pH was reduced from 7.7 to 6.7 in the presence of 2% CO_2, so bicarbonate was simultaneously reduced from 20 mM to 2 mM. During this external acidification pH_i fell steadily

Figure 9. Diagram of probable ionic mechanisms for pH_i regulation (acid extrusion) by snail, crayfish and leech neurones. The mechanism on the left is inhibited by SITS, that on the right is inhibited by amiloride in leech neurones.

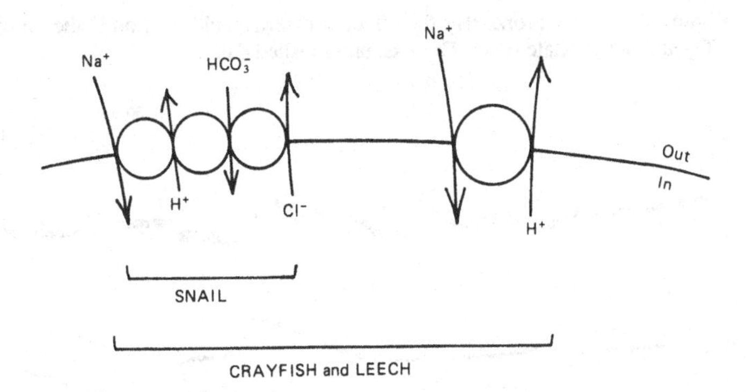

Figure 10. Snail neurone pH_i: the effect of external acidification in a bicarbonate-free superfusate. Except where shown the external pH was 8 (R.C. Thomas, unpublished data).

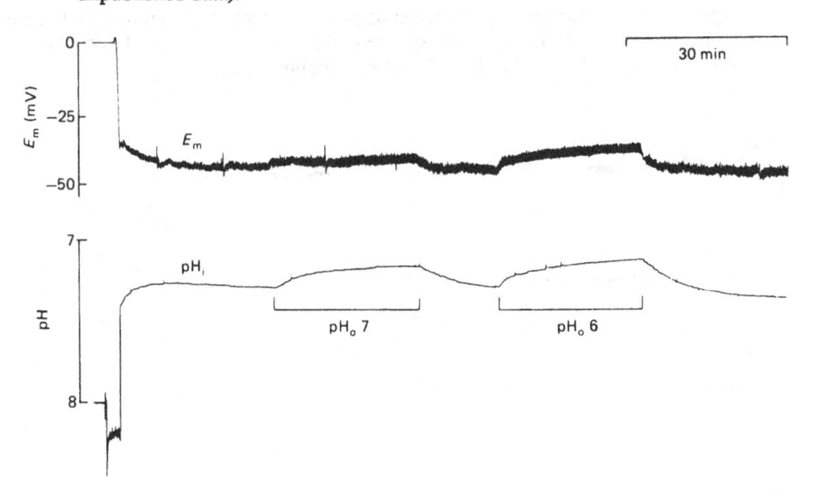

over the 20 min period, showing little sign of approaching a new steady state. From a number of experiments of this type it was clear that bicarbonate made pH_i fall faster when external pH was decreased.

Further investigation revealed that this was because the bicarbonate-dependent pH_i regulating system appeared to reverse quite easily. The evidence for this included finding an increase in internal Cl and a decrease in internal Na during the acidification, and the reduction of the acid influx by SITS. This last point is shown in Figure 12. In this experiment E_m was recorded by a K^+-sensitive microelectrode, so the calibrations are not very exact. Nevertheless it is clear that the extra pH_i decrease caused by exposure to pH 6.5 saline was largely eliminated by SITS, and somewhat reduced, especially when the lower buffering power is allowed for, by removing bicarbonate. Full details of the evidence on reversal is given in Evans & Thomas (1984).

Figure 11. Snail neurone pH_i: the effect of external acidification in the presence of CO_2 and bicarbonate (R.C. Thomas, unpublished data).

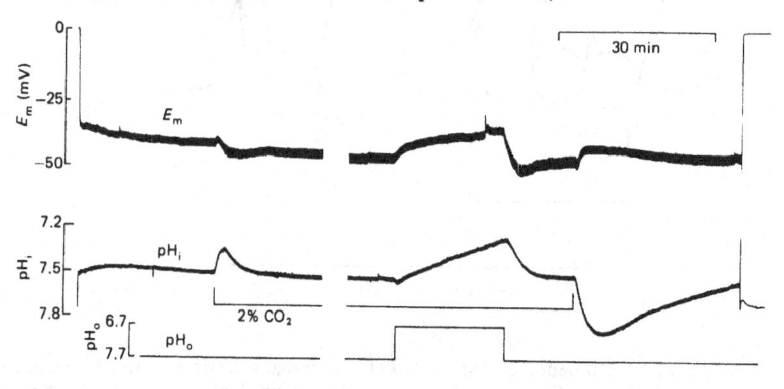

Figure 12. Snail neurone pH_i: the effect of bicarbonate and CO_2 removal and of the application of SITS, on the pH_i response to external acidification. Except where shown external pH was 7.5 (R.C. Thomas, unpublished data).

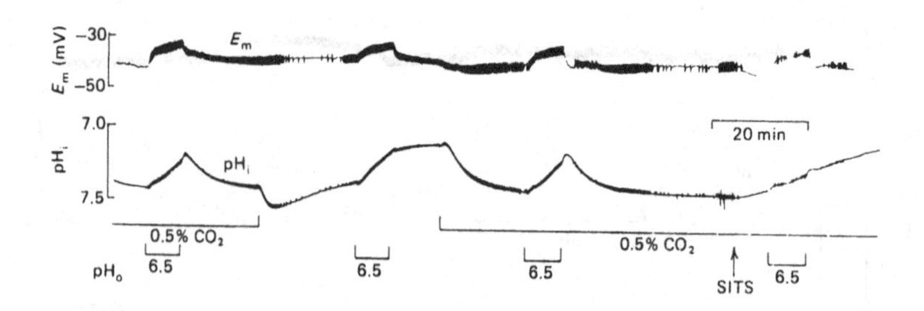

Conclusion

The results described above show that pH_i regulation depends on energy from the Na gradient being sufficient to drive H^+, HCO_3^- and Cl^- against their concentration gradients. Any change in H^+ and HCO_3^- gradient caused by extracellular acidification is likely to upset the pH_i regulating system by increasing the energy needed to extrude acid. Certainly relatively large extracellular pH decreases do cause pH_i to decrease.

My main take-home message must be that a stable, well-regulated pH_i depends absolutely on a constant extracellular pH. If an animal fails to keep its blood pH within normal limits, its pH_i regulating mechanisms will not be able to prevent pH_i changing as well.

References

Aickin, C.C. & Thomas, R.C. (1977). An investigation of the ionic mechamism of intracellular pH regulation in mouse soleus muscle fibres. *Journal of Physiology*, **273**, 295–316.

Boron, W.F. & De Weer, P. (1976). Intracellular pH transients in squid giant axons caused by CO_2, NH_3 and metabolic inhibitors. *Journal of General Physiology*, **67**, 91–112.

Evans, M.G. & Thomas, R.C. (1984). Acid influx into snail neurones caused by reversal of the normal pH_i-regulating system. *Journal of Physiology*, **346**, 143–54.

Moody, W.J. (1981). The ionic mechanism of intracellular pH regulation in crayfish neurones. *Journal of Physiology*, **316**, 293–308.

Roos, A. & Boron, W.F. (1981). Intracellular pH. *Physiological Reviews*, **61**, 296–434.

Schlue, W.R. & Thomas, R.C. (1985). A dual mechanism for intracellular pH regulation by leech neurones. *Journal of Physiology*, **354**, 327–38.

Thomas, R.C. (1977). The role of bicarbonate, chloride and sodium ions in the regulation of intracellular pH in snail neurones. *Journal of Physiology*, **273**, 317–38.

Thomas, R.C. (1978). *Ion-sensitive intracellular microelectrodes: how to make and use them*. London: Academic Press.

Thomas, R.C. (1984). Review Lecture: Experimental displacement of intracellular pH and the mechanism of its subsequent recovery. *Journal of Physiology*, **354**, 3–22.

Thomas, R.C. (1986). Intracellular pH. In *Acid-base balance*, ed. R. Hainsworth, pp. 50–74. Manchester: Manchester University Press.

CHRIS M. WOOD

The physiological problems of fish in acid waters

Introduction

The physiological effects of environmental acid stress on fish have been thoroughly reviewed in recent years (Muniz & Leivestad, 1980a; Fromm, 1980; Haines, 1981; Spry, Wood & Hodson, 1981; Brown, 1982; Leivestad, 1982; Wood & McDonald, 1982; McDonald, 1983a; Howells, Brown & Sadler, 1983; Howells, 1984); there are certainly not enough new data to justify yet another compendium. Instead, I will first summarize our current knowledge on the acute toxic mechanisms of pure acid stress to adult fish, and in so doing attempt to correct the widely held misconception that external water acidity must cause internal acidosis in the animal. By means of this brief summary, I hope to illustrate that external acidity has proven to be an exceptionally useful probe of normal physiological processes in freshwater teleosts. Secondly, I will describe some of our recent findings on the physiological responses to long term, low level acid stress, and acid-aluminium interactions, both of which may have greater ecological relevance than short term acid stress for ultimate survival in the wild.

Acute responses to pure acid stress

Background

Relatively short term depressions to pH = 4.0–4.5, usually as a result of snowpack melt in the spring, or highly acidic runoff in the summer and autumn, have often been observed in natural soft waters of both northern Europe and eastern North America (e.g. Jeffries, Cox & Dillon, 1979; Harvey *et al.*, 1981; Christophersen, Rustad & Seip, 1984; Marmorek *et al.*, 1985). Kills of adult fish have been documented in a matter of hours to days following the start of such acid surges (e.g. Leivestad & Muniz, 1976; Harvey & Lee, 1982; Muniz, 1984). While damage may involve additional factors such as aluminium, acidity itself is likely to be the dominant toxic agent under these acute conditions. In nature, such acid surges are almost exclusively a soft water problem (i.e. [Ca^{2+}] below 0.5, and generally less than 0.2 mequiv l^{-1}), because hard water catchments have sufficient bicarbonate alkalinity to neutralize precipitation acidity. Further laboratory and field data show that water [Ca^{2+}] is clearly protective against acid toxicity (e.g. Lloyd & Jordan, 1964;

Leivestad *et al.*, 1976; Wright & Snekvik, 1978; Graham & Wood, 1981; Brown, 1981, 1983). Unfortunately, as of 1982, only about 25% of all the physiological studies on fish under acute acid stress had been performed in soft water (Wood & McDonald, 1982). In view of calcium's well documented effects on membrane permeability and many other physiological processes, an understanding of its possible interaction with the mechanism(s) of acute acid toxicity seems essential.

Therefore, we have studied the physiology of the toxic syndrome and its relation to water [Ca^{2+}] in some detail using the rainbow trout (*Salmo gairdneri*) as a model. This species was selected because of its status as a standard reference animal in physiology and toxicology, and because of its relatively high sensitivity to acid stress (Grande, Muniz & Andersen, 1978). The fish were fitted with chronic, indwelling arterial and bladder catheters for the collection of blood and urine samples without disturbance. They were usually confined in darkened 'flux boxes' which could be operated as open flow-through or closed recirculating systems, the latter allowing measurements of ion and acidic equivalent exchanges between the animal and its pH-statted environment. The exposures were conducted in decarbonated waters of various defined composition at a mean pH *c.* 4.3, which approximates the seven day LC_{50} for *S. gairdneri* (Wood & Graham, 1981).

Blood acid–base and ionic status

Our initial experiments (McDonald, Höbe & Wood, 1980) showed that in soft water ([Ca^{2+})] = 0.3 mequiv l^{-1}), 50% mortality occurs over five days' exposure to pH *c.* 4.3 and that there are large equimolar depressions of plasma [Na^+] and [Cl^-], negligible blood acidosis, and only a very small accumulation of lactate. In contrast, hard water exposures ([Ca^2] = 1.9 mequiv l^{-1}) result in only 11% mortality, smaller overall ion losses from the plasma with [Na^+] falling much more than [Cl^-], and no lactate buildup. As lactate is a sensitive index of O_2 delivery problems, these results argue against the old idea of anoxia as the primary cause of death under these conditions (Westfall, 1945, Packer & Dunson, 1972) though it is clearly involved at environmentally unrealistic lower pH (<4.0; cf. Wood & McDonald, 1982) *and* sometimes at higher pH *when aluminum is present* (see below). Further experiments have shown that the influence of water hardness on the blood ion/acid-base response is simply a function of the water [Ca^{2+}], and not of other electrolytes which may vary with hardness (McDonald *et al.*, 1980; McDonald & Wood, 1981; Graham, Wood & Turner, 1982; McDonald, 1983b; C.M. Wood & S. Munger, unpublished data).
The summary in Figure 1 illustrates that blood metabolic acidosis and therefore pH depression are greatest at the highest water [Ca^{2+}], and are linearly reduced as [Ca^{2+}] falls. Thus, at about 0.2 mequiv l^{-1}, there is *no blood acidosis at all* after three days' exposure to pH = 4.3 in surviving fish, and at 0.05 mequiv l^{-1}, representative of many natural softwaters, there is a slight but significant *alkalosis* (see also Figure 2). In contrast, plasma ion depressions are least at the highest [Ca^{2+}], and become

greater as [Ca^{2+}] falls. The pattern is complex, with first Na^+ and then Cl^- loss increasing. At about 0.2 mequiv l^{-1}, the losses are equal, and by 0.05 mequiv l^{-1}, Cl^- loss slightly exceeds Na^+ loss. I will subsequently argue that death is associated with electrolyte loss, and that these ion and acid–base patterns result from fixed linkages dictated by the constraints of electrical neutrality and the strong ion difference concept (SID $\approx(Na^+ + K^+ - Cl^-)$; Stewart, 1978, 1983). A survey of other studies on salmonids in which blood acid–base status (very few) and plasma electrolytes (many) have been measured, during comparable acid stress at defined

Figure 1. The relationship between water calcium concentration and the extent of various acid-base and ionic disturbances in the arterial blood of rainbow trout exposed to pH *c*.4.3 for three days. *(a)* Change in arterial pH; *(b)* change in blood metabolic acid load; *(c)* change in plasma Na^+ concentration; *(d)* change in plasma Cl^- concentration. Means ± 1 SEM (n = 4–24). From McDonald, Hõbe & Wood (1980); McDonald (1983b); C.M. Wood & S. Munger (unpublished data).

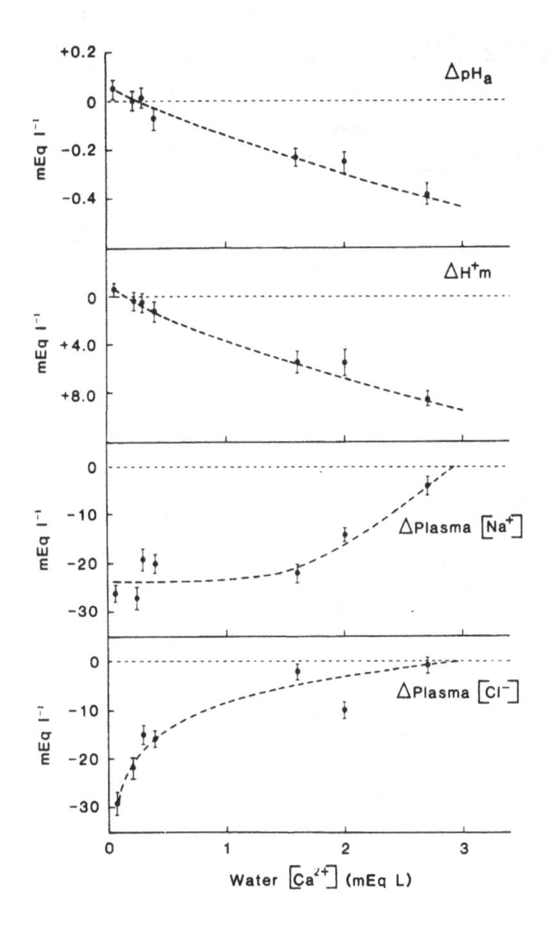

water [Ca^{2+}], indicates very good agreement with the relationships shown in Figure 1 (Leivestad & Muniz, 1976; Neville, 1979a,b; Leivestad, Muniz & Rosseland, 1980; Booth, Jansz & Holeton, 1982; Holeton, Booth & Jansz, 1983; Saunders *et al.*, 1983; Giles, Majewski & Hobden, 1984; Johnston *et al.*, 1984; Lacroix, 1985; Brown, Evans & Hara, 1986). It is particularly gratifying that the few measurements available on fish under acid stress in the wild, all at [Ca^{2+}] <0.2 mequiv l^{-1}, show Cl^- loss clearly in excess of Na^+ loss from the plasma, so that the unmeasured 'anion gap' (i.e. the difference between positive and negative electrolytes) has increased (Leivestad & Muniz, 1976; Leivestad *et al.*, 1976, 1980; Lacroix, 1985). This

Figure 2. The influence of three days' exposure to pH = 4.3 on extracellular and intracellular acid-base status in various tissues of the rainbow trout in waters of high and low [Ca^{2+}]. Means ± 1 SEM (n = 9–24). * indicates means significantly different (p <0.05) from respective control value (day 0). From C.M. Wood & S. Munger (unpublished data).

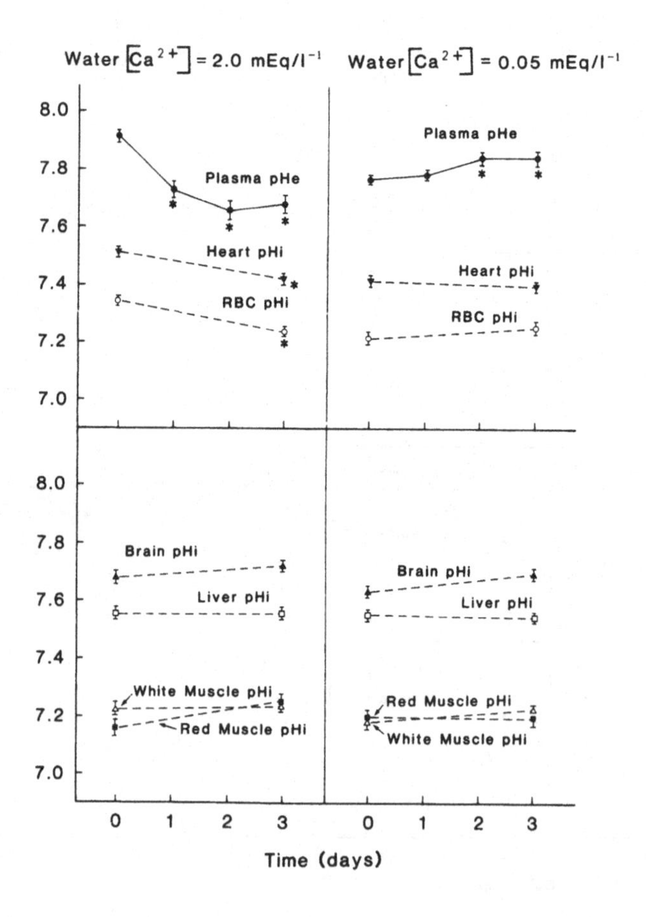

missing anion is probably HCO_3^-. Thus I suggest that during most acid surges in natural softwaters, the fish never experience metabolic acidosis, but rather unchanged internal pH or even metabolic alkalosis as plasma Cl^- losses equal or exceed Na^+ losses.

Tissue acid–base and ionic status

While extracellular patterns now seem relatively clear, the effects of environmental acidity on the pH of the various intracellular compartments (pH_i) has not been examined. Disturbances here might be far more damaging. We have addressed this problem by measuring the distribution of the weak acid ^{14}C–DMO (5,5-dimethyloxazolidine-2,4-dione; Waddell & Butler, 1959) in various tissues of trout in hard and very soft water. As expected, the blood plasma exhibits acidosis at high $[Ca^{2+}]$ and alkalosis at low $[Ca^{2}]$ during three days' acid exposure. However, there is no significant change in pH_i in any measured compartment in soft water (Figure 2). In hard water, the brain, liver, white muscle, and red muscle are well regulated in the face of extracellular acidosis, while the heart ventricle and red blood cells (RBC) do show small but significant decreases in pH_i. These results support the view that acid-base disturbance is unimportant under natural soft water conditions.

The tissues lose large amounts of ions, and in some studies gain significant quantities of water (Leivestad *et al.*, 1976; Neville, 1979b; McDonald & Wood, 1981; Fugelli & Vislie, 1982; Lee, Gerking & Jazierska, 1983; Giles *et al.*, 1984; Stuart & Morris, 1985). The relative intracellular reductions of Na^+, K^+ and Cl^- appear rather variable. As yet, nothing is known about possible effects of water $[Ca^{2+}]$ on these responses. Nevertheless, it seems reasonable to predict that tissue salt loss should follow, rather than precede, plasma loss, so the intracellular patterns should be dictated by the extracellular responses.

Branchial versus *renal responsibility for internal disturbances*

In freshwater fish, the two major sites of ion and acid-base regulation are gills and kidney (see Cameron, Heisler, this volume). The 'flux box' and bladder catheterization approach has been used to quantify their relative contributions to the observed internal effects (McDonald & Wood, 1981; McDonald, 1983b; McDonald, Walker & Wilkes, 1983; Wood, Wheatley & Hõbe, 1984; Wright & Wood, 1985). When the urine is collected and analysed separately, it is a relatively safe assumption that the remaining exchanges with the environment are almost entirely branchial. Cumulative measures over four days' exposure to pH *c*. 4.3 (Table 1) show that the gills are the dominant site of net flux in both hard and soft water, accounting for *c*. 80% of the major electrolyte losses, and all of the net acidic equivalent uptake ('H^+', i.e. the sum of titratable acid and ammonia fluxes, signs considered; Maetz, 1973). The kidney's contribution is negligible in soft water, probably because its

Table 1. *Cumulative net fluxes (μequiv kg^{-1}) of Na^+, K^+, Cl^- and 'H^+' (acidic equivalents) at the gills and kidney of rainbow trout, and their relative contributions to overall balance over four days' exposure to pH c.4.3 in hard ([Ca^{2+}] c. 2.0 mequiv l^{-1}; n = 10–14) vs soft water ([Ca^{2+}] = 0.22 mequiv l^{-1}; n = 4). Note the correspondence (underlined) between net H^+ uptake and the net SID flux ($Na^+ + K^+ - Cl^-$) at the gills in both media.*

	Gills		Kidney	
Soft water				
Na^+	–10,400	(–89%)	–1300	(–11%)
K^+	–2500	(–86%)	–400	(–14%)
Cl^-	–9500	(–89%)	–1150	(–11%)
H^+	+4850	(+114%)	–600	(–14%)
$Na^+ + K^+ - Cl^-$	–3400			
Hard water(a)				
Na^+	–12,550	(–87%)	–1950	(–13%)
K^+	–3900	(–74%)	–1350	(–26%)
Cl^-	–6500	(–69%)	–2900	(–31%)
H^+	+10,450	(+142%)	–3100	(–42%)
$Na^+ + K^+ - Cl^-$	–9950			

(a)Total fluxes are slightly overestimated in the hard water relative to the soft water group, as the means unavoidably include contributors from 4 non-surviving fish in the former only

From McDonald & Wood (1981), McDonald (1983b) and unpublished data.

volume output progressively falls as fluid shifts into the muscle ICFV, plasma oncotic pressure increases, and the osmotic gradient for water entry at the gills declines. In hard water, these effects are less severe and there is actually an initial diuresis. Thus the kidney accounts for a greater, though still small, fraction of electrolyte loss, and plays a significant role in acid–base compensation, excreting about 30% of the net 'H^+' entry at the gills through a combination of NH_4^+ and phosphate (titratable acid) output. This is therefore one of the relatively few circumstances where the teleost kidney plays an important role in net H^+ output (cf. Heisler, this volume).

The gill fluxes are clearly responsible for the [Ca^{2+}]-dependent patterns of blood ion and acid–base disturbance (Figure 1). Thus in soft water ([Ca^{2+}] = 0.22 mequiv l^{-1}), the approximately equal branchial losses of Na^+ and Cl^- (Table 1) explain their equal depressions in the blood plasma, and the small H^+ uptake explains the minimal blood acidosis. Conversely, in hard water ([Ca^{2+}] *c.* 2.0 mequiv l^{-1}), the excess of Na^+ over Cl^- loss and the large net H^+ uptake at the gills explain the corresponding blood patterns. K^+ is the other major electrolyte moving across the gills during acid exposure; K^+ losses across the gills largely originate from the ICFV and therefore have little influence on plasma [K^+], which actually increases slightly. Ca^{2+} fluxes are very small and little affected by low pH (Hōbe, Laurent & McMahon, 1984). An important feature of the branchial fluxes is the approximate equivalence between strong cation minus anion flux (SID $\approx Na^+ + K^+ - Cl^-$) and the net H^+ uptake, in both soft and hard water (Table 1). This again fits with the ideas of Stewart (1978, 1983).

Branchial exchange mechanisms

Unidirectional flux measurements with $^{22}Na^+$ and $^{36}Cl^-$ have shown that both influx and efflux components are affected, and net flux becomes generally negative, when water pH moves more than about one pH unit above or below circumneutral (Figure 3). Influx is more sensitive to acid than efflux. Effects on both become increasingly severe below pH *c.* 5.0. By pH *c.* 4.0, influx is virtually obliterated and efflux greatly stimulated, the latter accounting for the major portion of the net losses. These actions are immediate, occurring within minutes. The data of McWilliams & Potts (1978) and McWilliams (1980a,b, 1982a,b) on Na^+ fluxes in *S. trutta* are in general accord.

The inhibitory action of acid on Na^+ influx sheds some light on the controversies whether sodium uptake is via Na^+/NH_4^+ or Na^+/H^+ exchange, and whether branchial ammonia excretion is via Na^+/NH_4^+ exchange or NH_3 diffusion (cf. Cameron & Heisler, 1983). Acute exposure to pH *c.* 4.0 abolishes Na^+ influx in trout, but reduces ammonia efflux by only *c.* 30%; thus ammonia excretion can occur by both mechanisms (Wright & Wood, 1985). In the absence of Na^+ uptake, only NH_3 diffusion persists, driven by a measured, elevated P_{NH_3} gradient from blood to water. This situation provides an estimate of the diffusivity (D_{NH_3}) of the gills to NH_3. (Cameron & Heisler (1983) estimated a 50% higher value for D_{NH_3}, but their methodology did not eliminate the possibility of Na^+/NH_4^+ exchange.) Application of our D_{NH_4} value to other conditions, where the P_{NH_3} gradient and total ammonia excretion are known, allows estimation of the relative proportions of the latter moving as NH_3 and NH_4^+. In general these calculations show that Na^+/NH_4^+ exchange dominates over NH_3 diffusion under control, but not under acidic conditions. Furthermore, there is a linear, proportional relationship (though with considerable scatter) between NH_4^+ net flux and Na^+ influx (Figure 4), suggesting that the majority of Na^+ uptake occurs through Na^+/NH_4^+ exchange. This whole area remains controversial; consult Heisler (this volume) for an opposing view.

The blockade of Na^+ influx by acid is almost certainly due to a direct competition of H^+ with Na^+ for the transport sites and/or access channels to the carrier. Cl^- influx is usually considered an entirely separate process. Its inhibition by H^+ has never been adequately explained; possibilities include the depletion of HCO_3^- or OH^- for coupled exchange in the epithelial cells, access blockade or damage to the carrier, or the disquieting prospect that Na^+ and Cl^- transport are in some way linked. The massive stimulations of efflux in the face of obliterated influx can only result from increased passive diffusion, for by definition, the exchange diffusion components must be eliminated (Wood *et al.*, 1984). An increased diffusive permeability due to leaching of Ca^{2+} away from the paracellular channels in the branchial epithelium now

Figure 3. The influence of acute exposure (first 3 h) to water of different pH levels on branchial unidirectional and net fluxes of Na^+ and Cl^- in the rainbow trout (water [Ca^{2+}] *c.* 2.0 mequiv (l^{-1}). The control (acclimation) pH was 7.8–8.0. The upward-facing bars are influx, downward facing bars are efflux, and cross-hatched bars are net flux. Means ± 1 SEM ($n = 6-12$). Composite data from laboratories of C.M. Wood and D.G. McDonald.

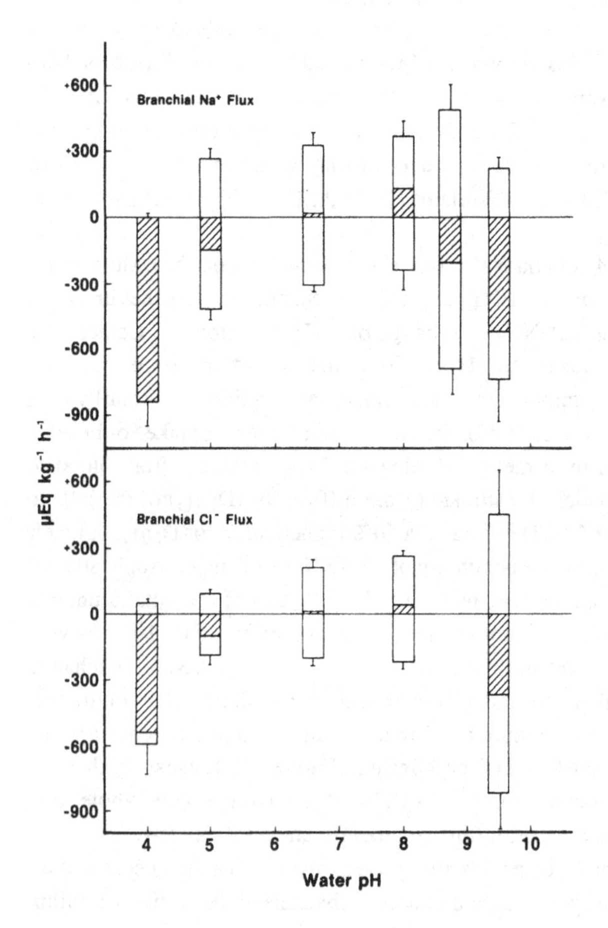

seems the most likely explanation (McWilliams, 1983; Marshall, 1985; McDonald, Reader & Dalziel, this volume).

McDonald *et al.* (1983) have studied time-dependent changes in branchial fluxes and the influence of water [Ca^{2+}] in great detail (Figure 5). After two days' acid exposure, surviving trout show a remarkable compensation in returning Na^+ and Cl^- efflux rates to control levels, or even below. Influx rates exhibit only a small recovery, so net fluxes, while greatly corrected, remain negative. Thus the overall balance, which was initially negative due to efflux stimulation, remains negative due to the persistent depression of influx. These adjustments may well be of hormonal origin, with prolactin and cortisol being the two most likely candidates (Wendelaar Bonga, van der Meij & Flik, 1984; Brown *et al.*, 1984, 1986). As before, there is a

Figure 4. The relationship in rainbow trout betwen branchial sodium influx (J_{in}^{Na+}) and branchial ammonium ion net flux ($J_{net}^{NH_4^+}$), as estimated from the measured P_{NH_3} gradient from blood to water (ΔP_{NH_3}), the diffusivity of the gills to NH_3 (D_{NH_3}), and the total ammonia excretion (J_{net}^{Amm}) by the equation given. Each point represents the mean of 4–9 fish under a different water pH treatment (water [Ca^{2+}] *c.*1.6 mequiv l^{-1}). From Wright & Wood (1985) and C.M. Wood (unpublished data).

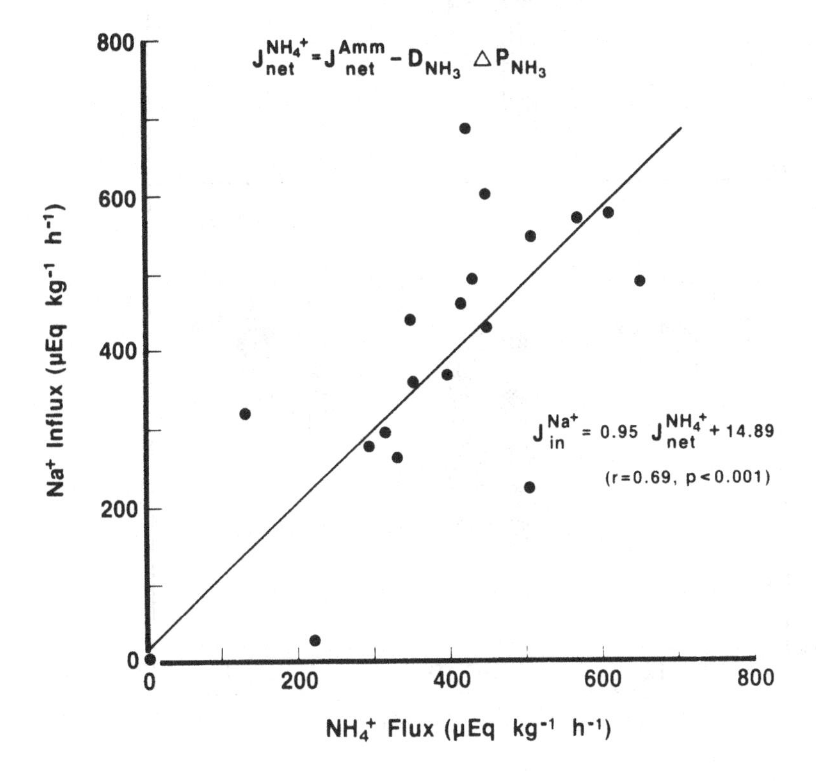

Figure 5. Changes in unidirectional and net Na^+, Cl^-, and net 'H^+' (acidic equivalent) fluxes across the gills of rainbow trout upon acute exposure to pH = 4.2 (first 4 h) and after 42 h of continuous exposure, in very hard water ($[Ca^{2+}]$) = 5.7 mequiv l^{-1}) and soft water ($[Ca^{2+}]$ = 0.24 mequiv l^{-1}). The upward-facing bars are influx, downward-facing bars efflux, and stippled bars net flux. Means ± 1 SEM (n = 4). * indicates means significantly different ($p < 0.05$) from respective control value. Redrawn from McDonald, Walker & Wilkes (1983).

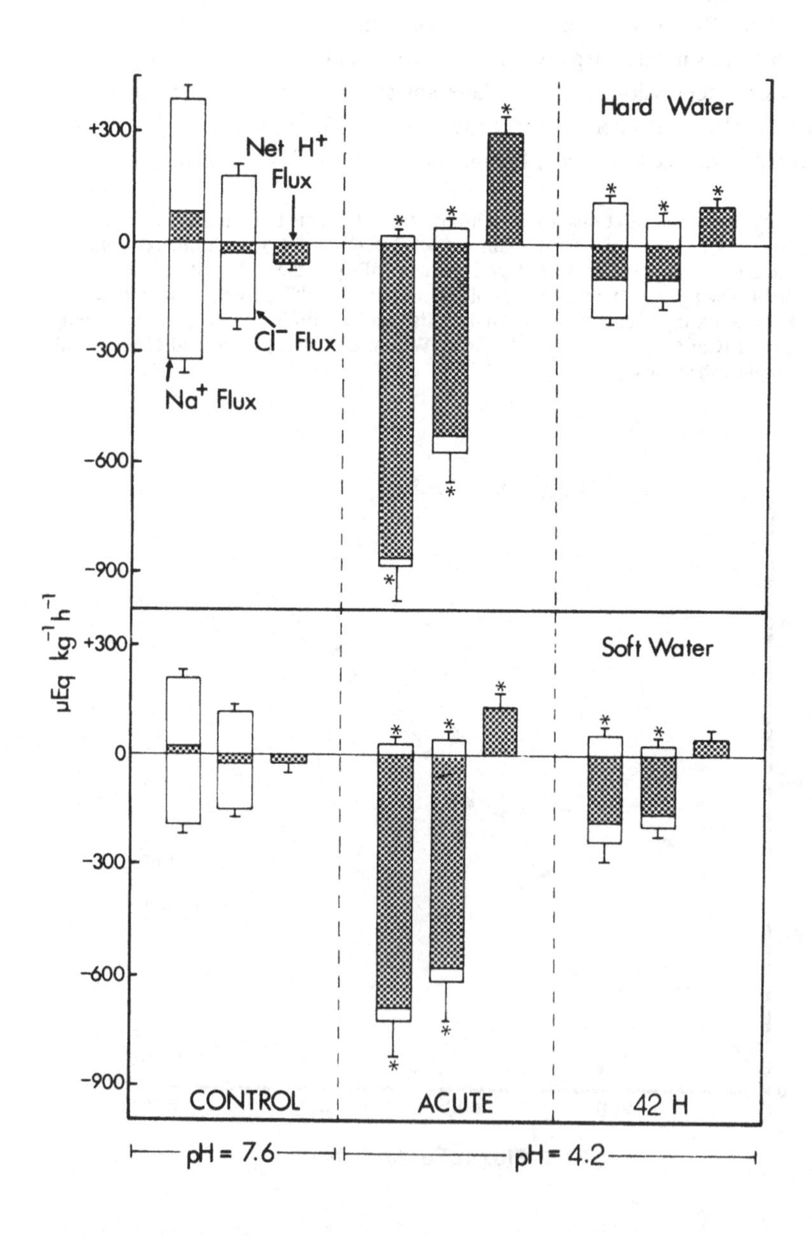

marked effect of water [Ca^{2+}] on the ratio of strong cation to strong anion loss, with the net SID flux corresponding well with the net H^+ flux (Figure 5). Initially this is due to a much lower Cl^- (relative to Na^+ + K^+) efflux in hard water, and after two days to a slightly better recovery of Na^+ influx.

The thesis I am developing is that the net H^+ flux across the gills must equal the SID flux. Additional empirical support comes from a study which followed the clearance of large acidic and basic equivalent loads across the gills of trout at neutral pH (Wood *et al.*, 1984). Here the only two strong electrolytes moving in significant quantities were Na^+ and Cl^-, and there was a linear, proportional relationship between net [Na^+ – Cl^-] flux and net H^+ flux over a wide range of positive and negative values (Figure 6). This correlation occurred irrespective of whether these fluxes were achieved through modifications of influx, efflux, or both components.

Figure 6. The relationship betwen the net flux of 'H^+' (acidic equivalents) and the net Na^+– net Cl^- (i.e. ≈ SID) flux across the gills of rainbow trout at neutral pH (water [Ca^{2+}] *c.* 1.6 mequiv l^{-1}). The trout were clearing large acid or base loads induced by experimental hyperoxia and return to normoxia. Individual values (48) for 12 fish under four different treatments are shown. Redrawn from Wood, Wheatly & Hõbe (1984).

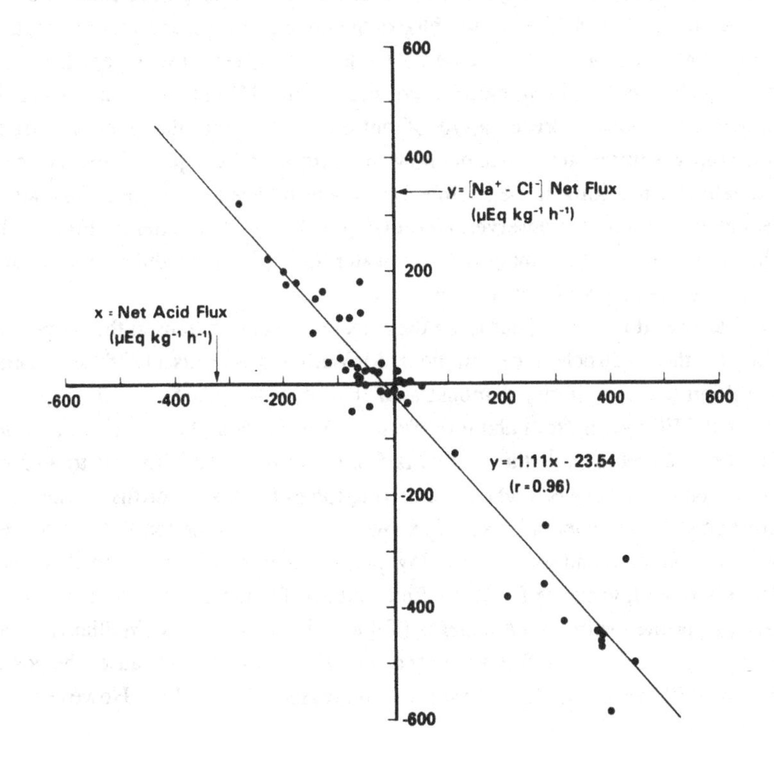

A simple gill model

Modern acid–base theory (Stewart, 1978, 1983) holds that solutions separated by membranes (e.g. the blood of the fish and the external water) can only interact in acid–base terms by processes which alter the values of their independent variables, which are the difference between strong cation and anion concentration (SID), the P_{CO_2}, and the total weak acid present (mainly protein in blood). In the acid stress situation, changes in the latter two are minor relative to those in SID. Viewed in this light, a net H^+ flux need not involve the physical movement of H^+ across the gills, for water is an infinite source or sink of acidic equivalents, and the acid gradient (a dependent variable) between the blood and the environment is immaterial to this flux. All that matters are the relative movements of strong cations and anions (*no matter how these occur*), for they will automatically cause a positive or negative net H^+ flux. Any acid–base disturbance which develops in the animal is therefore an unavoidable consequence of ionoregulatory disturbance.

The application of these ideas to the gill under acute acid stress is summarized in the simple model of Figure 7. Transcellular uptake processes for sodium (Na^+/NH_4^+, H^+, and self-exchange), chloride (Cl^-/HCO_3^-,OH^-, and self-exchange) and potassium (mechanism unknown; Eddy, 1985) occur in parallel to paracellular routes of diffusive loss. Acute acid exposure blocks the uptake of ions, and more importantly greatly stimulates their diffusive losses. Under high $[Ca^{2+}]$, low pH conditions, the leakage pathways are cation selective, constraining net H^+ uptake. Under low $[Ca^{2+}]$, low pH conditions representative of natural soft water, the overall diffusive permeability is further increased, but now anion permeability equals or exceeds cation permeability, resulting in negligible H^+ uptake or even excretion. This scheme adequately explains the observed blood (Figure 1) and flux patterns (Figures 3, 5; Table 1), but the mechanism(s) by which water $[Ca^{2+}]$ might modulate relative anion to cation losses at low pH remain unknown.

The answer does not appear to be the effect of changing transepithelial potential (TEP) on the electrochemical gradient. McWilliams & Potts (1978; see Potts & McWilliams, this volume) demonstrated that at water $[Ca^{2+}]$ *c.* 2.0 mequiv l^{-1}, branchial TEP moves from about 0 mV to +15 mV when pH falls to 4.3, while at $[Ca^{2+}]$ *c.* 0.2 mequiv l^{-1}, the change is from approximately –13 mV to +15 mV. Calculation of the net electrochemical driving forces for passive diffusion (i.e. TEP – Nernst equilibrium potential; Table 2) shows that they increase for Na^+ and decrease for Cl^- in both hard and soft water at low pH, and that in soft water, the changes are relatively more favourable for Na^+ efflux, and less favourable for Cl^- efflux. This is exactly opposite the observed patterns (Table 1; Figures 3, 5). McWilliams & Potts (1978) proposed that the diffusive entry of free H^+ ions at low pH causes the positive shift in TEP, which in turn drives the increased Na^+ efflux. However, their

Figure 7. A simple model of ion and acid–base exchanges across the gills of the rainbow trout and the effects of environmental acidity in hard water (high [Ca^{2+}]) and soft water (low [Ca^{2+}]). The symbol J represents flux.

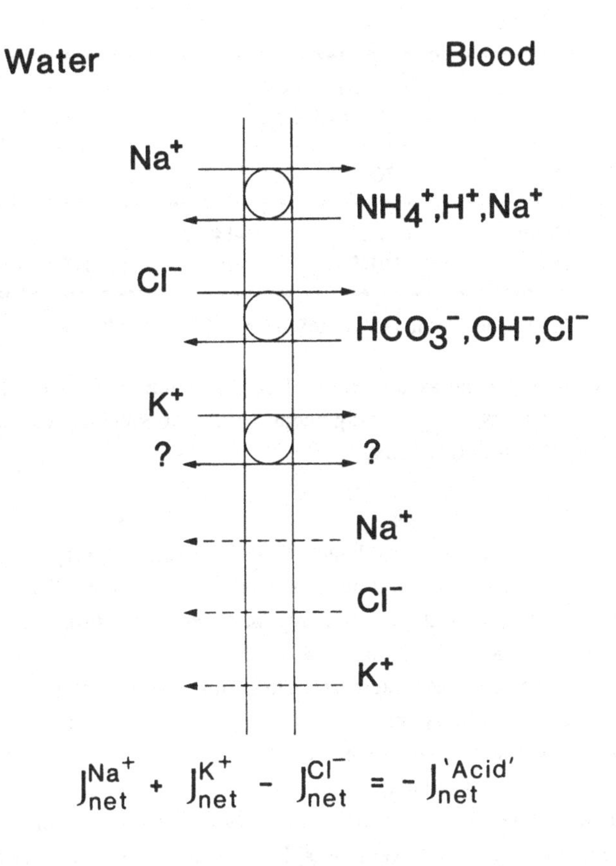

<u>High Calcium, Low pH</u>

$Na^+ + K^+$ Loss $>$ Cl^- Loss ➡ Net 'Acid Uptake'

<u>Low Calcium, Low pH</u>

$Na^+ + K^+$ Loss $\simeq$ Cl^- Loss ➡ $\sim$0 'Acid Flux'

Table 2. *Calculated net electrochemical driving forces (mV) for passive diffusion of* Na^+ *and* Cl^- *(i.e.* $F = TEP - \frac{RT}{zF} \ln \frac{[Na^+]int}{[Na^+]ext}$*) across the gills of trout under control and acid (pH c. 4.3) conditions in hard and soft water.*

	Hard water ($[Ca^{2+}]$ = 2.0 mequiv l^{-1})		Soft water ($[Ca^{2+}]$ = 0.2 mequiv l^{-1})	
	Na^+	Cl^-	Na^+	Cl^-
Control	+146.5	−142.5	+133.5	−155.5
Acid	+161.5	−127.5	+161.5	−127.5

Recalculated from data of McWilliams & Potts (1978).

arguments hinged on assumptions of constant branchial permeability to Cl^-, and diffusive H^+ entry as the driving agent, neither of which appear tenable in light of current results and SID theory.

The cause of death

The key parameters diagnostic of incipient mortality are plasma $[Na^+]$ and $[Cl^-]$. Once either or both fall more than 30% below normal (Na^+ *c.* 155, Cl^- *c.* 135 mequiv l^{-1}) in the rainbow trout, then death ensues within hours. The proximate mechanism is closely associated with this ionic dilution. While acidosis can be confidently dismissed as a cause of death during pure acid stress in natural soft water, this may not be entirely true in hard water (or when aluminium is involved – see below). In very hard water, the extent of acidosis (Figure1(*a*),(*b*)) may approach that occurring after exhaustive activity, where it has been implicated as a contributor to post-exercise lethality (Graham *et al.*, 1982; Wood, Turner & Graham, 1983; Cameron, this volume). Nevertheless, I have rarely seen hard water acid deaths in the absence of a 30% electrolyte loss; acidosis is likely to be a secondary factor at most.

It is doubtful that a 30% decrease in plasma NaCl levels could of itself cause death (by interfering with nerve or muscle function, for example), for anadromous salmonids experience similar dilution during the seawater to freshwater migration (Miles, 1971), and even over seasonal cycles within freshwater (Stuart & Morris, 1985). Indeed, they may tolerate much greater ion losses during longer term low level acid stress (see below). Rather, we have proposed that rapid ion losses during acid surges entrain haematological and fluid volume disturbances which ultimately kill the fish through circulatory failure (Milligan & Wood, 1982; Wood & McDonald, 1982). This theory has yet to be challenged, so some of the key points and evidence (Table 3) are reiterated here to stimulate debate.

Table 3. *Changes in various haematological, fluid volume, and cardiovascular parameters after three days' exposure to pH c. 4.3 in rainbow trout. Means ± 1 SEM.*

	Control ($n = 8-32$)	Acid – Day 3 ($n = 7-32$)
Haematocrit (%)	21.3 ± 2.9	47 3 ± 7.8[a]
Haemoglobin (g 100 ml^{-1})	5.95 ± 0.76	8.91 ± 0.95[a]
Plasma Protein (g 100 ml^{-1})	2.27 ± 0.08	3.46 ± 0.15[a]
Mean Cell [Haemoglobin] (g 100 ml^{-1}RBC)	29.43 ± 3.70	16.70 ± 1.48[a]
Mean [Haemoglobin] per RBC (pg $cell^{-1}$)	92.26 ± 9.45	67.49 ± 2.79[a]
RBC Count (10^{-4} mm^{-3})	65.6 ± 6.6	108.5 ± 13.1[a]
Reticulocyte Count (10^{-2} mm^{-3})	18.9 ± 5.4	83.6 ± 15.8[a]
Spleen Weight (g)	0.55 ± 0.07	0.34 ± 0.06[a]
Spleen [Haemoglobin] (g $spleen^{-1}$)	0.18 ± 0.04	0.06 ± 0.01[a]
Plasma Volume (ml kg^{-1})	43.5 ± 1.5	31.4 ± 1.7[a]
Whole Body ISFV (ml kg^{-1})	233.2 ± 8.9	169.9 ± 6.8[a]
Whole Body ECFV (ml kg^{-1})	273.5 ± 8.8	199.9 ± 6.8[a]
Whole Body ICFV (ml kg^{-1})	459.1 ± 9.4	524.8 ± 10.8[a]
Total Body Water (ml kg^{-1})	730.8 ± 4.5	731.1 ± 5.6
Muscle ICFV (ml kg^{-1} muscle)	690.0 ± 3.5	726.4 ± 4.5[a]
Arterial Blood Pressure (cm H_2O)	31.0 ± 1.0	39.2 ± 1.4[a]
Heart Rate (# min^{-1})	52.3 ± 2.2	62.9 ± 2.5[a]

[a] $p < 0.05$ from control.
From Milligan & Wood, 1982.

After three days' exposure to pH *c.* 4.3, surviving trout exhibit elevated heart rates and arterial blood pressures (Table 3); pharmacological evidence attributes these to catecholamine mobilization, a common stress response in fish (Mazeaud & Mazeaud, 1982). At the same time, there is a large shift of fluid from the extracellular (ECFV) to the intracellular fluid volume (ICFV), most of which can be found in the intracellular compartment of the white muscle mass; total body water does not change. This shift presumably occurs because of a dynamic imbalance: ions are lost more quickly from the ECF than from the ICF, entraining an osmotic flux of water in the opposite direction. As the ECFV contracts, both interstitial fluid volume (ISFV) and plasma volume fall by 27%. Haematocrit, haemoglobin, plasma protein, RBC, and reticulocyte (immature RBC) count all increase, while the mean cell haemoglobin concentration drops markedly. Thus the remarkable 125% increase in haematocrit has a complex origin. In order of importance, the three most important contributors are swelling of the RBCs due to the fall in plasma osmotic pressure, the reduction in plasma volume, and a discharge of stored RBCs from the spleen, again probably due to catecholamines. The elevated haematocrit, together with the higher plasma protein level, causes a doubling of blood viscosity. In combination with adrenergic stimulation, this thicker blood must undoubtedly contribute to increased peripheral resistance, arterial pressure, and cardiac workload at a time of severe hypovolaemia. The stage is set for circulatory collapse. The data of Table 1 are from *survivors only*. The few available measurements on fish just prior to death show haematocrits above 70%, up to three-fold elevations in viscosity, and blood volumes less than half the control level!

While this remains a theory rather than fact, more recent evidence adds additional support. Thus Lacroix (1985) found haematocrits as high as 70% in acid-stressed fish in the wild. Lee, Gerking & Jazierska (1983) showed that ion losses from the muscle ICFV do lag behind those from the ECFV, resulting in an expansion of muscle water. Stuart & Morris (1985) confirmed the contraction of whole body ECFV and reciprocal expansion of ICFV, while Giles *et al.* (1984) confirmed the increase in heart rate. McDonald (1983b) demonstrated that it is the *rapidity* of ion loss, rather than the absolute amount lost, which is the key to mortality; this supports the idea of a dynamic imbalance between ECFV and ICFV. Indeed, when salmonids lose ions more slowly under milder acid stress, they can tolerate plasma NaCl reductions well in excess of the 30% figure, probably through the mobilization of other osmo-effectors such as glucose and taurine (Fugelli & Vislie, 1982; Johnston *et al.*, 1984; Giles *et al.*, 1984). The remaining data needed to prove the theory are measurements of plasma catecholamines and continuous cardiovascular recordings prior to and during acute acid death.

Chronic responses to pure acid stress

Background

Chronic sublethal stress caused by pH levels (4.6–6.0) which are not acutely toxic may have more subtle, and ultimately more serious, effects on wild populations (Kelso *et al.*, 1986). It is important to know whether fish can acclimate to such conditions. Trout collected from naturally acidified waters show greater physiological resistance to acid challenge in the laboratory (McWilliams, 1980a, 1982, 1983; Brown, 1981), but it is unclear whether this results from acclimation, or simply from natural selection, for acid tolerance is a heritable trait (Leivestad *et al.*, 1976; Swarts, Dunson & Wright, 1978). In general, the results of laboratory tests on acclimation have been negative or equivocal, but few have been performed under environmentally realistic conditions (discussed in McWilliams, 1980a,b). The most convincing positive result is McWilliams' own study showing that *S. trutta* exposed to acidity for 42 d show a progressive recovery of plasma Na^+ and Cl^- levels, while Na^+ influx becomes more resistant to inhibition by lower pH levels. However the study was conducted in hard water ($[Ca^{2+}] = 1.0$ mequiv l^{-1}) under extremely mild acid stress (pH = 6.0).

Long-term studies in soft water

Recently, several long-term studies (one to three months) have been carried out on juvenile Atlantic salmon (*S. salar*) held in the laboratory under environmentally realistic pH levels (4.5–5.2) in natural soft water (Saunders *et al.*, 1983; Johnston *et al.*, 1984; Lacroix, Gordan & Johnston, 1985). The reported effects are severe, including retarded growth and development, reduced branchial ATPase activities, impaired ionoregulation, and failure of smoltification. Nevertheless, all studies indicate partial recovery during continued acid stress in those fish which survive. That this recovery is not complete agrees with the findings of Leivestad *et al.* (1976) that brook trout (*Salvelinus fontinalis*) held for one year at pH *c*.4.6 still exhibit significantly depressed [Na^+] and [Cl^-] in plasma.

To clarify this question, we have followed ionic and acid–base exchanges in individual rainbow trout exposed to pH *c*. 4.8 for three months in flowing soft water ([Ca^{2+}] = 0.05 mequiv l^{-1}). As predicted from previous short term studies (see above), the fish lose large, approximately equal amounts of Na^+ and Cl^- (Figure 8) over the first few weeks, mainly due to influx inhibition, and actually excrete small amounts of H^+. By week 4, positive net ion fluxes are achieved, and by week 12, a situation approximating control balance is restored through both recovery of influx and reduction of efflux components (Figure 8). We are not yet sure whether plasma ion levels are completely corrected, but brook trout held for a similar period at pH

Figure 8. The influence of three months' exposure to pH *c*.4.8 on Cl^- balance in rainbow trout held in flowing soft water ([Ca^{2+}] = 0.05 mequiv l^{-1}). The upward facing bars are influx, downward facing bars efflux, and stippled bars net flux. Means ± 1 SEM (n = 7–12). * indicates mean significantly different ($p < 0.05$) from respective control value. From C. Audet, S. Munger & C.M. Wood (unpublished data).

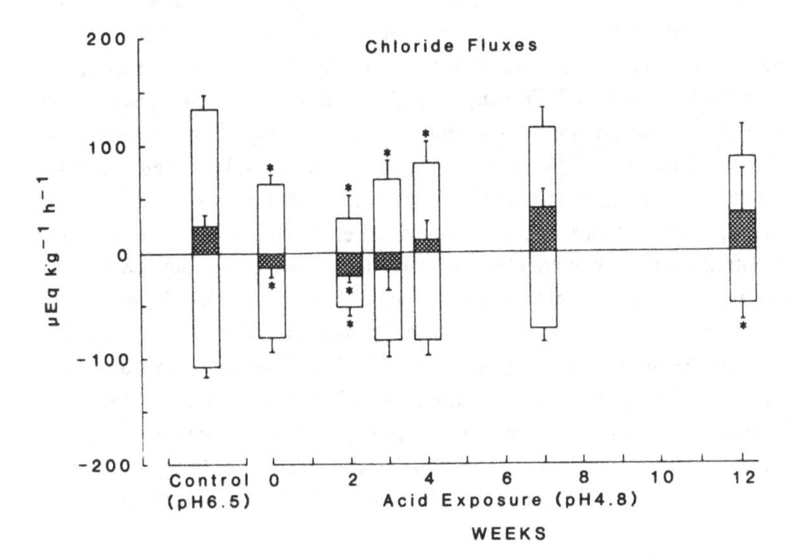

Table 4. *Plasma ions, osmolarity, and blood haemoglobin concentration in brook trout held for 10 weeks in flowing softwater ($[Ca^{2+}]$ = 0.025 mequiv l^{-1}) at various pH and aluminum levels. Means ±1 SEM (n = 6)*

pH	6.5	5.2	5.2	5.2
Aluminium (ppb)	0	0	75	150
Plasma Na^+ (mequiv l^{-1})	138.1 ± 1.1	141.0 ± 1.7	132.6 ± 1.9	141.3 ± 1.1
Plasma Cl^- (mequiv l^{-1})	130.0 ± 1.8	132.8 ± 1.2	127.5 ± 1.2	134.0 ± 0.7
Plasma osmolality (mosmol kg^{-1})	316.0 ± 5.5	323.3 ± 4.5	306.7 ± 1.6	313.2 ± 6.7
Blood haemoglobin (g 100 ml^{-1})	10.4 ± 0.4	10.5 ± 0.4	9.7 ± 0.4	9.9± 0.5

From Wood, MacDonald, Booth, Simons, Ingersoll & Bergman (1988).

c. 5.2 do show such compensation (Table 4). The mechanisms involved are unknown, but proliferation of chloride cells on the secondary lamellae (Leino & McCormick, 1984) and changes in endocrine status (Wendelaar Bonga *et al*., 1984; Brown *et al*., 1984) are attractive possibilities.

Responses to acid/aluminium stress

Background

Unfortunately, all of the preceding treatment may be a gross oversimplification with respect to many field situations. Aluminium is the third most abundant element in the earth's crust with a solubility which increases exponentially as pH falls below about 5.6. Its mobilization into acidified soft waters has been identified as the 'missing link' which explains why mortality in the field is often much greater than that in the laboratory at the same pH and $[Ca^{2+}]$ (Muniz & Leivestad, 1980a,b; Schofield & Trojnar, 1980; Baker & Schofield, 1982; Howells *et al*., 1983). The aqueous chemistry of aluminium is extremely complex. As a first approximation, Al bound to ligands (organic acids, fluoride, sulphate etc.) will be ignored as only inorganic monomeric Al seems to contribute to acute toxicity. However its speciation alone is complex and extremely pH-dependent: cationic hydroxides predominate around pH 5.0 ($Al(OH)_2^+$, $Al(OH)^{2+}$) but are largely replaced by free trivalent aluminium (Al^{3+}) by pH 4.0 (see McDonald *et al*., this volume, for details). At such low pH levels, Al^{3+} may actually protect against H^+ toxicity, perhaps because polyvalent cations exert a Ca^{2+}-like action in stabilizing branchial permeability (Schofield & Trojnar, 1980; Baker & Schofield, 1982). Al toxicity is greatest around pH 5.0, a level where acidity itself is not acutely lethal; this fact undoubtedly explains the many fish kills in the wild which have been observed in the pH *c*. 4.8–6.0 range (e.g. Grahn, 1980; Dickson, 1983; Henriksen, Skogheim &

Rosseland, 1984). As yet we are unsure whether this maximal toxicity around pH 5.0 is due to the predominance of the hydroxide species, or to loss of solubility and precipitation in the alkaline microenvironment of the branchial epithelium. These possibilities are not mutually exclusive, as McDonald *et al.* (this volume) point out.

Only a few studies have been reported on the physiology of acid/Al stress (Muniz & Leivestad, 1980a,b; Rosseland, 1980; Staurnes, Sigholt & Reite, 1984; Neville, 1985). In general these indicate more complex disturbances than during acid stress alone, with effects on O_2 and CO_2 exchange at the gills in addition to elevated losses of Na^+ and Cl^- from the plasma. However neither serial blood sampling nor ion exchange studies have yet been performed. In order to clarify the mechanism(s) of acid/Al toxicity, we have conducted flux and cannulation experiments in soft water using the brook trout (*Salvelinus fontinalis*) as a model, selected because of its great resistance to acid stress (Grande *et al.*, 1978). The methodology was generally similar to our previous work on rainbow trout, *with the important exception* that all experiments were run under one pass, flow-through conditions to minimize possible changes in Al concentration and/or speciation caused by the presence of the fish. Thus only net fluxes could be measured, using the Fick principle.

Acute responses

Al elevates net branchial losses of Na^+ and Cl^- above the levels due to acid stress alone (Figure 9). The effect becomes greater with increasing pH from 4.4 to 5.2, but is ameliorated by increasing $[Ca^{2+}]$ over the soft water range from 0.025 to 0.40 mequiv l^{-1} (see Figure 5 of McDonald *et al.*, this volume). The presence of 333 ppb [Al] (12.3 μmol l^{-1}) converts a sublethal acid stress (pH = 4.8) into a lethal one, killing most fish within 24–72 h (Figure 9). However, if trout survive the initial period of acute toxicity at this or lower Al levels, they restore positive ion balance over the next 7–10 d during continued exposure.

Based on these results, a standard challenge of pH = 4.8, [Al] = 333 ppb (which is well within the normal range of occurrence in acidified areas; Muniz & Leivestad, 1980a; Kelso *et al.*, 1986) has been employed in cannulation studies. Exposure to pH = 4.8 alone exerts negligible effects on most blood parameters, except for a small increase in arterial oxygen tension (Pa_{O_2}; Figure 10). However pH = 4.8 plus 333 ppb [Al] causes massive internal disturbances, some of which are shown in Figure 10. These means are only for the 43% of the fish which survived the first 42 h of exposure; changes in animals which died were more severe. In addition to the loss of plasma Na^+ and Cl^- and related haemoconcentration predicted by the flux experiments, these disturbances include a halving of Pa_{O_2}, reciprocal doubling of Pa_{CO_2}, and associated large decreases in O_2 saturation of the haemoglobin, increases in blood lactate, hyperventilation, and combined respiratory and metabolic acidoses. No one event clearly precedes another, so neither ionic nor gas exchange disturbance can be considered primary; I suggest that multiple toxic mechanisms (e.g. circulatory failure, tissue anoxia, acidosis) are involved.

In accord with previous studies (Grahn, 1980; Neville, 1985), we find that death is associated with an accumulation of large amounts of Al on the gills. Branchial mucification, oedema, and even lamellar fusion have also been reported under comparable conditions (Muniz & Leivestad, 1980a,b; Schofield & Trojnar, 1980). A present working hypothesis is that precipitation of aluminium hydroxides and/or organic ligand formation on the gill surface induces an inflammatory response which generally thickens and distorts the branchial epithelium, decreases its transcellular permeability to O_2 and CO_2, yet simultaneously increases paracellular channel permeability for ion loss. We will have to learn a great deal more about the branchial microenvironment before we can prove or disprove these ideas.

Chronic responses and acclimation

Acclimation to many other metals has been documented. The recovery of ion balance in surviving fish during continued exposures (Figure 9) suggests that this may also be true of Al. To test this possibility, we have examined brook trout held for long periods in flowing soft water at controlled [Al] (0, 75, 150 ppb), [Ca^{2+}] (0.025, 0.40 mequiv l^{-1}), and pH (5.2, 6.5). Data from selected treatments are summarized

Figure 9. Net flux rates of Na^+ in brook trout determined under constant flow-through conditions in soft water by the Fick principle. After control measurements *(C)* at pH = 6.5, the fish were subjected to 24 h exposure to pH 4.8 alone, followed by 240 h exposure to pH = 4.8 plus either 111 or 333 ppb [Al]. Means ± 1 SEM (*n* = 12 for first 24 h. *n* = 6 for each subgroup once Al was added). Data from individual fish are shown once mortality commenced. † indicates last measurement prior to death. Redrawn from Booth, McDonald, Simons & Wood (1988).

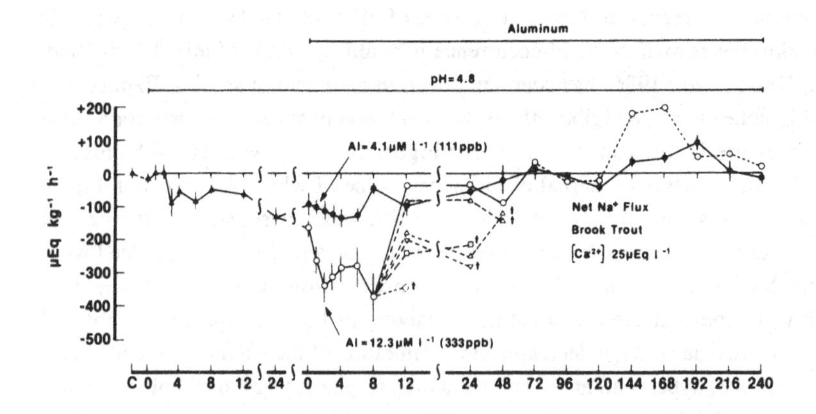

in Table 4. After 10 weeks' exposure, there is little substantive variation from control levels in plasma electrolytes, osmolality, and haematology in any treatment, suggesting that acclimation has occurred. This is not the result of selection, because all fish survived. To further test for acclimation, cannulated fish from various holding conditions were subjected to the standard challenge (pH = 4.8, [Al] = 333 ppb). The

Figure 10. Changes in oxygen tension (Pa_{O_2}), haemoglobin-bound oxygen content per unit haemoglobin (O_2 Hb^{-1}), carbon dioxide tension (Pa_{CO_2}) and plasma sodium and chloride levels in the arterial blood of brook trout subjected to either pH = 4.8 alone (●), [Ca^{2+}] = 0.025 mequiv l^{-1}) or pH = 4.8 plus [Al] = 333 ppb (0, [Ca^{2+}] = 0.40 mequiv l^{-1}). Means ± 1 SEM for the 9 (out of 9 tested) survivors in the pure acid group, and only for the 12 (out of 28 tested) survivors in the acid plus Al group. * indicates means significantly different ($p << 0.05$) from respective control value *(C)*. Redrawn from Wood, Playle, Simons Goss & McDonald (1988).

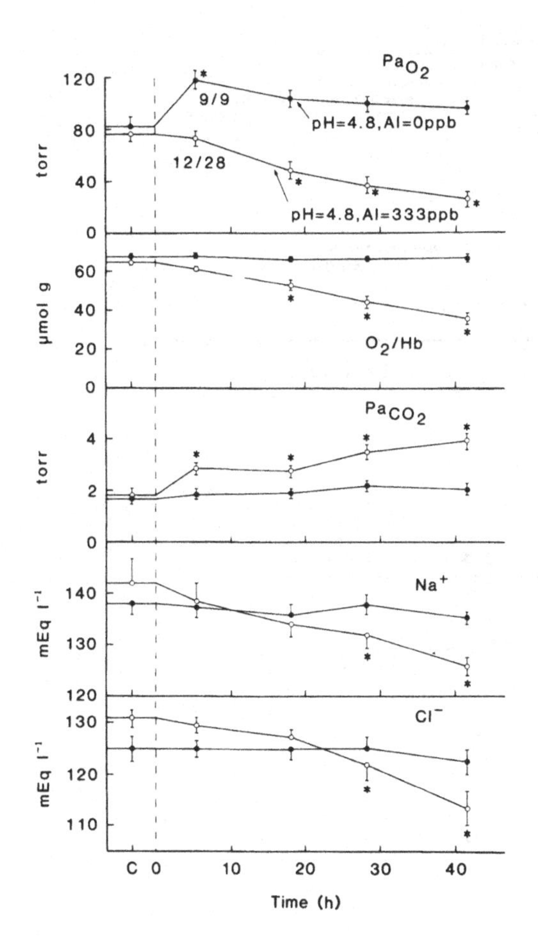

results (Figure 11) clearly demonstrate physiological acclimation in the traditional sense – increased tolerance of a more severe exposure. Trout pre-exposed for 10 weeks to pH = 5.2, [Al = 150 ppb] all survive the challenge for at least 66 h with minimal internal disturbance. In contrast, trout naïve to Al, whether or not they have been pre-exposed to acidity (pH = 5.2), exhibit the previously documented syndrome of ionoregulatory and gas exchange failure which results in rapid mortality. These

Figure 11. Changes in arterial oxygen tension (Pa_{O_2}), lactate concentration, and plasma chloride levels in the arterial blood of three groups of brook trout subjected to pH = 4.8 plus 333 ppb [Al] at [Ca^{2+}] = 0.025 mequiv l^{-1}. The fish were pre-exposed for 10 weeks to either pH = 5.2, Al = 150 ppb (●), pH = 5.2, [Al] = 0 ppb (Δ), or pH = 6.5, Al = 0 ppb (○). Means ± 1 SEM (n = 6–7 for each group). Data from individual fish are shown once mortality commenced. † indicates last measurement prior to death. * indicates mean significantly different ($p < 0.05$) from respective control value *(C)*. Figure redrawn from Wood, Simons, Mount & Bergman (1988).

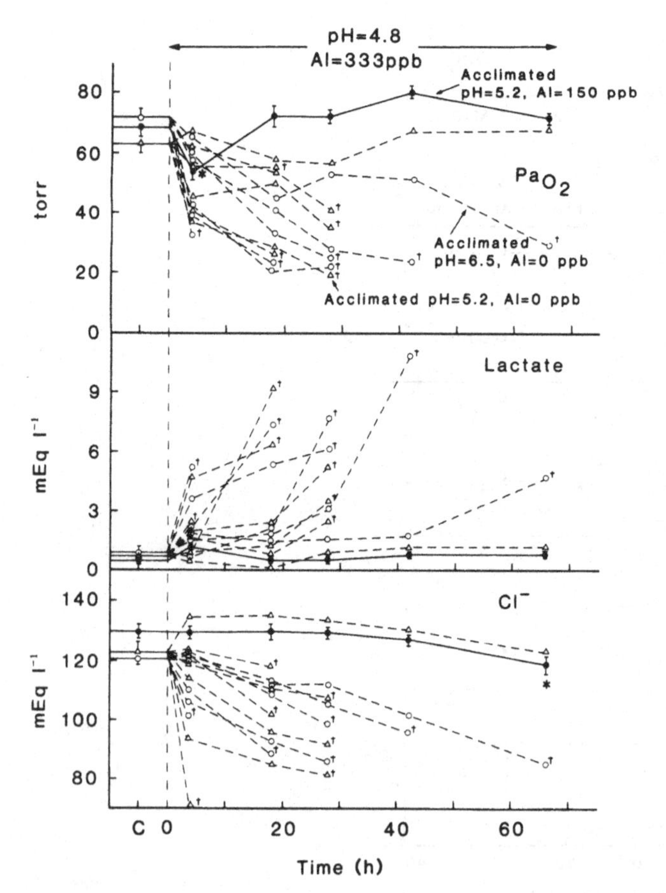

findings are in good agreement with a recent toxicity study on rainbow trout (Orr *et al.*, 1986).

Therefore, in the wild, fish chronically exposed to low levels of Al in mildly acidic soft water may have increased tolerance to short term increases in [Al] associated with acid surges. Morphological changes in the gills associated with chronic exposure to sublethal acid/Al levels have recently been described (Chevalier, Gauthier & Moreau, 1985; Karlsson-Norggren *et al.*, 1986a,b). These include detachment of the epithelium, deformation and fusion of lamellae, and a general swelling of the lamellae due to both epithelial and chloride cell hyperplasia. At the ultrastructural level, the chloride cells are grossly abnormal and contain Al precipitates in the cytoplasm. Within the framework of our current ideas, it is very difficult to see how such changes could provide the fish with increased resistance. The branchial surface and the effects of Al thereon should prove a rewarding field for future investigation.

Acknowledgements

I thank my co-workers cited herein and especially Drs D.G. McDonald and H.L. Bergman for their continuing collaboration. Original research reported here was supported by the NSERC strategic grants program in environmental toxicology, and by a contract ('Lake Acidification and Fisheries', RP–2346–01) from the Electric Power Research Institute, Environmental Assessment Department, through a subcontract from the University of Wyoming.

References

Baker, J.P. & Schofield, C.L. (1982). Aluminum toxicity to fish in acidic waters. *Water Air Soil Pollut.*, **18**, 289–310.

Booth, C.E., Macdonald, D.G., Simons, B.P. & Wood, C.M. (1988). The effects of aluminum and low pH on net ion fluxes and ion balance in the brook trout, (*Salvelinus fontinalis*). *Can. J. Fish. Aquat. Sci.*, **45**: in press.

Booth, J.A., Jansz, G.F. & Holeton, G.F. (1982). Cl^-, K^+, and acid–base balance in rainbow trout during exposure to, and recovery from, sublethal environmental acid. *Can. J. Zool.*, **60**, 1123–30.

Brown, D.J.A. (1981). The effects of various cations on the survival of brown trout, *Salmo trutta*, at low pHs. *J. Fish Biol.*, **18**, 31–40.

Brown, D.J.A. (1982). The effect of pH and calcium on fish and fisheries. *Water Air Soil Pollut.*, **18**, 343–51.

Brown, D.J.A. (1983). Effect of calcium and aluminum concentrations on the survival of brown trout (*Salmo trutta*) at low pH. *Bull. Envir. Contam. Toxicol.*, **30**, 582–7.

Brown, S.B., Eales, J.G., Evans, R.E. & Hara, T.J. (1984). Interrenal, thyroidal, and carbohydrate responses of rainbow trout (*Salmo gairdneri*) to environmental acidification. *Can. J. Fish. Aquat. Sci.*, **41**, 36–45.

Brown, S.B., Evans, R.E. & Hara, T.J. (1986). Interrenal, thyroidal, carbohydrate, and electrolyte responses in rainbow trout (*Salmo gairdneri*) during recovery from effects of acidification. *Can. J. Fish. Aquat. Sci.*, **43**, 714–18.

Cameron, J.N. & Heisler, N. (1983). Studies of ammonia in the rainbow trout: physico-chemical parameters, acid–base behaviour, and respiratory clearance. *J. exp. Biol.*, **105**, 107–25.

Chevalier, G., Gauthier, L. & Moreau, G. (1985). Histopathological and electron microscope studies of gills of brook trout, *Salvelinus fontinalis*, from acidified lakes. *Can. J. Zool.*, **63**, 2062–70.

Christophersen, N., Rustad, S. & Seip, H.M. (1984). Modelling streamwater chemistry with snowmelt. *Phil. Trans. R. Soc. Lond.* B, **305**, 427–39.

Dickson, W. (1983). Liming toxicity of aluminium to fish. *Vatten*, **39**, 400–4.

Eddy, F.B. (1985). Uptake and loss of potassium by rainbow trout (*Salmo gairdneri*) in freshwater and dilute seawater. *J. exp. Biol.*, **118**, 277–86.

Fromm, P.O. (1980). A review of some physiological and toxicological responses of freshwater fish to acid stress. *Envir. Biol. Fish.*, **5**, 79–93.

Fugelli, K. & Vislie, T. (1982). Physiological response to acid water in brown trout (*Salmo trutta* L.). Cell volume regulation in heart ventricle tissue. *J. exp. Biol.*, **101**, 71–82.

Giles, M.A., Majewski, H.S. & Hobden, B. (1984). Osmoregulatory and hematological responses of rainbow trout (*Salmo gairdneri*) to extended environmental acidification. *Can. J. Fish. Aquat. Sci.*, **41**, 1686–94.

Graham, M.S. & Wood, C.M. (1981). Toxicity of environmental acid to the rainbow trout: interactions of water hardness, acid type and exercise. *Can. J. Zool.*, **59**, 1518–26.

Graham, M.S., Wood, C.M. & Turner, J.D. (1982). The physiological responses of the rainbow trout to strenuous exercise: interactions of water hardness and environmental acidity. *Can. J. Zool.*, **60**, 3153–64.

Grahn, O. (1980). Fish kills in two moderately acid lakes due to high aluminum concentration. In *Proc. Int. Conf. Ecological Impact of Acid Precipitation*, ed. D. Drabløs & A. Tollan, pp. 310–11. Oslo–Ås: SNSF Project.

Grande, M., Muniz, I.P. & Anderson, S. (1978). The relative tolerance of some salmonids to acid waters. *Verh. Int. Ver. Limnol.*, **20**, 2076–84.

Haines, T.A. (1981). Acidic precipitation and its consequences for aquatic ecosystems: a review. *Trans. Am. Fish. Soc.*, **110**, 669–707.

Harvey, H.H. & Lee, C. (1982). Historical fisheries changes related to surface water pH changes in Canada. In *Acid Rain/Fisheries*, ed. R.E. Johnson, pp. 45–55. Bethesda: American Fisheries Society.

Harvey, H.H., Pierce, R.C., Dillon, P.J., Kramer, J.R. & Whelpdole, D.M. (1981). Acidification in the Canadian aquatic environment: scientific criteria for assessing the effects of acidic deposition on aquatic ecosystems. *National Research Council of Canada, Assoc. Comm. on Sci. Criteria for Env. Quality, NRCC Report no. 18475*, 369 pp.

Henriksen, A., Skogheim, O.K. & Rosseland, B.O. (1984). Episodic changes in pH and aluminium-speciation kill fish in a Norwegian salmon river. *Vatten*, **40**, 255–63.

Hõbe, H., Laurent, P. & McMahon, B.R. (1984). Whole body calcium flux rates in freshwater teleosts as a function of ambient calcium and pH levels: A comparison between the euryhaline trout, *Salmo gairdneri* and stenohaline bullhead, *Ictalurus nebulosus*. *J. exp. Biol.*, **113**, 237–52.

Holeton, G.F., Booth, J.H. & Jansz, G.F. (1983). Acid–base balance and Na^+ regulation in rainbow trout during exposure to, and recovery from, low environmental pH. *J. exp. Zool.*, **228**, 21–32.

Howells, G.D. (1984). Fishery decline: mechanisms and predictions. *Phil. Trans. R. Soc. Lond.* B, **305**, 529–47.

Howells, G.D., Brown, D.J.A. & Sadler, K. (1983). Effects of acidity, calcium and aluminium on fish survival and productivity – a review. *J. Sci. Food. Agric.*, **34**, 559–70.

Jeffries, D.S., Cox, C.M. & Dillon, P.J. (1979). Depression of pH in lakes and streams in central Ontario during snowmelt. *J. Fish. Res. Bd Can.*, **36**, 640–6.

Johnston, C.E., Saunders, R.L., Henderson, E.B., Harman, P.R. & Davidson, K. (1984). Chronic effects of low pH on some physiological aspects of smoltification in Atlantic salmon (*Salmo salar.*) *Can. Tech. Rep. Fish. Aquat. Sci.*, 1294, iii + 7 pp.

Karlsson-Norrgren, L., Dickson, W., Ljunberg, O. & Runn, P. (1986a). Acid water and aluminium exposure: gill lesions and aluminium accumulation in farmed brown trout, *Salmo trutta* L. *J. Fish. Diseases*, **9**, 1–8.

Karlsson–Norrgren, L., Bjorklund, I., Ljunberg, O. & Runn, P. (1986b). Acid water and aluminium exposure: experimentally induced gill lesions in brown trout, *Salmo trutta* L. *J. Fish. Diseases*, **9**, 11–25.

Kelso, J.R.M., Minns, C.K., Grey, J.E. & Jones, M.L. (1986). Acidification of surface waters in eastern Canada and its relationship to aquatic biota. *Can. Spec. Publ. Fish. Aquat. Sci.*, **87**, 42 pp.

Lacroix, G.L. (1985). Plasma ionic composition on the Atlantic salmon (*Salmo salar*), white sucker (*Catostomus commersoni*), and alewife (*Alosa pseudoharengus*) in some acidic rivers of Nova Scotia. *Can. J. Zool.*, **63**, 2254–61.

Lacroix, G.L., Gordon, D.J. & Johnston, D.J. (1985). Effects of low environmental pH on the survival, growth, and ionic composition of postemergent Atlantic salmon (*Salmo salar*). *Can. J. Fish. Aquat. Sci.*, **42**, 768–75.

Lee, R.M., Gerking, S.D.& Jazierska, B. (1983). Electrolyte balance and energy mobilization in acid-stressed rainbow trout, *Salmo gairdneri*, and their relation to reproductive success. *Envir. Biol. Fish.*, **8**, 115–23.

Leino, R.L. & McCormick, J.H. (1984). Morphological and morphometrical changes in chloride cells of the gills of *Pimephales promelas* after chronic exposure to acid water. *Cell Tiss. Res.*, **236**, 121–8.

Leivestad, H. (1982). Physiological effects of acid stress on fish. In *Acid Rain/Fisheries*, ed. R.E. Johnson, pp. 157–64. Bethesda: American Fisheries Society.

Leivestad, H. & Muniz, I.P. (1976). Fish kill at low pH in a Norwegian river. *Nature (Lond.)*, **259**, 391–2.

Leivestad, H., Hendrey, G., Muniz, I.P. & Snekvik, E. (1976). Effects of acid precipitation on freshwater organisms. In *Impact of Acid Precipitation on Forest and Freshwater Ecosystems in Norway*, ed. F.H. Braekke, pp. 87–111. SNSF Project FR6/76.

Leivestad, H., Muniz, I.P. & Rosseland, B.O. (1980). Acid stress in trout from a dilute mountain stream. In *Proc. Int. Conf. Ecological Impact of Acid Precipitation*, ed. D. Drabløs & A. Tollan, pp. 318–9. Oslo–Ås: SNSF Project.

Lloyd, R. & Jordan, D.H.M. (1964). Some factors affecting the resistance of rainbow trout (*Salmo gairdneri* Richardson) to acid waters. *Int. J. Air Water Pollut.*, **8**, 393–403.

Maetz, J. (1973). Na^+/NH_4^+, Na^+/H^+ exchanges and NH_3 movement across the gills of *Carassius auratus*. *J. exp. Biol.*, **58**, 255–75.

Marmorek, D.R., Cunningham, G., Jones, M.L. & Bunnell, P. (1985). Snowmelt effects related to acidic precipitation: a structured review of existing knowledge and current research activities. *LRATP Workshop no. 3*. Downsview:Atmospheric Environment Service.

Marshall, W.S. (1985). Paracellular ion transport in trout opercular epithelium models osmoregulatory effects of acid precipitation. *Can. J. Zool.*, **63**, 1816–22.
McDonald, D.G. (1983a). The effects of H^+ upon the gills of freshwater fish. *Can. J. Zool..*, **61**, 691–703.
McDonald, D.G. (1983b). The interaction of calcium and low pH on the physiology of the rainbow trout, *Salmo gairdneri*. I. Branchial and renal net ion and H^+ fluxes. *J. exp. Biol.*, **102**, 123–40.
McDonald, D.G., Höbe, H. & Wood, C.M. (1980). The influence of calcium on the physiological responses of the rainbow trout, *Salmo gairdneri*, to low environmental pH. *J. exp. Biol.*, **88**, 109–31.
McDonald, D.G. & Wood, C.M. (1981). Branchial and renal acid and ion fluxes in the rainbow trout, *Salmo gairdneri*, at low environmental pH. *J. exp. Biol.*, **93**, 101–18.
McDonald, D.G., Walker, R.L. & Wilkes, P.R.H. (1983). The interaction of environmental calcium and low pH on the physiology of the rainbow trout, *Salmo gairdneri* II. Branchial ionoregulatory mechanisms. *J. exp. Biol.*, **102**, 141–55.
McWilliams, P.G. (1980a). Effects of pH on sodium uptake in Norwegian brown trout (*Salmo trutta*) from an acid river. *J. exp. Biol.*, **88**, 259–67.
McWilliams, P.G. (1980b). Acclimation to an acid medium in the brook trout *Salmo trutta. J. exp. Biol.*, **88**, 269–80.
McWilliams, P.G. (1982a). A comparison of physiological characteristics in normal and acid exposed populations of the brown trout *Salmo trutta. Comp. Biochem. Physiol.*, **72A**, 515–22.
McWilliams, P.G. (1982b). The effects of calcium on sodium fluxes in the brown trout, *Salmo trutta*, in neutral and acid media. *J. exp. Biol.*, **96**, 439–42.
McWilliams, P.G. (1983). An investigation of the loss of bound calcium from the gills of the brown trout, *Salmo trutta*, in acid media. *Comp. Biochem. Physiol.*, **74A**, 107–16.
McWilliams, P.G. & Potts, W.T.M. (1978). The effects of pH and calcium concentrations on gill potentials in the brown trout, *Salmo trutta. J. comp. Physiol.*,**126**, 277–86.
Miles, H.M. (1971). Renal function in migrating adult coho salmon. *Comp. Biochem. Physiol.*, **38A**, 787–826.
Milligan, C.L. & Wood, C.M. (1982). Disturbances in hematology, fluid volume distribution, and circulatory function associated with low environmental pH in the rainbow trout, *Salmo gairdneri. J. exp. Biol.*, **99**, 397–415.
Muniz, I.P. (1984). The effects of acidification on Scandinavian freshwater fish fauna. *Phil. Trans. R. Soc. Lond.* B, **305**, 517–28.
Muniz, I.P. & Leivestad, H. (1980a). Acidification-effects on freshwater fish. In *Proc. Int. Conf. Ecological Impact of Acid Precipitation*, ed. D. Drabløs & A. Tollan, pp. 84–92. Oslo–Ås: SNSF Project.
Muniz, I.P. & Leivestad, H. (1980b). Toxic effects of aluminum on the brown trout, *Salmo trutta* L. In *Proc. Int. Conf. Ecological Impact of Acid Precipitation*, ed. D. Drabløs & A. Tollan, pp. 320–1. Oslo–Ås: SNSF Project.
Neville, C.M. (1979a). Sublethal effects of environmental acidification on rainbow trout (*Salmo gairdneri*). *J. Fish. Res. Bd Can.*, **36**, 84–7.
Neville, C.M. (1979b). Influence of mild hypercapnia on the effects of environmental acidification on rainbow trout (*Salmo gairdneri*). *J. exp. Biol.*, **83**, 345–9.
Neville, C.M. (1985). Physiological response of juvenile rainbow trout, *Salmo gairdneri*, to acid and aluminum – prediction of field responses from laboratory data. *Can. J. Fish. Aquat. Sci.*, **42**, 2004–19.

Orr, P.L., Bradley, R.W., Sprague, J.B. & Hutchinson, N.J. (1986). Acclimation-induced change in toxicity of aluminum to rainbow trout (*Salmo gairdneri*). *Can. J. Fish. Aquat. Sci.*, **43**, 243–6.
Packer, R.K. & Dunson, W.A. (1972). Anoxia and sodium loss associated with the death of brook trout at low pH. *Comp. Biochem. Physiol.*, **41A**, 17–26.
Rosseland, B.O. (1980). Physiological responses to acid water in fish. 2. Effects of acid water on metabolism and gill ventilation in brown trout, *Salmo trutta* L., and brook trout, *Salvelinus fontinalis* Mitchell. In *Proc. Int. Conf. Ecological Impact of Acid Precipitation*, ed. D. Drabløs & A. Tollan, pp. 348–9. Oslo–Ås: SNSF Project.
Saunders, R.L., Henderson, E.B., Harmon, P.R., Johnston, C.E. & Eales, J.G. (1983). Effects of low environmental pH on smolting of Atlantic salmon (*Salmo salar*) *Can. J. Fish. Aquat. Sci.*, **40**, 1203–11.
Schofield, C.L. & Trojnar (1980). Aluminum toxicity to brook trout (*Salvelinus fontinalis*) in acidified waters. In *Polluted Rain*, ed. T.Y. Toribara, M.W. Miller & P.E. Morrow, pp. 341–66. New York: Plenum Press.
Spry, D.J., Wood, C.M. & Hodson, P.V. (1981). The effects of environmental acid on freshwater fish with particular reference to the softwater lakes in Ontario and the modifying effects of heavy metals. A literature review. *Can. Tech. Rep. Fish. Aquat. Sci.*, **999**, 145 pp.
Stewart, P.A. (1978). Independent and dependent variables of acid–base control. *Respir. Physiol.*, **33**, 9–26.
Stewart, P.A. (1983). Modern quantitative acid–base chemistry. *Can. J. Physiol. Pharmacol.*, **61**, 1444–61.
Staurnes, M., Sigholt, T. & Reite, O.B. (1984). Reduced carbonic anhydrase and Na–K-ATPase activity in gills of salmonids exposed to aluminium-containing acid water. *Experientia*, **40**, 226-7.
Stuart, S. & Morris, R. (1985). The effects of season and exposure to reduced pH (abrupt and gradual) on some physiological parameters in brown trout (*Salmo trutta*). *Can. J. Zool.*, **63**, 1078–83.
Swarts, F.A., Dunson, W.A. & Wright, J.E. (1978). Genetic and environmental factors involved in increased resistance of brook trout to sulfuric acid solutions and mine acid polluted waters. *Trans. Am. Fish. Soc.*, **107**, 651–77.
Waddell, W.J. & Butler, T.C. (1959). Calculation of intracellular pH from the distribution of 5,5-dimethyl-2,4-oxazolidine dione (DMO): application to skeletal muscle of the dog. *J. Clin. Invest.*, **38**, 720–9.
Wendelaar Bonga, S.E., van der Meij, J.C.A. & Flik, G. (1984). Prolactin and acid stress in the teleost *Oreochromis* (formerly *Sarotherodon*) *mossambicus*. *Gen. Comp. Endocrinol.*, **55**, 323–32.
Westfall, B.A. (1945). Coagulation film anoxia in fishes. *Ecology*, **26**, 283–7.
Wood, C.M. & McDonald, D.G. (1982). Physiological mechanisms of acid toxicity to fish. In *Acid Rain/Fisheries*, ed. R.E. Johnson, pp. 197–226. Bethesda: American Fisheries Society.
Wood, C.M., McDonald, D.G., Booth, C.E., Simons, B.P., Ingersoll, C.G. & Bergman, H.L. (1988). Physiological evidence of acclimation to acid/aluminum stress in adult brook trout (*Salvelinus fontinalis*). 1. Blood composition and net sodium fluxes. *Can. J. Fish. Aquat. Sci.*, **45**: in press.
Wood, C.M., Playle, R.C., Simons, B.P., Goss, G.A. & McDonald, D.G. (1988). Blood gases, acid–base status, ions, and hematology in adult brook trout (*Salvelinus fontinalis*) under acid/aluminum exposure. *Can. J. Fish. Aquat. Sci.*. **45**: in press.
Wood, C.M., Simons, B.P., Mout, D.R. & Bergman, H.L. (1988). Physiological evidence of acclimation to acid/aluminum stress in adult brook trout (*Salvelinus*

fontinalis). 2. Blood parameters by cannulation. *Can. J. Fish. Aquat. Sci.* **45**: in press.

Wood, C.M., Turner, J.D. & Graham, M.S. (1982). Why do fish die after severe exercise? *J. Fish. Biol.*, **22**, 189–201.

Wood, C.M., Wheatly, M.G. & Höbe, H. (1984). The mechanisms of acid–base and ionoregulation in the freshwater rainbow trout during environmental hyperoxia and subsequent normoxia. III. Branchial exchanges. *Respir. Physiol.*, **55**, 175–92.

Wright, P.A. & Wood, C.M. (1985). An analysis of branchial ammonia excretion in the freshwater rainbow trout: effects of environmental pH change and sodium uptake blockade. *J. exp. Biol.*, **114**, 329–53.

Wright, R.F. & Snekvik, E. (1978). Acid precipitation: chemistry and fish populations in 700 lakes in southernmost Norway. *Verh. Int. Ver. Limnol.*, **20**, 765–75.

J.H.D VANGENECHTEN, H. WITTERS AND
O.L.J. VANDERBORGHT

Laboratory studies on invertebrate survival and physiology in acid waters

Introduction

Although most attention has been focused on the impoverishment of fish populations in acidified lakes, numerous ecological studies have shown that phytoplankton, zooplankton and benthic invertebrates have decreased in diversity in recently acidified waters. There is ample evidence to suggest that the pH is a major variable in determining the distribution of species, although recent work includes other parameters, like the Ca^{2+} content and the metal concentrations into the list of distribution-limiting factors. Very elegant work in this respect has been performed by Økland and Økland (Økland, J., 1980; Økland, K.A., 1980; Økland & Økland, 1980) who studied in detail the gastropod fauna in acid-polluted and non-polluted lakes in Norway. The literature describing *in situ* studies of the pH influence on invertebrate distribution, mortality and diversity, being large and quite impressive, will not be quoted in detail in this text, and the reader is advised to study the reports of the SNSF Project in Norway (Leivestad *et al.*, 1976; Overrein *et al.*, 1980), or the review paper of Sutcliffe & Hildrew (this volume) covering the subject.

Laboratory experiments on mortality of freshwater fauna reveal clearly that there exists a critical pH value below which survival is significantly reduced. This critical value is species-dependent but it is also determined by the composition of the testwater. The concentration of sodium, calcium and (heavy) metals can markedly influence the rate of mortality at low pH. An overall summary of pH tolerance tests together with a detailed description of the ambient water quality composition would show an almost inexhaustible list of invertebrate species and LD_{50} values. Therefore we will discuss studies that reveal indications on the physiological mechanisms of tolerance or sensitivity.

Physiological tolerance on the other hand is certainly not the only factor determining the survival of aquatic organisms. Availability of food and the presence or absence of predator organisms are two other crucial factors.

Several papers are presented in this volume concerning the mechanisms of action of acid pH on the physiology of fishes. This area of research has been studied extensively in recent years and mostly use was made of animal species which appeared to be very sensitive to low pH and to elevated metal concentrations. Our research in this area of acid stress physiology has been conducted with a different approach. The physiological processes of interaction of acid pH and elevated metal concentrations with ion regulation have been studied in freshwater animals which survive extremely acidic and metal-rich environments, e.g. the waterbugs *Corixa dentipes* and *C. punctata* are captured in Belgian acid lakes with pH around 3.8 and Al concentrations often exceeding 2 mg l^{-1}. We studied the kinetics of Na^+ and Cl^- flux (influx and outflux) at different H^+ concentrations together with elevated aluminium concentrations in the ambient water. There have been few studies in this area of research using invertebrate species and, thus, progress is slow and use is often made of models borrowed from research in fish physiology to understand observed phenomena. In the present paper, we shall discuss some new ideas and expand the existing hypotheses with some of our more recent findings.

Laboratory studies on invertebrate survival in conditions of acid stress

Numerous efforts have been made to define the mortality rate for different freshwater invertebrates at conditions of acid pH. In these efforts use is made mostly of daphnids, insect larvae and gammarids. In an attempt to review all these data one is immediately concerned with the complexity of the problem. It hardly seems justified to investigate the survival in laboratory conditions using only freshly prepared solutions of carefully defined chemical composition, as natural conditions are seldom mimicked in this way in the laboratory. This is especially true when the combined toxicity of low pH and elevated metals are investigated. One can hardly imagine a natural water in which no organic complexation of metals occurs. In these survival experiments one thus has to take into account the toxicity of the different metal species which can be present. This holds also for the physiological studies where metals are used.

Survival of daphnids has been studied by Havas and co-workers (Havas, 1985; Havas & Likens, 1985a,b). *Daphnia magna* died rapidly below pH 5.0 but addition of Al (1.02 mg l^{-1}) temporarily ameliorated the high H^+ toxicity. Aluminium appeared to be harmful or beneficial to *D. magna* depending on pH and on the Al concentration in the water. Maximum Al toxicity and maximum Al bioaccumulation were observed at pH 6.5. These laboratory data support field surveys where *D. magna* is not known to occur in acidic lakes, and is typically found in alkaline waters (Potts & Fryer, 1979). *D. catawba* on the other hand appeared to be more tolerant of both Al and H^+ than *D. magna* (Havas & Likens, 1985a). Mortality, attributable to Al, occurred only at pH 6.5 for *D. catawba* at the highest Al concentration tested (1.02 mg l^{-1} Al). The pH tolerance and reproduction of *D. pulex*, another important crustacean, was tested by Walton *et al.*, (1982). A 96 h acute test suggested a threshold effect at pH 4.2. Further studies however, using chronic tests and measurements of reproduction failure indicated that pH 5 would be a more realistic value of the critical limit for population maintenance in nature.

Based on mortality rates, crustaceans appear to be considerably more sensitive to low pH than insect larvae (Havas & Hutchinson, 1983). Laboratory studies revealed that insect larvae were also sensitive to aluminium (Havas & Likens, 1985a), but it was concluded that the species tested, *Chaoborus punctipennis* and *Chironomus anthrocinus*, should be able to tolerate Al concentrations in excess of those which now occur in recently acidified oligotrophic lakes.

Together with the studies on survival, some authors have made concomitant measurements of physiological parameters (Havas *et al.*, 1984; Havas & Likens, 1985b). The data are used to explain the observed mortalities and therefore give supplementary information on the mode of toxic action. Some of these data will be discussed in detail in the appropriate section of this chapter, but it is obvious that both H^+ and Al interfere with Na^+ regulation.

Several experiments on the mortality of gammarids at different ambient pH are published (Brehm & Meijering, 1982; Meinel *et al.*, 1985). But no concomitant physiological measurements were made. *Gammarus fossarum* was shown to be highly sensitive to acid environmental pH and was seen to become damaged below pH 6.0 in low buffered water. It was further found that in between pH 4 and 4.5, survival could be increased by using higher concentrations of sodium in the water, other cations (K^+, Mg^{2+} or Ca^{2+}) apparently had no ameliorating influence.

In all the experiments where different external conditions are applied to study survival or mortality at acid pH, the ameliorating effect of increased ionic (mostly NaCl) concentrations is acknowledged. It therefore seems appropriate to assume that the principal mode of toxic action of low pH has to be sought in a failure of ionic regulation. In the next section we will study in detail the effects of a low environmental pH and of elevated aluminium concentrations, on the overall Na^+ and Cl^- exchange and balance in freshwater invertebrates.

The influence of acid pH on NaCl exchange in aquatic invertebrates

The harmful influence of acid pH on Na^+ and Cl^- exchanges in fish gills is very well documented nowadays. The fish gill epithelium, being the primary site of ion exchange and of gas exchange between the water and the blood, represents a large thin-walled ectodermal surface area in intimate contact with the external medium. It is not surprising that fish species which need this epithelial surface for gas exchange suffer heavily from acid stress. Among invertebrates, however, several groups from different phyla are air breathers or have evolved an interesting respiratory system known as 'plastron' respiration taking advantage of air bubbles which are carried on dorsal or ventral body parts. Waterbugs, who colonize acid habitats are known to use this type of respiration.

In the discussion that follows, the majority of data come from studies on waterbugs (Insecta: Hemiptera, Heteroptera). The uptake of ions like Na^+ and Cl^- in aquatic insects is carried out by specialized cells which are arranged mostly in transport epithelia. Despite anatomical variations, all such systems seem to possess a cell type which is often referred to as a 'chloride cell'. These cells are rich in mitochondria and they are believed to be responsible for ion exchange. The interested reader is referred to the excellent review on this subject by Komnick (1977).

Interesting data on acid stress are further available for crayfish, but as these will be described in detail by McMahon & Stuart (this volume), we will not discuss them here. Occasionally, lowering the pH of the water has been used to study the nature of the counter-ion in ion exchange experiments using crustaceans (Shaw, 1960) or aquatic insect larvae (Stobbart, 1967). pH effects on ion exchange mechanisms have been described recently in daphnids (Havas *et al.*, 1984; Havas & Likens, 1985b), but no data are available for the majority of other phyla.

pH effects on Na^+ influx in aquatic insects: the waterbugs Corixa dentipes and C. punctata

In contrast to many other aquatic insects (Ephemeroptera or Plecoptera), the representatives of the Hydrocorisae (Heteroptera) normally live in water throughout their whole lives. They do not possess any anatomical structure which resembles a gill-like organ but they do possess ion transporting cells called 'chloride cells' in the integument of the larvae as well as that of the adults (Komnick, 1977; Komnick & Schmitz, 1977). In adult *C. punctata* these chloride cells are located on the legs and on the front of the head, the latter holding about 50% of the total number. These authors, using different techniques (histochemistry, autoradiography, Cl^- influx measurements) suggest that these cells are the sites of cutaneous osmoregulatory ion absorption.

If we assume that the basic physiological principles of ion exchange in ion uptake cells are the same in vertebrates and invertebrates, we may describe ion uptake at the apical membrane of the chloride cell in corixids as follows. Sodium uptake at the external side of the uptake cell is accomplished by a carrier-mediated Na^+/H^+ or Na^+/NH_4^+ exchange mechanism whereas chloride uptake is made possible by a Cl^-/HCO_3^- exchange mechanism. The affinity of the carrier being high for Na^+ at the external side of the membrane, is high for H^+ at the cytosolic side. When the ambient H^+ concentration is increased by lowering the pH, one may expect a competition between H^+ and Na^+ at the external side of the carrier. Eventually one may expect a decreased sodium influx. Lowering the pH of the ambient water was used by Shaw (1960) when he studied the counter-ion for Na^+ uptake in *Astacus pallipes*. Lowering the pH from 6 to 5 for example depressed Na^+ influx by about 40%, whereas at pH 4 the influx hardly attained 25% of control values. Stobbart (1967) studied the influence of different cations on ion exchange in the larvae of the insect *Aëdes aëgypti*. In his studies, lowering the pH significantly reduced Na^+ influx whereas the maximum Na^+ influx was measured at a pH of about 10. These data on invertebrates served as valuable information to define the counter ion in Na^+ ion transport mechanisms, but it was not until the mid-seventies when alarming reports on lake acidification became widely known, that scientists became systematically interested in acid-stress physiology *per se*. But even then, research efforts were almost exclusively directed towards fish physiology.

The effects of different pH levels on the kinetics of Na^+ influx in the waterbugs *C. dentipes* and *C. punctata*, two inhabitants of acid boglakes in Belgium were studied by our group (Vangenechten *et al.*, 1979a,b). Na^+ influx in these waterbugs exhibits enzyme-substrate saturation kinetics, which can be described by a Michaëlis–Menten relation (Figure 1*(a)*). This behaviour can be regarded as proof of the existence of a saturable Na^+ uptake carrier mechanism as explained before. The two parameters describing this behaviour: K_m (the external concentration at which the carrier is 50% saturated) and V_{max} (the maximum influx rate) are clearly pH-dependent within certain ranges. In between pH 4.5 and 9, V_{max} does not change and ranges from 1.67 to 1.76 μmol Na^+ ml^{-1} haemolymph h^{-1} whereas under these conditions K_m ranges from 0.20 mmol Na^+ l^{-1} to 0.48 mmol Na^+ l^{-1} in *C. punctata*. It seems appropriate at this stage to discuss the dispersal of both species of waterbugs in regard to the Na^+ and the H^+ concentrations of their natural habitats. The individuals of both species which were used in the experiments, have been captured in bogpools with pH values around pH 3.9 and with Na^+ concentrations in between 0.2 and 0.6 mmol Na^+ l^{-1}. But whereas *C. dentipes* has never been found in lakes with a pH above 5.5 in Belgium, *C. punctata* has a more ubiquitous occurrence and can be found even in saline waters up to 3.25 ‰ Cl^- and pH values up to pH 9.5 (R. Bosmans, personal communication). Figure 1*(a)* shows that extremely acidic conditions (pH values below 3.5, which occasionally occur in these acid boglakes in Belgium) increase the maximum influx rate for sodium (V_{max}). This increase equals 53% in *C. punctata* and 86% in *C. dentipes*, compared with V_{max} at pH 6. This can be explained by assuming an increase in the number of active sodium carriers or in the pumping frequency of the carriers. But despite this increased V_{max}, Na^+ influx at pH 3 is still lower than at pH 6 when the external Na^+ concentration remains below 1.4 mmol Na^+ l^{-1} for *C. punctata* and below 0.95 mmol Na^+ l^{-1} in *C. dentipes* (Figure 2).

Let us consider the consequences of these laboratory data for the ion uptake processes in nature. Under natural acidic conditions (pH 3.5–3.8) in bogpools (0.2–0.6 mmol Na^+ l^{-1}) the Na^+ carriers in both animal species will be 20% to 42% saturated. The possibility of increasing V_{max} warrants a Na^+ influx at pH 3.5 of 62 to 88% of the value at pH 6 for *C. dentipes* and of 54 to 77% in *C. punctata*. The

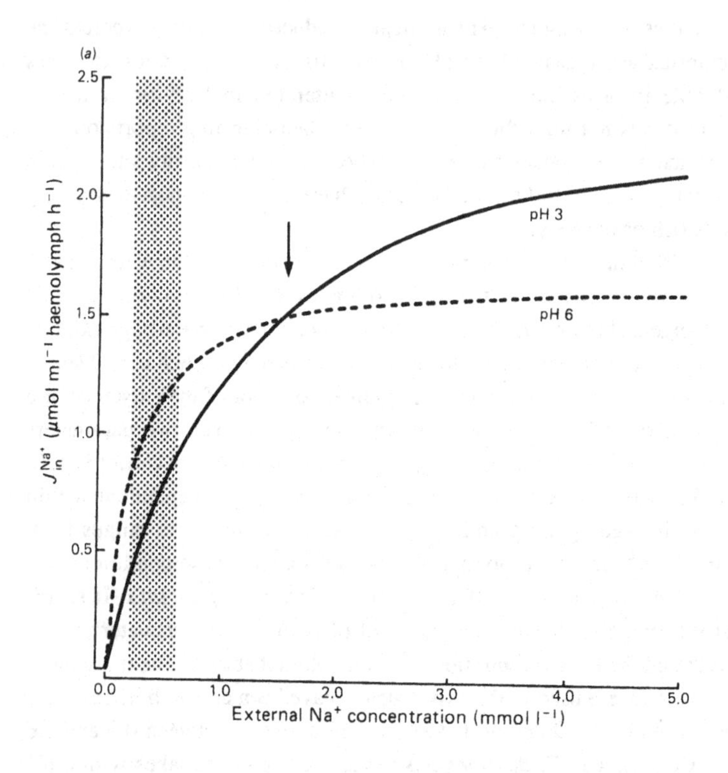
(a)
2.5
2.0
1.5
1.0
0.5
J in Na+ (μmol ml⁻¹ haemolymph h⁻¹)
pH 3
pH 6
0.0
1.0
2.0
3.0
4.0
5.0
External Na+ concentration (mmol l⁻¹)

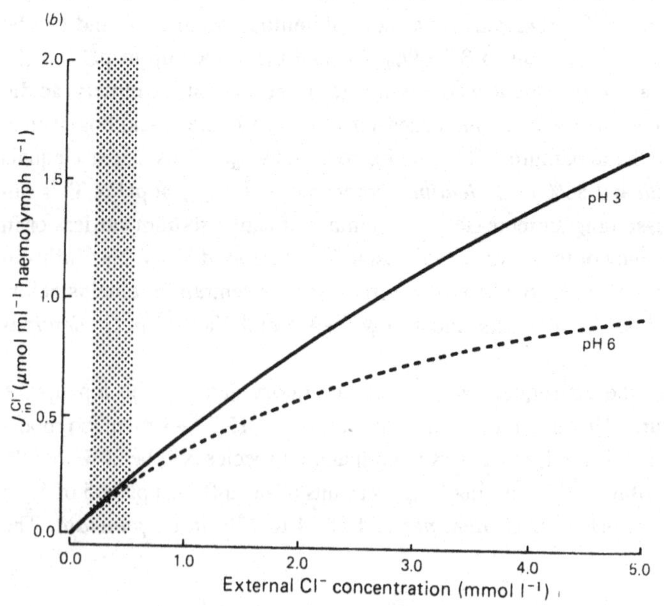
(b)
2.0
1.5
1.0
0.5
0.0
J in Cl⁻ (μmol ml⁻¹ haemolymph h⁻¹)
pH 3
pH 6
0.0
1.0
2.0
3.0
4.0
5.0
External Cl⁻ concentration (mmol l⁻¹)

influence of the pH on the Na^+ influx is thus relatively low. The Na^+ influx of course is only part of all the processes involved in maintenance of the body Na^+ concentration. The influence of pH on total Na^+ balance will be discussed in one of the following sections.

At acid pH the competition between Na^+ and H^+ binding on the transport carrier ultimately must lead to an increased H^+ uptake into the animal. In fish, McWilliams & Potts (1978) observed a large inward diffusion of H^+ immediately following acidification of the external medium, producing a positive shift in the measured transepithelial potential. A decrease in blood pH has been noticed in fishes as a consequence of this increased H^+ permeability (McDonald *et al.*, 1980; McDonald &

Figure 2. Difference of Na^+ influx in *C. dentipes* and *C. punctata* between pH 6 and pH 3 expressed as a percentage of Na^+ influx at pH 6. Note that Na^+ influx at pH 3 exceeds influx at pH 6 if external Na^+ concentration exceeds 0.95 mmol Na^+ l^{-1} in *C. dentipes* and 1.4 mmol Na^+ l^{-1} in *C. punctata* (recalculated using data published by Vangenechten *et al.*, 1979a). The shaded area represents the Na^+ concentration in the natural water.

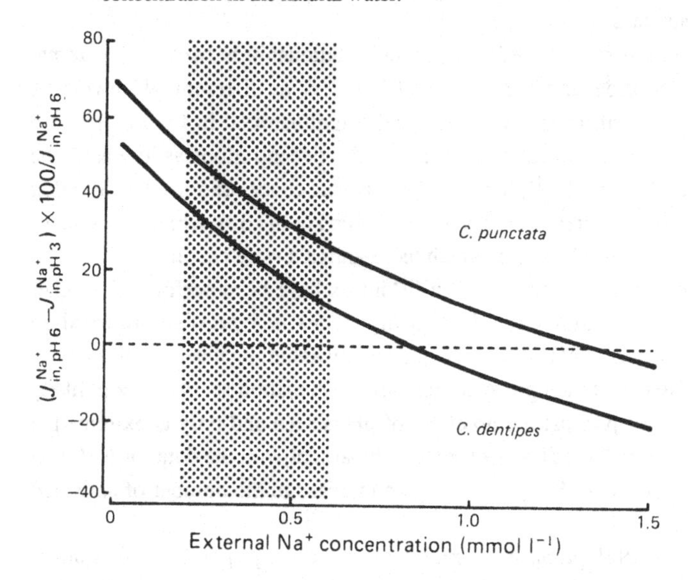

Figure 1 (Facing page). *(a)* The relationship between the external concentration of Na^+ and Na^+ influx at pH 3 and pH 6 in *Corixa punctata*. Note the intersection of both curves at an external Na^+ concentration of 1.4 mmol l^{-1} (arrow). The shaded area represents the Na^+ concentration in the natural environment. Drawn using data from Vangenechten *et al.*, 1979b. *(b)* Relationship between the external Cl^- concentration and the Cl^- influx in *C. punctata* at pH 3 and at pH 6. The shaded area represents the Cl^- concentration of the natural environment. Drawn using data from Vangenechten *et al.*, 1980.

Table 1. *Haemolymph pH values ± 95% confidence limits in* C. dentipes *and* C. punctata *kept for four days in a NaCl solution of 0.5 mmol l^{-1} at pH 6 and at pH 3*

	Corixa dentipes	*Corixa punctata*
pH 6	7.04 ± 0.09	6.89 ± 0.13
pH 3	6.97 ± 0.22	6.88 ± 0.10

Wood, 1981; Ultsch *et al.*, 1981) but seems to depend on the external Ca^{2+} concentration. In waterbugs, however, a stay for four days in a NaCl solution of 0.5 mmol l^{-1} at pH 3 did not change the haemolymph pH as measured using microelectrodes *in vivo* (Table 1) (Vangenechten, unpublished observations).

pH effects on Cl^- *influx in aquatic insects: the waterbugs* Corixa dentipes *and* C. punctata

Vangenechten *et al.* (1980) studied the relationship between the external concentration of chloride and the rates of Cl^- influx at different pH levels. A comparison of the Cl^- influx at pH 3 and at pH 6 is illustrated in Figure 1*(b)*. No significant difference in Cl^- influx was found at Cl^- concentrations as low as in the natural habitat (0.2–0.5 mmol l^{-1}), but at higher external concentrations the Cl^- influx is increased at pH 3. No increase however was noticed at pH 3.6, which approaches the natural conditions (pH 3.5–3.8) in which these animals were caught.

Few experiments on the effects of pH on Cl^- influx have been performed in other freshwater invertebrates. Stobbart (1967) using *Aëdes aëgypti* also observed an increase in Cl^- influx in acid pH which he explained by assuming a decreased external competition with HCO_3^- at the uptake carrier. This explanation seems to be unlikely in our experiments, as up to pH 3.6 no effect of pH is seen and only at extreme pH conditions an increase in Cl^- influx is observed. In parallel with the data for Na^+, one could explain the increase in V_{max} for Cl^- by an increase in the number of activated Cl^- carriers.

The affinity of the Na^+ pump in *C. punctata* in the range pH 4.5–9.0 is about 0.2–0.5 mmol $Na^+ l^{-1}$ whereas the affinity for the Cl^- pump in the same pH region ranges from 2 to 4 mmol $Cl^- l^{-1}$. This difference in affinity may be the result of an evolutionary adaptation of the Na^+ transport mechanism to conditions of low Na^+ combined with high, competitive H^+ concentrations. In *A. aëgypti*, for example, Stobbart (1967) calculates the affinity for Na^+ to range from 0.5 to 0.6 mmol $Na^+ l^{-1}$ and for Cl^- from 0.2 to 0.5 mmol $Cl^- l^{-1}$.

The effects on NaCl exchange and balance in aquatic insects: the waterbugs Corixa dentipes *and* C. punctata

It is well known nowadays, that acid pH leads to a severe disturbance of Na^+ and Cl^- balance in fish. The interested reader is referred to the specialized papers covering the subject in this volume. Whereas Na^+ and Cl^- influx seem to be largely decreased, drastic increases in Na^+ and Cl^- losses are the main mechanisms responsible for the observed decline in the blood Na^+ and Cl^- concentrations. It therefore seemed of interest to study the net Na^+ and Cl^- fluxes in waterbugs at acid pH.

Na^+ influx in *C. punctata* decreased by about 50% after 24 h of acid stress (pH 3), the Na^+ loss was not changed, resulting in a net Na^+ loss of about 0.56 μmol Na^+ ml^{-1} haemolymph h^{-1} (Figure 2*(a)*). Over a period of 24 h this net Na^+ loss would equal about 13 mmol Na^+ l^{-1}. Such a decrease in haemolymph sodium would reduce its concentration from about 114 mmol l^{-1} to 101 mmol l^{-1}. We have never been able in any experiment to detect such a decrease. On the contrary, we observed a significant increase in haemolymph sodium after 24 h to 127 mmol l^{-1} (Vangenechten & Vanderborght, 1980). In another experiment (Vangenechten, 1980) we compared the haemolymph Na^+ concentration in *C. dentipes* after a stay for seven days in pH 6 (131 mmol Na^+ l^{-1} haemolymph) and pH 3 (130 mmol Na^+ l^{-1} haemolymph). It is clear that these waterbugs are capable of maintaining their haemolymph Na^+ concentration, it is therefore postulated that other body compartments compensate for the observed net loss. It is known for *C. dentipes,* for example, that about 70% of the total body sodium is contained in the tissue compartment (Vangenechten *et al.*, 1979a). Over long periods of time one may assume that a new balance between influx and outflux will be achieved, although even after a stay of 50 h at pH 3 the measured sodium influx was still lower than at pH 6 and was comparable to the influx values in animals subjected to acute (one to two hours) acid stress (Vangenechten, unpublished observations).

It was noticed in our experiments with waterbugs (H. Witters *et al.*, unpublished results) that systematically smaller volumes of haemolymph samples could be taken from animals at low pH. This, together with the observed increase in haemolymph sodium concentration at pH 3, which could be the result of a decreased haemolymph water content, led us to investigate the ECFV (Extracellular Fluid Volume) in these animals. In salmonids, for example, a 27% decrease of the ECFV is observed after a three day acidification period (Milligan & Wood, 1982). Our experiments with waterbugs did not reveal any change in ECFV between animals at pH 6 and animals which stayed for 13 d at pH 3.

Chloride flux in *C. punctata* after a stay for three days at acid pH is compared with flux values at control pH in Figure 3*(b)* (Vangenechten & Vanderborght, 1980). The four-fold increase in Cl^- influx at pH 3 is almost totally compensated by the increase

Figure 3. Na^+ influx, Na^+ outflux and net Na^+ flux (shaded area) in *C. punctata* at pH 3 and pH 6 *(a)* (Vangenechten & Vanderborght, 1980) and during conditions of elevated Al concentrations *(c)* (Witters *et al.*, 1984b). Cl^- influx, Cl^- outflux and net Cl^- flux (shaded area) in *C. punctata* at pH 3 and at pH 6 *(b)* (Vangenechten & Vanderborght, 1980). Mean values ± SE. * denotes a significant difference ($p < 0.05$) compared with control values.

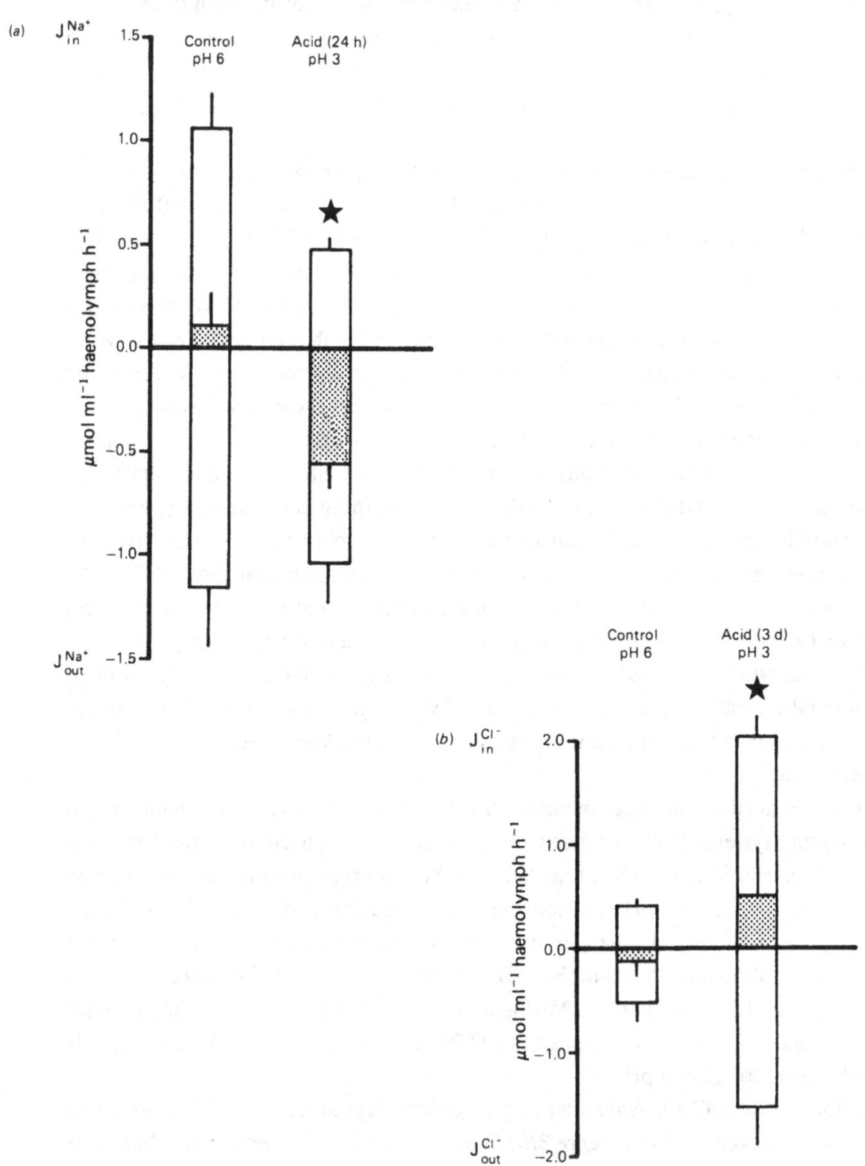

in Cl^- loss, yielding a positive net Cl^- flux which was not significantly different from zero. The haemolymph Cl^- concentration was accordingly not changed and averaged 64 mmol Cl^- l^{-1} at pH 6 and 69 mmol Cl^- l^{-1} at pH 3. In *C. dentipes*, a stay for seven days at pH 3 did not change the haemolymph Cl^- concentration (84 mmol Cl^- l^{-1} at pH 6 compared with 86 mmol Cl^- l^{-1} at pH 3).

pH effects on Na^+ and Cl^- exchange and balance in freshwater crustaceans

Potts & Fryer (1979) studied the effects of pH on the sodium balance in two cladocerans *Daphnia magna* and *Acantholeberis curvirostris*. They found that *A. curvirostris* retains sodium better in acid waters than does *D. magna*. In the former, the rate of sodium loss at pH 3 is lower than at pH 7 while in the latter, the rate of loss at pH 3 is four times that at pH 7. Sodium influx on the other hand was depressed in acid waters to a value of about 25% of the influx at pH 7 in *A. curvirostris* and the depression was even more pronounced in *D. magna*. In this respect Na^+ balance of *A. curvirostris* in acid water is comparable to the Na^+ regulation of the waterbugs described above.

Havas, Hutchinson & Likens (1984) used *D. magna* to study the Na^+ flux in soft (78 μmol Na^+ l^{-1}) and hard water (0.87 mmol Na^+ l^{-1}) at different acidities. No difference was found in Na^+ influx in the pH range 4.5–8.0 in hard waters, whereas in soft waters Na^+ influx seemed to be decreased below pH 5.0. The rate of sodium outflux in hard water increased significantly below pH 4.5. In soft water this increased Na^+ outflux was measurable even at pH 4.5. The net loss of Na^+ in soft water was due to both an increase in Na^+ outflux and a decrease in Na^+ influx, whilst in hard water the effect was primarily on Na^+ outflux. Both species tested, *D. magna* and *D. middendorffiana*, showed problems with Na^+ regulation below pH 5.5 in soft

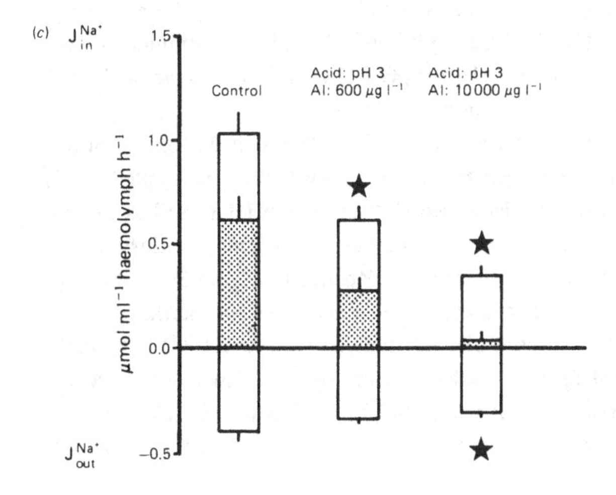

water and below pH 4.5 in hard water, which indicates that they are considerably more sensitive to low pH in soft water than in hard water. Havas & Likens (1985b) in another set of experiments, using the same *D. magna,* found a drastic decrease (from about 350 mmol Na^{+} mg^{-1} animal to about 60 mmol Na^{+} mg^{-1} animal) of the Na^{+} content at pH 4.5 within less than five hours. They further found a significant correlation in *D. magna* between the Na^{+} content and the morbidity. The decreased body Na^{+} content was the result of a decreased Na^{+} influx (to about 27% of control values) and an increased Na^{+} loss from about 100 mmol Na^{+} mg^{-1} h^{-1} at pH 6.5 to about 150 mmol Na^{+} mg^{-1} h^{-1} at pH 4.5. The Na^{+} balance was calculated to change from a positive value (gain) of about 1 nmol Na^{+} mg^{-1} h^{-1} at pH 6.5 to a net loss value of 125 nmol Na^{+} mg^{-1} h^{-1} at pH 4.5.

The influence of combined stress on the physiology of aquatic invertebrates: high acidity and elevated aluminium concentrations

Earlier studies on the toxicity of acid waters assumed that the elimination of biota from acidified lakes and rivers was due to elevated concentrations of hydrogen (H^{+}) ions *per se.* More recent studies, however, have pointed out that high concentrations of metals – mostly aluminium which is leached from the soil by the action of acid percolating groundwater – are toxic in combination with the high acidity. Whereas relatively few studies on the physiology of high acidity have been performed using invertebrates, those on the toxicity of aluminium are still more scarce. Havas & Likens (1985b) studied the mortality together with the Na^{+} balance in *D. magna* at elevated Al concentrations. Both H^{+} and Al interfered with Na^{+} regulation: at pH 6.5, Al (1.02 mg Al l^{-1}) decreased Na^{+} influx by 46% and increased Na^{+} outflux by 25%, which led to a net loss of Na^{+}. At pH 4.5, Na^{+} influx was significantly inhibited (by 73%) compared with the reference pH 6.5 treatment in the absence or in the presence of Al (0.02 mg l^{-1} or 1.02 mg l^{-1}). Aluminium, on the other hand, decreased Na^{+} outflux (by 31%) at pH 4.5, which reduced the net loss of Na^{+} and temporarily prolonged survival of the daphnids.

The waterbug *C. punctata,* which lives in acid boglakes with Al concentrations often exceeding 2 mg l^{-1} (Vangenechten *et al.*, 1984), seems very well adapted to this situation. In *C. punctata* aluminium decreased Na^{+} influx at pH 3 as well as at pH 4 (Witters *et al.*, 1984a). The decrease was concentration dependent as illustrated in Figure 4. The Na^{+} outflux however was little Al-dependent (Figure 3*(c)*) although extreme Al concentrations of 10 mg l^{-1} significantly reduced the Na^{+} outflux. The net Na^{+} gain which was observed at control conditions (pH 6 without Al) is strongly diminished at pH 3 with 600 μg Al l^{-1} and even more so at pH 3 with 10 mg Al l^{-1} (Witters, unpublished observation). But a stay for 20 h in pH 3 with 10 mg Al l^{-1} did not decrease the haemolymph Na^{+} concentration nor did a stay for as long as 13 d.

Even after 13 d, the haemolymph Na^+ concentration was not influenced by low pH nor by the elevated Al concentration of 10 mg l^{-1}.

Inter and intra species differences in sensitivity to acid stress

The influence of pH on Na^+ balance in two cladocerans (Potts & Fryer, 1979), *D. magna*, an inhabitant of calcium-rich alkaline waters with a high level of inorganic salts and *A. curvirostris* which normally is found in acid peaty waters of low salt content, has been discussed extensively in an earlier section. Na^+ balance in both species showed a different sensitivity for pH. *A curvirostris* especially retains Na^+ far better in acid waters than *D. magna* and herewith illustrates the physiological basis of the observed difference in their ecological distribution.

Within aquatic insects, Vangenechten *et al.* (1979a,b) and Vangenechten & Vanderborght (1980) compared two species of waterbugs, *Corixa dentipes* and *C. punctata*, in their reaction to acid pH. In Belgium, the former is found only in acid ion-poor boglakes, the latter on the contrary has a more ubiquitous distribution and is captured in alkaline brackish waters as well as in acid boglakes. In the experiments

Figure 4. Na^+ influx in *C. punctata* at pH 3 and at pH 4 in bogwater with different concentrations of added aluminium (drawn using data from Witters *et al.*, 1984a).

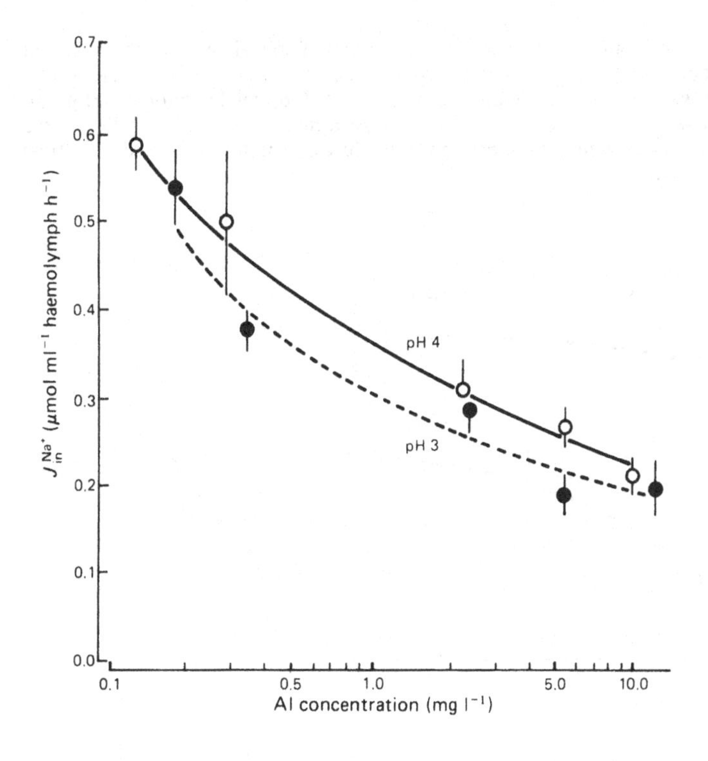

where both species were used, they were captured simultaneously in the same acid bogpool (pH around 3.8). In different respects the influence of acid pH on Na^+ influx is seen to be worse for *C. punctata*, the ubiquitous species, than in *C. dentipes*, the typical acid pool inhabitant. The latter displayed a larger increase in V_{max} at acid pH (86% compared with 55%) (Vangenechten *et al.*, 1979b), which is believed to be an indication of a reaction to overcome the increased competition between H^+ and Na^+ at the uptake carrier in conditions of acid pH. Moreover, the difference between influx at pH 3 and pH 6 was less in *C. dentipes* than in *C. punctata* (Figure 2) at least in conditions of low external Na^+ concentration as is noticed in acid pools. Unfortunately no comparison in total Na^+ balance has been made for both species.

Witters *et al.* (1984b and unpublished observations) report experiments of pH effects on Na^+ influx in *C. punctata* captured in physico-chemically differing surface waters. Within the limits of the experiments, the physiological tolerance of both groups did not differ. The response to acid pH however was significantly greater for the animals from the non-acid pool (pH 7.6) where a decrease of 50% in Na^+ influx was observed after a stay for 75 h at pH 4 compared to an insiginificant decrease in the animals from the acid pool (pH 4.1) (Figure 5). In the former animals, an increased haemolymph Na^+ content was found after the treatment, a phenomenon

Figure 5. Na^+ influx in *C. punctata*. One group of animals was caught in an acid ion-poor boglake (pH 4.1, conductivity = 100 μmho cm^{-1}) whereas the second group was caught in a circumneutral pool (pH 7.6, conductivity = 600 μmho cm^{-1}). Measurements of Na^+ influx were performed after a stay for 75 h in the bogpool water with the respective pH as illustrated in the figure (adapted from Witters *et al.*, 1984b).

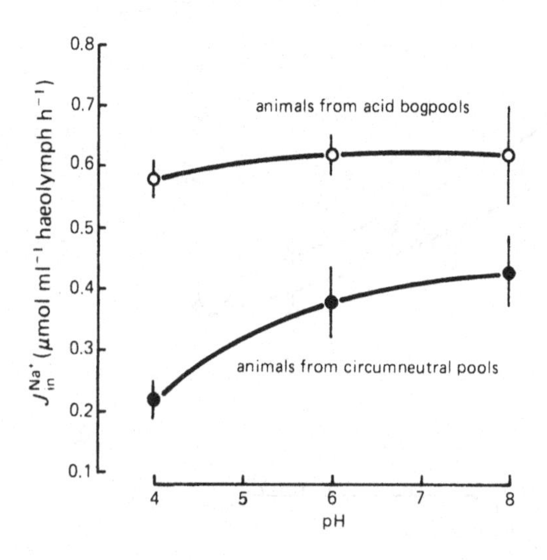

which was also observed to occur in animals from the acid pool but only after a stay in water with pH 3.0. The combination of acid pH (pH 4) and elevated Al concentrations (up to 10 mg l^{-1}) decreased the Na^+ influx further with 0.3 μmol Na^+ ml^{-1} haemolymph h^{-1} in acid pool animals and with 0.1 μmol Na^+ml^{-1} haemolymph h^{-1} in the other animals herewith indicating a decrease of 50% in both animal groups. These experiments thus illustrate that both groups of animals react differently to conditions of applied acidity (pH 4). The animals from non-acid pools experience an acid stress reaction which is comparable to the reaction of the acid-pool animals only at much lower pH (pH 3.0). It was never tested however, whether these differences between both batches of animals remain after a longer stay in acid water.

Conclusions

The pH regime, together with the concentrations of various (heavy) metals in the surface waters, has a profound efffect on the distribution of many invertebrates. Many field surveys have clearly pointed out that not only fishes are pH sensitive but invertebrates also suffer from the increased acidity. As was shown in this text, laboratory data on survival of invertebrates exist and some of these data are also explained by concomitant observations of changes in physiological parameters especially Na^+ regulation. Experiments in which the ion regulatory ability is tested in different invertebrate species under conditions of acid pH and elevated Al concentrations reveal that the influence of the ambient water quality is largely correlated with the known environmental distribution of the animals. Species which are found in acid waters display less disturbance in their Na^+ regulation than closely related species which do not live in acid waters. Even within the same species (*C. punctata*) a strongly different reaction on Na^+ influence was noticed at pH 4 when animals from an acid bog were compared with animals from a circumneutral pond. Species living in acid waters seem to display some adaptation of their Na^+ influx mechanism towards high concentrations of H^+. During conditions of acid pH a significant activation of the Na^+ carriers in the integument of insects (corixids) was observed which gives rise to a relatively high net Na^+ uptake even when competition with H^+ is severe. It would be interesting to investigate how the Na^+ influx mechanism is activated in these animals because this could contain clues which would enable us to understand the difficulties faced by freshwater fishes when adapting their influx mechanism to acid pH.

References

Brehm, J. & Meijering, M.P.D. (1982). Zur Säure-Empfindlichkeit ausgewählter süsswasser Krebse (Daphnia und Gammarus, Crustacea). On the sensitivity to low pH of some selected crustaceans (Daphnia and Gammarus, Crustacea). *Arch. Hydrobiol.*, **95**, 17–27.

Havas, M. (1985). Aluminum bioaccumulation and toxicity to *Daphnia magna* in soft water at low pH. *Can. J. Fish. Aquat. Sci.*, **42**, 1741–8.

Havas, M. & Hutchinson, T.C. (1983). Effect of low pH on the chemical composition of aquatic invertebrates from tundra ponds at the Smoking Hills, NWT, Canada. *Can. J. Zool.*, **61**, 241–9.

Havas, M. & Likens, G.E. (1985a). Toxicity of aluminum and hydrogen ions to *Daphnia catawba, Holopedium gibberum, Chaoborus punctipennis,* and *Chironomous anthrocinus* from Mirror Lake, New Hampshire. *Can. J. Zool.*, **63**, 114–9.

Havas, M. & Likens, G.E. (1985b). Changes in ^{22}Na influx and outflux in *Daphnia magna* (Straus) as a function of elevated Al concentrations in soft water at low pH. *Proc. Natl. Acad. Sci. USA*, **82**, 7345–9.

Havas, M., Hutchinson, T.C. & Likens, G.E. (1984). Effect of low pH on sodium regulation in two species of *Daphnia*. *Can. J. Zool.*, **62**, 1965–70.

Komnick, H. (1977). Chloride cells and chloride epithelia of aquatic insects. *Int. Rev. Cytol.*, **49**, 285–329.

Komnick, H. & Schmitz, M. (1977). Cutane Chloridaufnahme aus hypo-osmotischer Konzentration durch die Chloridzellen von *Corixa punctata*. *J. Insect Physiol.*, **23**, 165–73.

Leivestad, H., Hendrey, G., Muniz, I.P. & Snekvik, E. (1976). Effects of acid precipitation on freshwater organisms. In *Impact of Acid Precipitation on Forest and Freshwater Ecosystems in Norway*, ed. F.H. Braekke, pp. 86–111. Research Report. Research Report FR6/76, SNSF project, NISK, 1432 Aas-NLH, Norway.

McDonald, D.G. & Wood, C.M. (1981). Branchial and renal acid and ion fluxes in the rainbow trout, *Salmo gairdneri*, at low environmental pH. *J. exp. Biol.*, **93**, 101–18.

McDonald, D.G., Hôbe, H. & Wood, C.M. (1980). The influence of calcium on the physiological responses of the rainbow trout, *Salmo gairdneri*, to low environmental pH. *J. exp. Biol.*, **88**, 109–31.

McWilliams, P.G. & Potts, W.T.W. (1978). The effects of pH and calcium concentrations on gill potentials in the brown trout, *Salmo trutta*. *J. comp. Physiol.*, **126**, 277–86.

Meinel, W., Matthias, U. & Zimmermann, S. (1985). Okophysiologische Untersuchungen zur Säuretoleranz von *Gammarus fossarum* (Koch). Ecophysiological studies on acid tolerance of *Gammarus fossarum*. *Arch. Hydrobiol.*, **104**, 287–302.

Milligan, C.L. & Wood, C. (1982). Disturbances in haematology, fluid volume distribution and circulatory function associated with low environmental pH in the rainbow trout, *Salmo gairdneri*. *J. exp. Biol.*, **99**, 397–415.

Økland, J. (1980). Environment and snails (Gastropoda): studies of 1000 lakes in Norway. In *Proc. Int. Conf. Ecological impact of acid precipitation*, ed. D. Drabløs & A. Tollan, pp. 322–3. Oslo–Ås: SNSF Project.

Økland, K.A. (1980). Mussels and crustaceans: studies of 1000 lakes in Norway. In *Proc. Int. Conf. Ecological impact of acid precipitation*, ed. D. Drabløs & A. Tollan, pp. 224–5. Oslo–Ås: SNSF Project.

Økland, J. & Økland, K.A. (1980). pH levels and food organisms for fish: studies of 1000 lakes in Norway. In *Proc. Int. Conf. Ecological impact of acid precipitation*, ed. D. Drabløs & A. Tollan, pp. 226–7. Oslo–Ås: SNSF Project.

Overrein, L.N., Seip, H.M. & Tollan, A. (1980). *Acid precipitation – effects on forest and fish. Final report of the SNSF project 1972–80*. Research Report FR19/80, SNSF project, NISK, 1432 Aas–NLH, Norway, 175 pp.

Potts, W.T.W. & Fryer, G. (1979). The effects of pH and salt content on sodium balance in *Daphnia magna* and *Acantholeberis curvirostris* (Crustacea: Cladocera). *J. comp. Physiol.*, **129**, 289–94.

Shaw, J. (1960). The absorption of sodium by the crayfish *Astacus pallipes* Lereboullet. III. The effect of other cations in the external solution. *J. exp. Biol.*, **37**, 548–56.
Stobbart, R.H. (1967). The effect of some anions and cations upon the fluxes and net uptake of chloride in the larva of *Aëdes aegypti* (L.) and the nature of the uptake mechanism for sodium and chloride. *J. exp. Biol.*, **47**, 35–57.
Ultsch, G.R., Ott, M.E. & Heisler, N. (1981). Acid-base and electrolyte status in carp (*Cyprinus carpio*) exposed to low environmental pH. *J. exp. Biol.*, **93**, 65–80.
Vangenechten, J.H.D. (1980). Fysico-chemisch onderzoek van de verzuring in Kempische oppervlaktewaters en invloed van de zuurtegraad op de ionenregeling van waterwantsen. Physico-chemistry of the acidification process in Belgian surface waters and the effects of environmental acidity on the ionic regulation in waterbugs (Hemiptera, Heteroptera). Ph.D. Thesis, University of Antwerp, Belgium.
Vangenechten, J.H.D., Van Puymbroeck, S. & Vanderborght, O.L.J. (1979a). Basic physiological data relative to ionic regulation in two waterbugs: *Corixa dentipes* (Thoms.) and *Corixa punctata* (Illig.) (Hemiptera, Heteroptera). *Comp. Biochem. Physiol.*, **64A**, 523–9.
Vangenechten, J.H.D., Van Puymbroeck, S. & Vanderborght, O.L.J. (1979b). Effect of pH on the uptake of sodium in the waterbug *Corixa dentipes* (Thoms.) and *Corixa punctata* (Illig.) (Hemiptera, Heteroptera). *Comp. Biochem. Physiol.*, **64A**, 509–21.
Vangenechten, J.H.D., Van Puymbroeck, S. & Vanderborght, O.L.J. (1980). Effect of pH on the chloride uptake and efflux in two waterbugs (Insecta, Hemiptera) from acid freshwaters. *Comp. Biochem. Physiol.*, **67A**, 85–90.
Vangenechten, J.H.D., Van Puymbroeck, S. & Vanderborght, O.L.J. (1984). Acidification in Campine boglakes. In *Proc. Symp. on Belgian research on acid deposition and the sulphur cycle*, ed. O. Vanderborght, pp. 251–62. Mol, Belgium: SCK/CEN.
Vangenechten, J.H.D. & Vanderborght, O.L.J. (1980). Effect of acid pH on sodium and chloride balance in an inhabitant of acid freshwaters, the waterbug *Corixa punctata* (Illig.) (Insecta, Hemiptera). *Proc. Int. Conf. Ecological impact of acid precipitation*, ed D. Drabløs & A. Tollan, pp. 342–3. Oslo–Ås: SNSF Project.
Walton, W.E., Compton, S.M., Allan, J.D. & Daniels, R.E. (1982). The effect of acid stress on survivorship and reproduction of *Daphnia pulex* (Crustacea: Cladocera). *Can. J. Zool.*, **60**, 573–9.
Witters, H., Vangenechten, J.H.D., Van Puymbroeck, S. & Vanderborght, O.L.J. (1984a). Interference of aluminium and pH on the Na-influx in an aquatic insect *Corixa punctata* (Illig.). *Bull. Envir. Contam. Toxicol.*, **32**, 575–9.
Witters, H., Vangenechten, J.H.D., Van Puymbroeck, S.& Vanderborght, O.L.J. (1984b). The effect of pH and aluminium on the Na-balance in an aquatic insect *Corixa punctata* (Illig.). In *Proc. Symp. on Belgian research on acid deposition and the sulphur cycle*, ed. O. Vanderborght, pp. 287–98. Mol, Belgium: SCK/CEN.

BRIAN R. MCMAHON AND
SALLY A. STUART

The physiological problems of crayfish in acid waters

Introduction

The devastating effects of acid precipitation (resulting from the conversion of atmospheric oxides of sulphur (SO_2) and nitrogen (NO_x) to sulphuric and nitric acid), on lake fauna and flora have been well documented (Likens & Borman, 1974; Leivestadt *et al.*, 1976; Harvey & Lee, 1982). Losses of freshwater fish populations from such affected areas have been recorded (see, for example, Leivestadt & Muniz, 1976; Harvey, 1979, 1982) and possibly because of their great economic value many physiological studies have concentrated on the effects of acid exposure on fish (reviews by Wood & McDonald, 1982; McDonald, 1983; Wood, this volume). Decreases in invertebrate populations have also been reported (Abrahamsson, 1972; Almer *et al.*, 1974; France, 1983, 1985), but this has stimulated rather few studies on physiological disturbances resulting from acid exposure in invertebrate animals. Interestingly, the few studies which have been undertaken have shown invertebrates to be more tolerant to acid exposure than fish, or more accurately, than those few, mostly salmonid, fish on which physiological studies have concentrated.

A second common feature of the fish studies is that they have mostly used sulphuric acid (H_2SO_4) as the acid stressor. While H_2SO_4 precipitation still remains a most serious problem, in recent years emission of SO_2 have declined while those of NO_x have continued to rise. Despite this, few studies of the physiological effects of nitric acid (HNO_3) pollution on aquatic animals have been carried out. Thirdly, almost all previous investigations on the physiological effects of acid exposure have concentrated on acute (here defined as one to five days) exposures to acid water. Chronic effects of acid exposure as well as possible mechanisms of compensation, and mechanisms of recovery following temporary acid shock have all been largely ignored.

These are, however, important areas of concern. NO_x forms a substantial fraction of acid emissions (Haines, 1981), acid precipitation is a chronic rather than acute problem in many areas and invertebrates form an essential part of the food chain in potentially acid stressed areas. The present review thus concentrates on work on chronic physiological and other effects of increased ambient acidity (using both sulphuric and nitric acids), on aquatic invertebrates. This review will largely concentrate on two genera of crayfish, *Orconectes* and *Procambarus*. *Orconectes* sp. occurs commonly in areas of Ontario and the Northeastern United States and thus this species is perhaps currently more likely to be threatened by acid precipitation than *Procambarus* sp. which have a more southerly distribution in North America.

Tolerance of acute acid exposure

Initial studies on *Orconectes rusticus* and *Procambarus clarki* demonstrated marked tolerance to sulphuric acid exposure (96 h LC_{50} = 3.1 mM H_2SO_4 (pH 2.5) and 1.6 mM H_2SO_4 (pH 2.8) respectively (Figure 1, Morgan & McMahon, 1982; McMahon & Morgan, 1983). More recent evidence suggests that greater differences in toxicity may occur between crayfish species. *Orconectes propinquus* exposed to sulphuric acid (pH 4.0, 0.1 mM) in soft water showed 28% mortality at five days and

Figure 1. Toxicity of sulphuric acid to two species of crayfish *Orconectes rusticus* and *Procambarus clarki*. Toxicity is reported as the LC_{50} at 24–96 h, quantified in terms of molar concentration as well as pH.

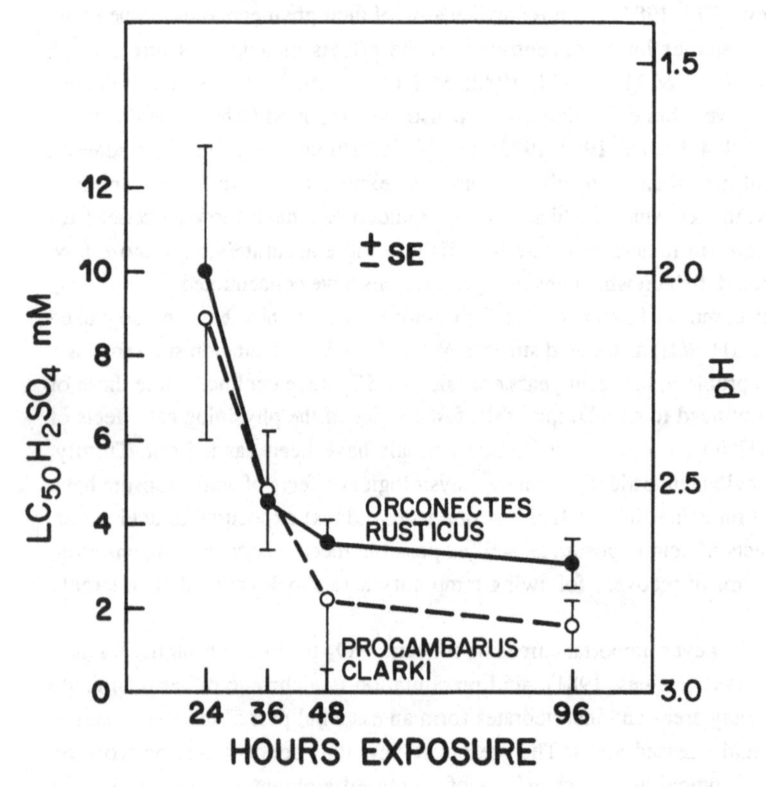

58% mortality after 12 d (Wood & Rogano, 1986). Whilst rigorous comparison is not possible due to differences in mass, acid strength, water type and time betwen these tests, *O. propinquus* is clearly less tolerant than the other species tested. Nonetheless, in general, crayfish are substantially more tolerant than either the salmonid or cyprinid fishes studied thus far (Ultsch *et al.*, 1982; McDonald & Wood, 1981; Hōbe *et al.*, 1984; Wood, 1988, this volume). These figures, however, must be interpreted with care since these are short term laboratory studies on adult, non-moulting animals. Baker (1979) and France (1983) show greater sensitivity for juveniles of *Orconectes rusticus* and *O virilis* respectively, while Malley (1980) showed reduction of tolerance associated with moulting. Thus natural populations of *Orconectes* sp. are likely to be substantially less tolerant and France (1983) has shown reduction of female reproductive success and reduction in carapace rigidity in *O. virilis* and Abrahamsson (1972) has shown mortality of *Astacus astacus*, in 'naturally' stressed populations at 0.003 mM (pH 5.6). Additionally France & Graham (1985) showed decline in resistance to infection in an acid-stressed crayfish population.

Physiological effects of acute acid exposure

It is necessary here to define the terms acute and chronic. In the current review we will define acute as being a period of one to seven days and chronic as being from 7–60 d maintained acid exposure. These are substantially longer periods than those used in the majority of work on fish (see, for example, McDonald, 1983) but were chosen as being more meaningful in time of natural acid exposure levels.

Measurements of physiological changes in acute sublethal exposure to sulphuric acid have been carried out both on fish and on crayfish. Responses of several species of salmonid fish are reviewed by Wood & McDonald (1982) and McDonald (1983), while a lesser volume of work on cyprinid and other fishes is described in Ulsch *et al.*, 1981; Hōbe *et al.*, 1984. Additionally pertinent data for fish species are reviewed by Wood, 1986. Briefly, the effects of acid exposure in fish involve disruption of blood acid–base status, and blood and muscle ion status and at higher acid levels, possibly disruption of mechanisms of blood O_2 uptake and supply. The relative contributions of ionic and acid base disruption vary with the degree of hardness (amounts of Ca^{2+} and possibly Na^+ and other ions) in the water used. Animals treated in hard water tend to show a greater degree of blood acid–base disturbance, while animals exposed in soft water show relatively less acid–base disturbance but a greater degree of disruption of blood ionic status (McDonald, 1983; Hōbe *et al.*, 1984). Since the most acid threatened areas are those with relatively soft water a majority of studies have concentrated on this aspect.

Physiological responses of several crayfish to a similar acute acid stress have been investigated. These include *Orconectes rusticus* (Morgan & McMahon, 1982; Wood & Rogano, 1986; *Astacus astacus* (Jarvenpaa *et al.*, 1983); *Procambarus clarki*

(McMahon & Morgan, 1983; Patterson, 1983) and *Orconectes propinquus* (Wood & Rogano, 1986). The majority of these studies have used a H^+ concentration of approximately 0.1 mM (pH = 4.0) and sulphuric acid and are thus easily comparable.

Changes in acid–base status

In *Procambarus clarki* exposed to sublethal levels of sulphuric acid in hard water (H^+ = 0.16 mM, pH = 3.8, 1/10 LC_{50}) initial responses included development

Figure 2. Acid–base disturbance in *Procambarus clarki* haemolymph during and following four days' exposure to hard water acidified to pH 3.8 with sulphuric acid (0.16 mM). Post-branchial (arterial) haemolymph *(a)* pH; *(b)* Total CO_2 concentration ([CO_2]) *(c)* carbon dioxide partial pressure (Pa_{CO_2}). Following a two day acclimation period in decarbonated hard water, animals were transferred to water which had been acidified to pH 3.8 and subsequently decarbonated. After four days' acid exposure animals were transferred to water at normal pH for recovery. Mean data for acid-treated animals are shown as dashed lines. Solid lines denote data from 'control' animals, treated similarly but not acid-exposed. Vertical bars at each data point show 1 SEM. Asterisks show points of significant difference from initial (day –2) values. From Morgan & McMahon (1982).

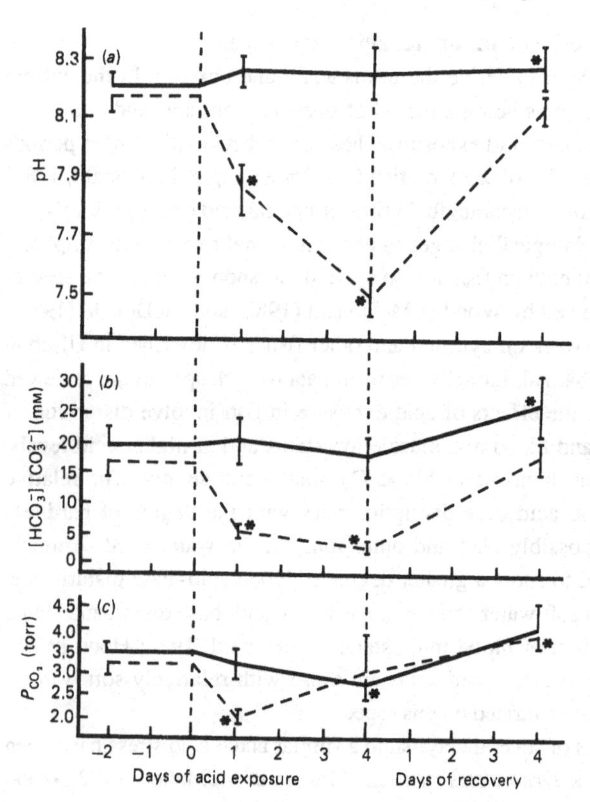

of marked haemolymph acidosis with $[H^+]_p$ increasing significantly from 6.92 to 14.13 nM H^+ (pH decrease from 8.16 to 7.86) by 24 h exposure and to 32.0 nM H^+ (pH 7.495) by 96 h (Figure 2, Morgan & McMahon, 1982). These high pH values are characteristic of this species, especially in hard water. At 24 h this acidosis is associated with reduction of both P_{CO_2} and C_{CO_2}. This presumably results from both the change in the carbonate equilibrium and from CO_2 washout associated with the hyperventilation which occurs initially in acid exposure (Patterson, 1983). The hyperventilation is transient and haemolymph P_{CO_2} returns to control levels by 96 h acid exposure (Figure 2). Total CO_2 remained significantly depressed throughout acid

Figure 3. Disturbance in *Procambarus clarki* haemolymph ion status during, and following, four days' exposure to hard water acidified to pH 3.8 with sulphuric acid (0.16 mM). Post-branchial (arterial) haemolymph contents of *(a)* sodium, *(b)* chloride, *(c)* potassium, *(d)* magnesium, *(e)* calcium. Experimental protocol as described in legend of Figure 2. Data for acid-treated animals are shown as dashed lines. Solid lines denote data from 'control' animals, treated similarly but not acid-exposed. Vertical bars at each data point show 1 SEM. Asterisks show points of significant difference from initial (day −2) values. From Morgan & McMahon (1982).

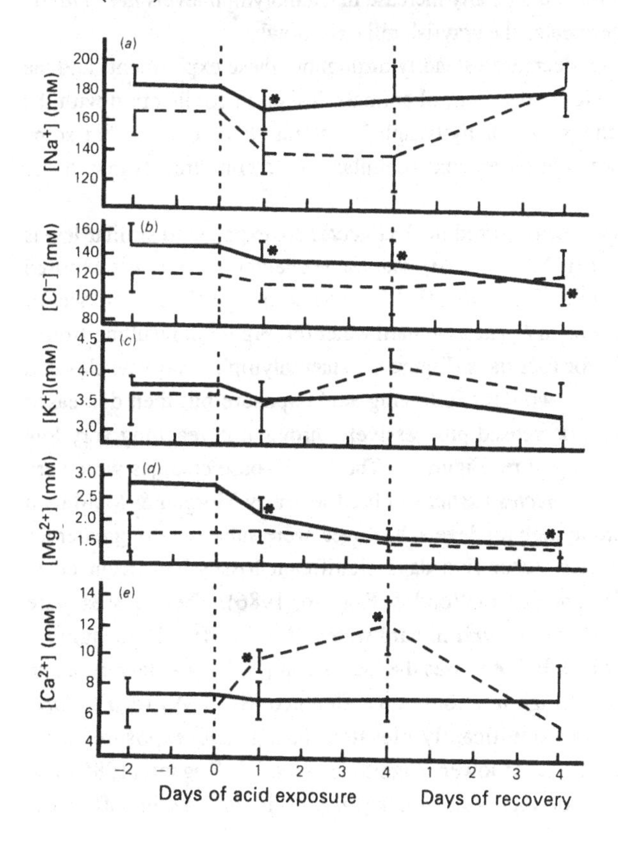

exposure. Essentially similar responses were observed in *Orconectes rusticus* exposed to a similar level of H_2SO_4 in hard water, except that the acidosis and depression of haemolymph [CO_2] were smaller and P_{CO_2} increased rather than decreased initially.

Changes in haemolymph ion status

Procambarus clarki exposed to sulphuric acid in hard water showed minimal changes in haemolymph ion status over a four day exposure period (Figure 3). Only $[Ca^{2+}]_p$ changed significantly, increasing dramatically at both one day and further at the four day measurement points. This maintenance of haemolymph ion levels in acid is the more remarkable since the levels of most haemolymph ions in control animals generally decreased (significantly for $[Mg^{2+}]_p$ and $[Cl^-]_p$) throughout the experimental period, possibly associated with starvation. Similar changes in haemolymph acid base and ion status occurred in hard water exposed *O. rusticus* (with the exception that no change in haemolymph P_{CO_2} occurred (Morgan & McMahon, 1982)). Although ambient SO_4^{2-} level increased five-fold during acidification this was not reflected by any increase in haemolymph levels in *P. clarki*, suggesting that this ion permeates the crayfish gill only slowly.

Haemolymph osmolarity decreased steadily throughout these experiments and the decrease was greater in acid-treated control animals. Taken in conjunction with the absence of significant change in haemolymph ion status this suggests that some reallocation of fluid between intra- and extracellular fluid compartments may occur during acid exposure.

Somewhat similar responses occurred in *Orconectes* sp. exposed to similar levels of acid in soft (Ca^{2+} 0.2 mM, Na^+ 0.2 mM) water. *Orconectes propinquus* exposed for 5–12 d to sulphuric acid (0.1 mM, pH 4.0) showed a significant acidosis of similar magnitude to that seen in *P. clarki* in hard water (cf. Figure 2) resulting from a progressive decline in pH for four days (Figure 4). Haemolymph P_{CO_2} was elevated slightly but significantly for two days following acid exposure but then decreased progressively, while [CO_2] decreased progressively throughout, reaching very low levels (<1 mM) after 12 d exposure (Figure 4). These acid–base changes were very similar to those seen for *Orconectes rusticus* in hard water by Morgan & McMahon (1982). Decreases in haemolymph ion levels, however, were substantially greater for *O. propinquus* in soft water. After five days significant losses had occurred in haemolymph [Na^+], [K^+] and [Cl^-] (Wood & Rogano, 1986). These losses were greater than those recorded for *P. clarki* in hard water ($[Na^+]_p$ −46:−25 mequiv l^{-1} and $[Cl^-]_p$ −32:−4 mequiv l^{-1}). In both cases the increase in $[Na^+]_p$ was larger than in $[Cl^-]_p$ indicating possible change in strong ion difference (SID). As for *P. clarki* [Ca^{2+}] in haemolymph was significantly elevated during acid exposure in *O. propinquus* but the increase was of lower magnitude. Wood & Rogano (1986) also measured decreases in haemolymph ions in *O rusticus*, acid-stressed in soft water.

The results (Figure 5) are similar to those of *O. propinquus*, showing marked and progressive reduction in $[Na^+]_p$ throughout five days exposure, but different in that $[K^+]_p$ showed a significant increase, and in the magnitude of increase in $[Ca^{2+}]_p$. These results largely confirmed preliminary data for *O. rusticus* (Morgan & McMahon, 1982).

Comparable data for *P. clarki* exposed to sulphuric acid in soft water are available only at one and seven days (Figure 6; S.A. Stuart & B.R. McMahon, unpublished

Figure 4. Disturbance in haemolymph acid–base status in soft water acclimated *Orconectes propinquus* exposed for 5–12 d to sulphuric acid (0.1 mM, pH 4.0). Post-branchial (arterial) *(a)* pH, *(b)* bicarbonate content, *(c)* carbon dioxide partial pressure (Pa_{CO_2}). The vertical dashed line indicates transfer to acidified soft water. Close symbols illustrate data for acid-treated animals, open symbols data for control animals treated similarly but not acid-exposed. Vertical bars at each data point show 1 SEM. Asterisks denote significant difference from initial (day 0) values in either vase. From Wood & Rogano (1986).

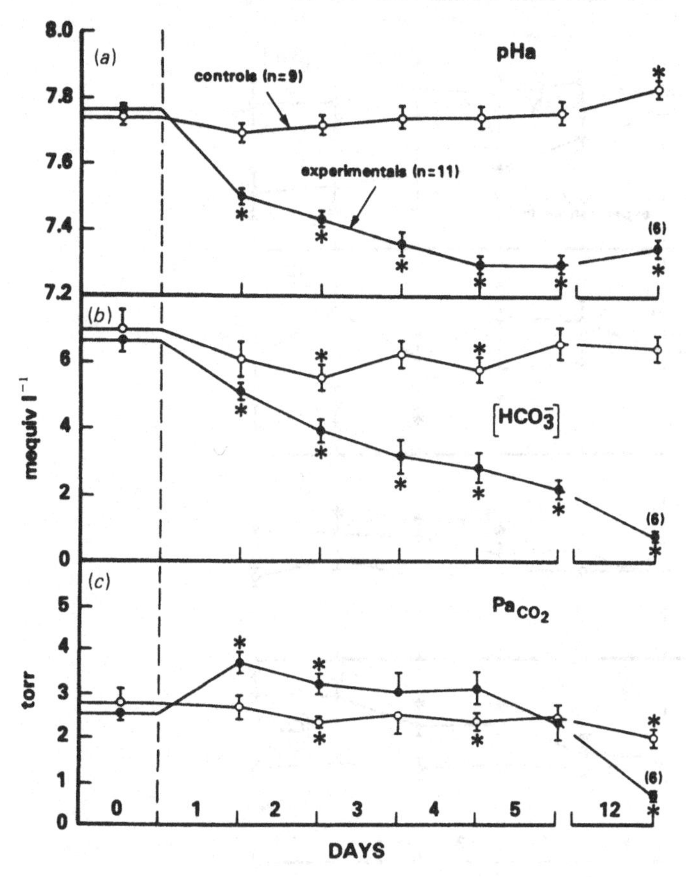

data). Haemolymph of *P. clarki* showed progressive acidosis at both one and seven days, although the rate of decline was reduced after day one. The magnitude of this haemolymph acidosis was similar in terms of the final pH but marginally less in terms of H^+ than occurred for this species in hard water. Also in contrast with the results in hard water, PCO_2 increased initially and remained elevated at seven days, while no significant decrease in $[CO_2]$ occurred. The degree of acidosis and depression of HCO_3^- are substantially less than observed in *O. propinquus* in soft water (Wood & Rogano, 1986).

Figure 5. Disturbance in haemolymph ion status in *Orconectes rusticus* exposed for five days to soft water acidified to pH 4.0 with sulphuric acid. Post-branchial (arterial) haemolymph *(a)* sodium, *(b)* chloride, *(c)* potassium, *(d)* calcium content. Closed symbols illustrate data from acid treated animals, open symbols data for control animals treated similarly but not acid-exposed. Vertical bars at each data point show 1 SEM. Asterisks denote significant difference from initial (day 0) values in either case. From Wood & Rogano (1986).

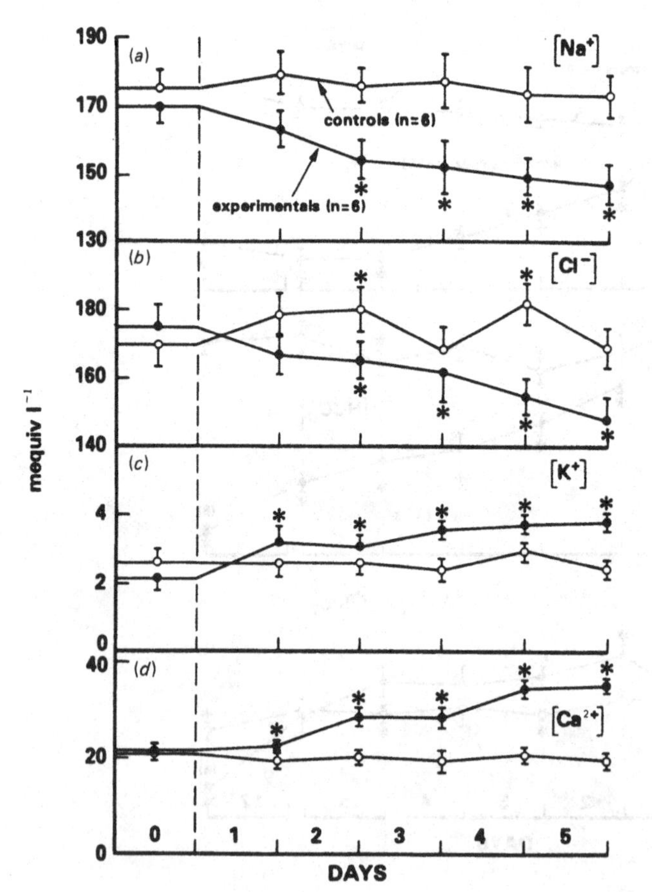

Changes in haemolymph ion status in soft water acid-exposed *Procambarus clarki*, however, are completely different from results for any other crayfish in soft water (Figures 7, 8; S.A. Stuart & B.R. McMahon, unpublished data). In this species no significant change in measured ion occurred in the first seven days of acid exposure.

Figure 6. Acid–base disturbance in *Procambarus clarki* haemolymph during and following chronic (60 d) exposure to soft water acidified to pH 4.0 with sulphuric acid (0.1 mM). Post-branchial (arterial) haemolymph *(a)* pH, *(b)* carbon dioxide partial pressure (Pa_{CO_2}). *(c)*Total CO_2 concentration [CO2]). Following acclimation in soft water, animals were transferred to water which had been acidified to pH 4.0. After 60 d acid-exposed animals were transferred to water at normal pH and monitored over a subsequent 35 d recovery period. Mean data for acid-treated animals are shown as dashed lines. Solid lines denote data from 'control' animals, treated similarly but not acid-exposed. Vertical bars at each data point show 1 SEM Asterisks show points of significant difference from control values at that interval. Note X-axis scale is discontinuous between 21 and 60 d acid exposure and between 7 and 28 d recovery.

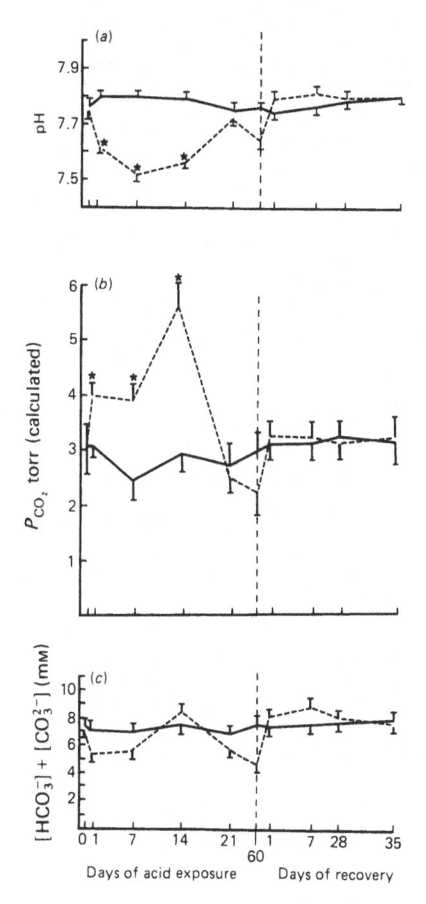

The exception was a significant two-fold elevation in [SO_4^{2-}] indicating that some permeability to this ion may occur in the crayfish gill in soft water.

The differences in toxicity between crayfish species suggested above is apparently borne out in these short term physiological experiments, to the extent that

Figure 7. Disturbance in *Procambarus clarki* haemolymph cation status during and following chronic (60 d) exposure to soft water acidified to pH 4.0 with sulphuric acid (0.1 mM). Post-branchial (arterial) haemolymph *(a)* sodium, *(b)* potassium, *(c)* calcium content. Following acclimation in soft water, animals were transferred to water which had been acidified to pH 4.0. After 60 d acid exposure animals were transferred to water at normal pH and monitored over a subsequent 35 d recovery period. Mean data for acid treated animals are shown as dashed lines. Solid lines denote data from 'control' animals, treated similarly but not acid exposed. Vertical bars at each data point show 1 SEM. Asterisks show points of significant difference between acid-treated and control values at that interval. Note X-axis scale is discontinuous betwen 21 and 60 d acid exposure and between 7 and 28 d recovery.

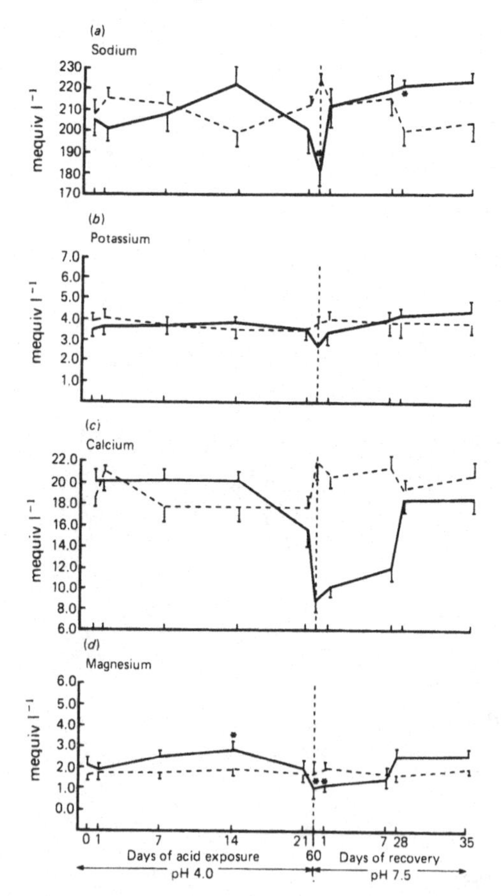

O. propinquus, the only species which showed significant mortality in acute exposure, also showed both greater ionic loss and greater acid–base disturbance. It should be noted here that the individual *O. propinquus* tested were smaller than either *O. rusticus* or *P. clarki*. Hypoxic exposure apparently increased both acid–base and ionic disturbance resulting from exposure to acid waters in *A. astacus* (Jarvenpaa *et al.*, 1983).

Taken overall these results show both similarity and difference to those reported for salmonid and other fish species (Jarvenpaa *et al.*, 1983). For *Orconectes* sp. they are similar in that a greater ionic depletion is observed during acid exposure in soft as opposed to hard water. This, however, is clearly not true for *P. clarki*. The results are

Figure 8. Disturbance in *Procambarus clarki* haemolymph anion status during and following chronic (60 d) exposure to soft water acidified to pH 4.0 with sulphuric acid (0.1 mM). Post-branchial (arterial) haemolymph *(a)* chloride, *(b)* sulphate, *(c)* nitrate. Following acclimation soft water, animals were transferred to water which had been acidified to pH 4.0. After 60 d acid exposure animals were transferred to water at normal pH and monitored over a subsequent 35 d recovery period. Mean data for acid-treated animals are shown as dashed lines. Solid lines denote data from 'control' animals, treated similarly but not acid-exposed. Vertical bars at each data point show 1 SEM. Asterisks show points of significant difference between acid-treated and control values at that interval. Note X-axis scale is discontinuous between 21 and 60 d acid exposure and between 7 and 28 d recovery.

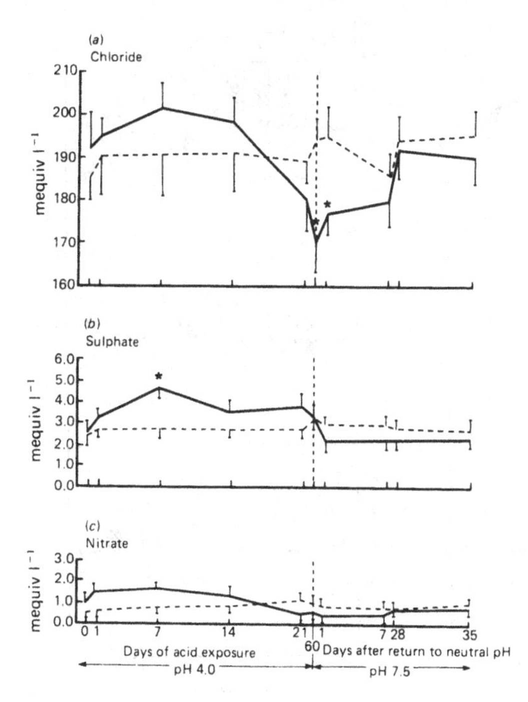

strikingly different, however, in that marked acid–base disturbance occurs in all crayfish exposed to H_2SO_4 in soft water exposed *Orconectes* sp. thus more closely resemble those of trout and cyprinid fish in hard water (Hõbe *et al.*, 1983; McDonald & Wood, 1983). This (as noted by Wood & Rogano, 1986) may result from the much greater Ca^{2+} levels available in crayfish or may be due to the continued presence of Ca^{2+} at the gill (i.e. in cuticle) which reduces the permeability changes resulting from acid exposure in this species.

Only the two most recent studies (McMahon & Stuart, 1985; S.A. Stuart & B.R. McMahon, unpublished data; Wood & Rogano, 1986) have attempted to isolate the mechanisms by which the acid–base and ionoregulatory changes observed occur.

Figure 9. Efflux and influx of *(a)* titratable acidic equivalents and ammonium ions, *(b)* sodium ions, *(c)* chloride ions, to and from *Orconectes propinquus* prior to, during and in recovery from exposure to soft water at neutral pH or acidified to pH 4.0 with sulphuric acid. Positive and negative going histogram bars indicate the magnitude of influx and efflux respectively. Shaded areas show the extent and direction of net flux, i.e. the sum (signs considered) of efflux and influx. The histogram set on the right presents data for acid-treated animals. Data for control animals are included on the left to facilitate comparison. Bars at the top of each histogram indicate 1 SEM. From Wood & Rogano (1986).

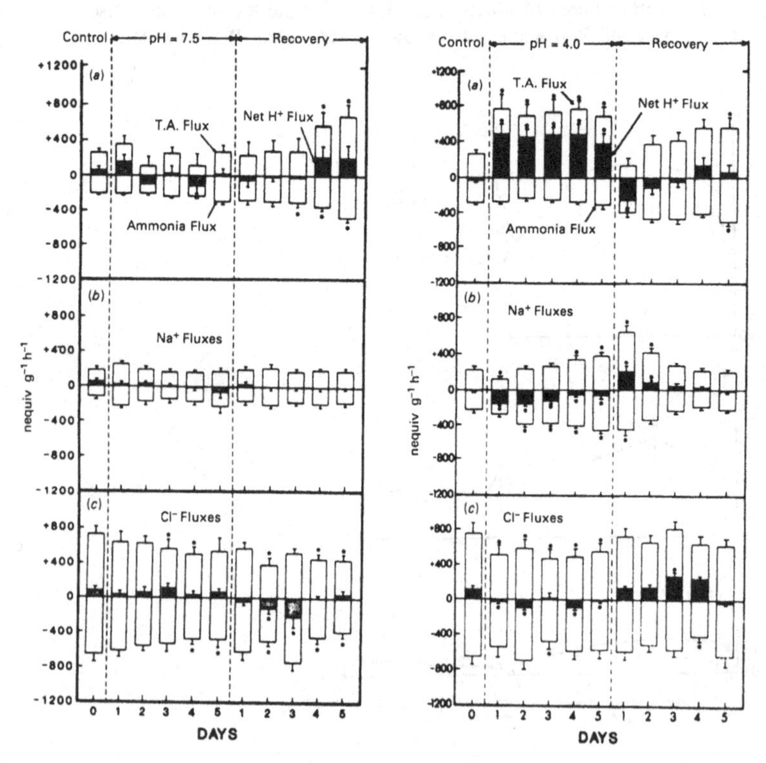

Wood & Rogano (1986) measured influx and efflux of ions across the body surface in *O. propinquus* exposed for five days to sulphuric acid in soft water. In water of neutral pH *O. propinquus* exhibited a slight efflux of protons (or influx of base) and while Cl^- and Na^+ showed substantial influx and efflux components these and other ions were clearly in balance across the gills since no significant net flux could be determined (Figure 9*(a)*). In acid exposure a marked net influx of protons occurred. This was associated (Figures 9, 10) with net efflux of Na^+, K^+, Ca^{2+}, Cl^-, and SO_4^{2-} confirming the loss of haemolymph ions for most species in soft water outlined above. The influx of protons, presumably occurring at the gills, is equal throughout the acid exposure period, yet accumulation of H^+ in haemolymph remains stable after 24 h (Figure 9*(b)*). Wood & Rogano (1986) conclude that within 24 h most of the protons (acidic equivalents) entering the animal are effectively removed from the haemolymph either by ionic exchange with the tissue or by dissolution of Ca^{2+} from the skeletal compartment (or other $Ca(CO_3)$ stores). The elevated haemolymph [K^+] lends support to the involvement of K^+/H^+ exchange with the tissues while similar trends for haemolymph [Ca^{2+}] and Ca^{2+} efflux support the latter contention: presumably both systems are involved. As has been pointed out previously (Morgan & McMahon, 1982) dissolution of skeletal $Ca(CO_3)$ can be only a

Figure 10. Net flux of *(a)* potassium, *(b)* calcium ions, *(c)* sulphate ions, to and from *Orconectes propinquus* prior to, during and in recovery from exposure to soft water at neutral pH or acidified to pH 4.0 with sulphuric acid. Positive or negative going histogram bars indicate net influx or efflux respectively. The histogram set on the right presents data for acid-treated animals. Data for control animals are included on the left to facilitate comparison. Bars at the top of each histogram indicate 1 SEM. From Wood & Rogano (1986).

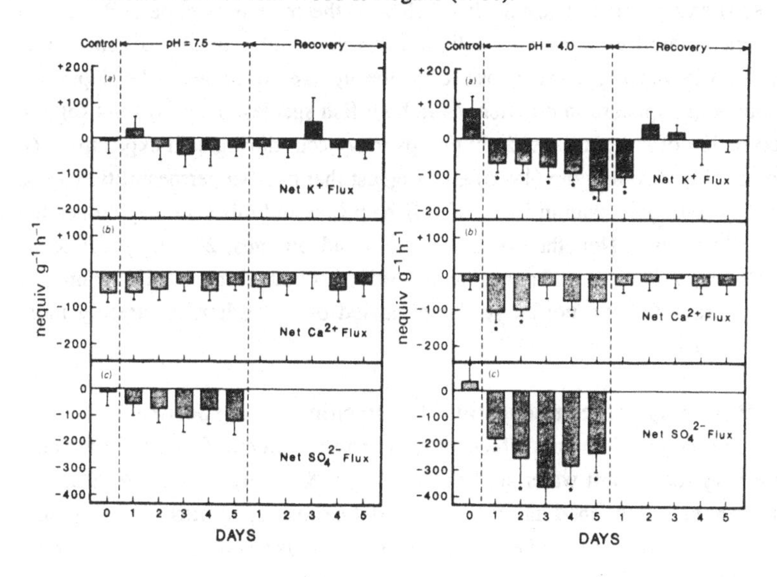

temporary measure since eventually skeletal weakening and problems of hardening following moulting will result. Indications of both problems, i.e. reduction in skeletal rigidity and problems in postmoult calcification, are already evident in the literature for naturally acid-stressed *Orconectes* populations (Malley, 1980; France, 1983, 1985).

For both Na^+ and Cl^- flux, the net decrease during acid exposure was initially associated with decreased influx, presumably due to inhibition of the inwardly directed pumps by the increase in external acidity. While this inhibition is maintained for Cl^-, Na^+ influx recovers quickly and, in fact, has increased by four to five days acid exposure (Figure 9*(b)*) to levels significantly above control. Sodium efflux, however, increases progressively throughout acid exposure in *O. propinquus*. Thus, although there is a progressive reduction in net efflux, this does not allow a return to net Na^+ balance and the animals continue to lose sodium to the environment throughout. Similar losses occur for K^+, Ca^{2+} and Cl^-. A small, but measurable SO_4^{2-} influx in control conditions, and a marked and significant efflux of SO_4^{2-} during acid exposure, again confirm the permeability of this ion across the crayfish gill in soft water. Since the gills of the sucker are also clearly permeable to SO_4^{2-} previous assumptions as to the impermeability of gills to this ion and therefore the absence of effects of the SO_4^{2-} ion on internal ionic balance must be seriously questioned.

Wood & Rogano (1986) again note a general similarity between these flux data for crayfish in soft water and those reported for the trout in hard water. Some differences nonetheless occur. Whereas inhibition of active Na^+ uptake pumps was maintained throughout acid exposure in both trout and sucker (McDonald & Wood, 1983; Höbe *et al.*, 1984) and passive efflux declined slowly, the reverse is apparently true in crayfish since inhibitory effects on sodium influx are quickly overcome, but cannot keep up with increasing passive loss. Chloride uptake continues to be depressed throughout acid exposure in crayfish, as in both fish species. These results suggest that acclimation of at least the sodium pumps may occur during acid exposure in *O. propinquus*. Wood & Rogano (1986) also suggest that passive permeability may be lower in *O. propinquus* than in (salmonid) fish species and this also may reduce ion loss in the long term. Nonetheless, despite these advantages, 28% *O. propinquus* could not survive five day acid exposure and 54% succumbed in 12 d. Unfortunately similar flux studies have not as yet been carried out on other, apparently more resistant, crayfish species.

Physiological compensation to chronic acid exposure

Chronic physiological effects of acid exposure on crayfish are known only from two very recent soft water studies (McMahon & Stuart, 1985; S.A. Stuart & B.R. McMahon, unpublished data) detailing 60 d exposures to nitric and sulphuric acids respectively. The effects of exposure to the two acids are substantially different.

Mortality

Crayfish in these studies were kept in somewhat crowded conditions and showed some mortality in control and experimental groups. To 21 d no significant difference occurred between acid-exposed and control crayfish. Subsequently mortality doubled in the acid-treated groups reaching 44–50% at 60 d exposure. Following return to neutral water mortality continued at an equal rate in both groups, reaching 24–30% in control and 52–62% in the acid treated groups at 100 d. Both the magnitude of, and time course for, mortality were essentially similar in both HNO_3 and H_2SO_4 treated groups.

Physiological effects of chronic exposure to sulphuric acid

Acid–base status

The significant acidosis resulting from exposure of *P. clarki* to H_2SO_4 in soft water (above) continued for 14 d acid exposure (Figure 6). The rate of decline was greatest over the first 24 h and peak levels of acidosis occurred at seven days. Haemolymph pH then gradually increased and returned to control levels between 14 and 21 d acid exposure. P_{CO_2} levels were also significantly elevated during the period of the acidosis reaching maximum at 14 d declining to control levels by 21 d and continuing to decline for the last 46 d of the acid-exposure period (Figure 6).

Ion status: haemolymph

Estimates of ICFV and ECFV (Cl^-/K^+ space) indicate that a 12% decrease in ECFV occurs during the first 24 h of acid exposure, suggesting a shift of fluid out of the extracellular compartment. The occurrence of haemoconcentration at this time is confirmed by increase in both haemolymph $[Cu^{2+}]$ and haemolymph protein concentration (both components of the respiratory carrier molecule haemocyanin). At least in part, the fluid lost from the haemolymph may enter the intracellular compartment, which increases 2.2% over the same time period (Table 1). Despite this fluid loss and the more prolonged acidosis (Figure 6) few significant changes in haemolymph ion concentration accompanied initial acid exposure (Figures 7, 8). The anticipated 12% increase in ion levels resulting from haemoconcentration was observed only in $[Cu^{2+}]_p$ and $[NO_3^-]_p$ and in $[SO_4^{2-}]_p$ indicating that the loss of extracellular fluid may have masked an initial loss of haemolymph ions. Following seven days exposure measured haemolymph ion levels increased initially, reaching a peak or plateau by 7–14 d. In the absence of further significant change in haemolymph pH, or of further increase in $[SO_4^{2-}]_p$, after 14 d all haemolymph ion contents decreased (Figures 7, 8) until all ion contents, except $[SO_4^{2-}]_p$ and $[NO_3^-]_p$, were significantly depressed by 60 d acid exposure. ECFV and ICFV remained essentially unchanged over day acid values indicating maintained loss of extracellular fluid (Table 1).

Table 1. *Calculated values of muscle extracellular (ECFV) and intracellular (ICFV) fluid volume during acid exposure*

Ambient pH	Days Spent at pH	HNO_3 ECFV[a]	HNO_3 ICFV[a]	H_2SO_4 ECFV[a]	H_2SO_4 ICFV[a]
7.5–8.0	Mean over 21 d	0.145	0.675	0.162	0.678
4.0	1	0.099	0.721	0.143	0.683
4.0	7	0.094	0.736	–	–
4.0	14	0.099	0.741	–	–
4.0	21	0.094	0.736	0.148	0.549
4.0→7.5	7	0.138	0.692	0.158	0.678

(a)Measured in l kg wet wt^{-1}.

Ion status in muscle

Ion concentrations $[X^+]_m$ were also measured in muscle samples at one day and 21 d acid exposure. Since these contained some ions originating in the ECFV the original values are not representative of intracellular ion levels $[X^+]_i$. The latter, however, were estimated using the following equation.

$$[X^+]_i = \frac{[X^+]_m\text{-muscle extracellular ion content}}{\text{total tissue water ECFV}}$$

(*cf.* McDonald & Wood, 1981). Muscle extracellular ion content was taken as the product of haemolymph ion content $[X^+]_p$ and muscle ECFV. Muscle ECFV was estimated as the Cl^-, K^+ space, calculated using the equations given in Bedford & Leader (1977). Although this method does not give absolute values for intracellular ion contents, it is sufficiently accurate to display major trends (Bedford & Leader, 1977; Houston & Mearow, 1979).

In initial (one day) exposure only $[Na^+]_i$ increased significantly (Table 2). Intracellular $[Cl^-]$ (one day) increased and $[K^+]_i$, $[Ca^{2+}]_i$ and $[Mg^{2+}]_i$ decreased but not significantly. At 21 d exposure measured changes in all muscle ion levels were almost completely opposite in direction to those measured in the first seven days, $[Ca^{2+}]_i$, $[Mg^{2+}]_i$ and $[K^+]_i$ increased to levels significantly above control, $[Na^+]_i$ and $[Cl^-]_i$ returned almost to control levels. The overall changes in intracellular ions at 21 d, however, are relatively slight and apparently do not approach lethal levels.

Carapace ion levels

The predominant ion in the carapace was $[Ca^{2+}]$ which comprised 97.5% of the total. $[Mg^{2+}]_c$ at 1.5%, $[Na^+]_c$ at 1% and $[K^+]_c$ at 0.2% were the largest minor constituents. The small amounts of Na^+ and K^+ possibly represent tissue and body

Table 2. *Intramuscular ion contents (Mequiv kg cell H_2O) in Procambarus clarki prior to, during and following chronic (21 d) exposure to H_2SO_4*

	Control $\bar{X}$ 21 d period	Acid Treated		Return to Neutral pH
Na^+	6.3	18.2[a]	7.9	6.9
K^+	129.8	120.6	155.3[a]	139.1
Ca^{2+}	13.82	11.99	26.0[a]	18.9
Mg^{2+}	27.3	26.9	34.6[a]	30.9
Cl^-	16.4	18.6	17.2	17.3
Cu^{2+}	0.36	0.56	0.34	0.37

[a] Significant elevation over control.

fluids incorporated into or adhering to the carapace at time of sampling. At one day of acid exposure $[Ca^{2+}]_c$ and $[Na^+]_c$ were reduced but not significantly, $[Mg^{2+}]_c$ was significantly elevated. By 21 d exposure $[Ca^{2+}]_c$ had decreased significantly (10%) but no other carapace ion level was significantly different from control.

Nitric acid

Since nitric acid is a very poorly studied acid pollutant and its physiological effects are very poorly known, both acute and chronic effects will be examined in detail for comparison with chronic effects of H_2SO_4.

Acute responses

Haemolymph acid–base status

Physiological responses to nitric acid exposure have been reported only for *Procamburus clarki* (Stuart & McMahon, 1984; McMahon & Stuart, 1985). At 24 h HNO_3 exposure a significant acidosis occurs, however, unlike the situation for soft water H_2SO_4, the acidosis is associated with significant reduction in both haemolymph P_{CO_2} and $[CO_2]$. By two days HNO_3 exposure haemolymph pH had recovered slightly, but remained significantly depressed, both P_{CO_2} and $[CO_2]$ had increased and were no longer significantly lower than control values (Figure 11).

Haemolymph ion status

Associated changes in haemolymph ion status (Figures 12, 13) included increase in $[NO_3^-]_p$, $[SO_4^{2-}]_p$, and $[K^+]_p$ and decrease in $[Na^+]_p$ and $[Cl^-]_p$, none of these trends, however, were significant at this time.

Estimates of muscle extracellular fluid (ECFV) and intracellular fluid volume (ICFV) indicated that after 24 h acid exposure ECFV decreased by 32% while ICFV increased by 6.8%, suggesting that a shift in water from ECF to ICF (Table 1) accompanied initial nitric acid exposure. Confirmation of a loss of water from haemolymph was noted by increase in both oxygen capacity and haemolymph protein at one day's acid exposure (Stuart & McMahon, 1984). This 'haemoconcentration' was reflected in increase in only $[SO_4^{2-}]_p$ and $[NO_3^-]_p$. A decreasing trend in

Figure 11. Acid–base disturbance in *Procambarus clarki* haemolymph during and following chronic (60 d) exposure to soft water acidified to pH 4.0 with nitric acid (0.1 mM). Post-branchial (arterial) haemolymph *(a)* pH, *(b)* carbon dioxide partial pressure (Pa_{CO_2}), *(c)* total CO_2 concentration ($[CO_2]$). Following acclimation in soft water, animals were transferred to water which had been acidified to pH 4.0. After 60 d acid exposure animals were transferred to water at normal pH and monitored over a subsequent 35 d recovery period. Mean data for acid-treated animals are shown as dashed lines. Solid lines denote data from 'control' animals, treated similarly but not acid-exposed. Vertical bars at each data point show 1 SEM. Asterisks show points of significant difference from control values at that interval. Note X-axis scale is discontinuous betwen 21 and 60 d acid exposure and between 7 and 28 d recovery.

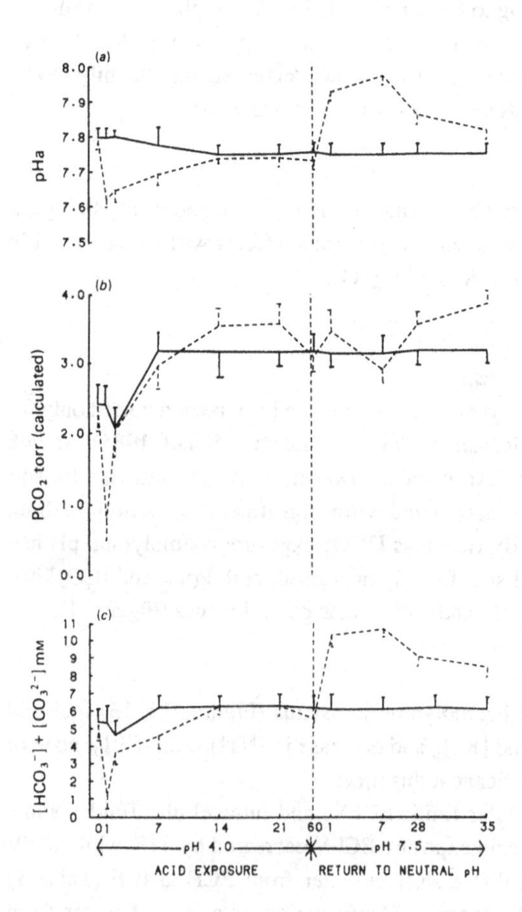

the other haemolymph ions measured suggests that initial ion loss from haemolymph was more extensive than appears in Figures 12 and 13, but that this was masked by a loss of fluid from the haemolymph compartment.

Figure 12. Disturbance in *Procambarus clarki* haemolymph cation status during and following up to chronic (60 d) exposure to soft water acidified to pH 4.0 with nitric acid (0.1 mM). Post-branchial (arterial) haemolymph *(a)* potassium, *(b)* calcium, *(c)* magnesium, *(d)* sodium content. Following acclimation in soft water, animals were transferred to water which had been acidified to pH 4.0. AFter 60 d acid-exposure animals were transferred to water at normal pH and monitored over a subsequent 35 d recovery period. Mean data for acid-treated animals were transferred to water at normal pH and monitored over a subsequent 35 d recovery period. Mean data for acid-treated animals are shown as dashed lines. Solid lines denote data from 'control' animals, treated similarly but not acid-exposed. Vertical bars at each data point show 1 SEM. Asterisks show points of significant difference between acid-treated and control values to the interval. Note X-axis scale is discontinuous between 21 and 60 d acid exposure and between 7 and 28 d recovery.

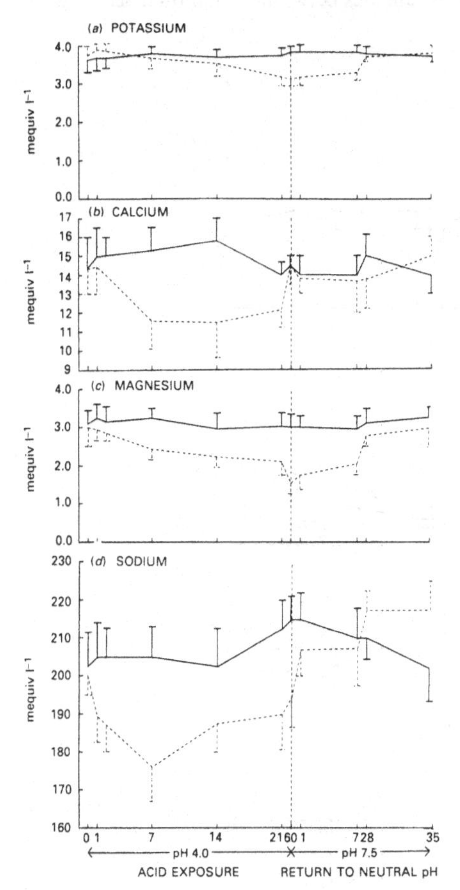

Muscle ion levels

Despite the apparently increased muscle water content (Table 1) only $[K^+]_i$ and $[Cl^-]_i$ decreased at one day's acid exposure and only the decrease in $[Cl^-]_i$ was significant. No significant change occurred in $[Mg^{2+}]_i$, while $[Ca^{2+}]_i$ and $[Na^+]_i$ increased significantly (Figure 14). These data are basically similar for $[K^+]_i$,

Figure 13. Disturbance in *Procambarus clarki* haemolymph anion status during and following chronic (60 d) exposure to soft water acidified to pH 4.0 with nitric acid (0.1 mM). Post-branchial (arterial) haemolymph *(a)* sulphate, *(b)* nitrate, *(c)* chloride. Following acclimation in soft water, animals were transferred to water which had been acidified to pH 4.0 and subsequently decarbonated. After 60 d acid exposure animals were transferred to water at normal pH and monitored over a subsequent 35 d recovery period. Mean data for acid-treated animals are shown as dashed lines. Solid lines denote data from 'control' animals, treated similarly but not acid-exposed. Vertical bars at each data point show 1 SEM. Asterisks show points of significant difference between acid-treated and control values at that interval. Note X-axis scale is discontinuous between 21 and 60 d acid exposure and between 7 and 28 d recovery.

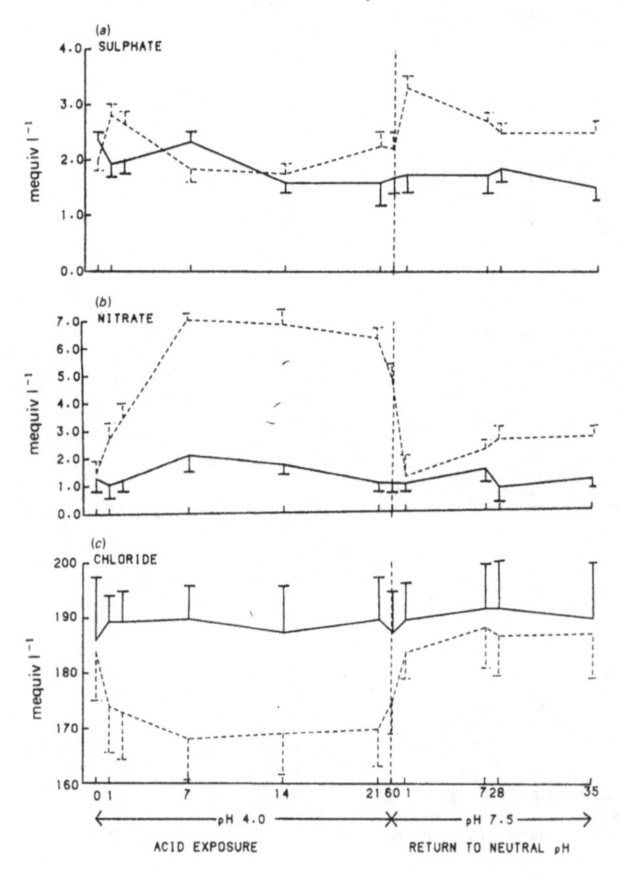

$[Mg^{2+}]_i$ and $[Na^+]_i$ to those observed at one day's exposure to sulphuric acid, (Figures 7, 8) excepting that the increase in $[Na^+]_i$ was of considerably greater magnitude. $[Ca^{2+}]_i$ was increased and $[Cl^-]_i$ decreased significantly, changes of opposite direction from those reported for H_2SO_4. To this point the data confirm that initial exposure to nitric acid causes a small acidosis which persists to 48 h exposure.

Figure 14. Changes in muscle intracellular ion concentrations prior to, during and following a 60 d exposure to soft water acidified to pH 4.0 with nitric acid. *(a)* calcium, *(b)* chloride, *(c)* sodium, *(d)* magnesium, *(e)* potassium content expressed as mequiv kg^{-1} cell water. Following acclimation in soft water, animals were transferred to water which had been acidified to pH 4.0. After 60 d acid exposure animals were transferrd back to water at neutral pH monitored over a subsequent 35 d recovery period. Mean data for acid-treated animals are shown as dashed lines. Solid lines denote data from 'control' animals, treated similarly but not acid-exposed. Vertical bars at each data point show 1 SEM. Asterisks show points of significant difference between acid-treated and control values at that interval. Note X axis scale is discontinuous between 21 and 60 d acid exposure and between 7 and 28 d recovery.

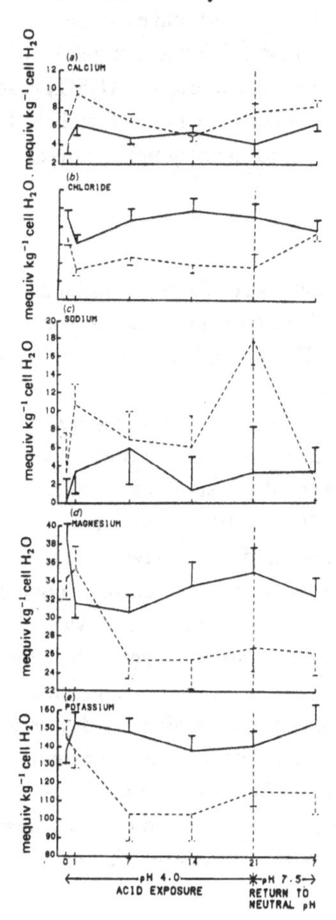

The acidosis is smaller than that observed for sulphuric acid exposure in either hard or soft water. Acute sublethal effects of nitric acid (0.2 mM, pH 4.0) in soft water are initially similar, but not identical, to the effect of sulphuric acid in hard water in the same species (cf. Figures 2 and 3; 9 and 10). Considerable exchange between intra– and extracellular ion pools may accompany the onset of acidosis as part of the cells compensatory responses to either increase in the acidity or decrease in the osmolarity of its haemolymph environment and could cause large change in intracellular SID and possibly internal acid base status.

Chronic responses

As with sulphuric acid exposure haemolymph acid–base balance is restored despite maintained acid exposure. Despite the apparently greater changes in haemolymph ion and water levels, recovery of acid–base levels was more rapid in HNO_3 than H_2SO_4. After seven days' exposure pH was not significantly reduced below control levels and at 14 d pH, P_{CO_2} and $[CO_2]$ were at control levels and showed no further significant change in 60 days of maintained acid exposure (Figure 11). Haemolymph acid–base status thus recovers completely despite continuing nitric acid exposure in *P. clarki*. In contrast, fish exposed to similar levels of H_2SO_4 in soft water show no change in acid–base status but nonetheless die within a few days. It is thus of considerable interest to observe what happens to haemolymph and tissue ion levels during chronic exposure.

Haemolymph ion status

Following the shift from ECF to ICF which occurred in the initial 24 h of acid exposure no further changes occurred in ECFV and only minor increase occurred in ICFV (Table 1) during 21 d of acid exposure. This variable was not determined during acid exposure after this point.

No significant change in haemolymph content of any ion had resulted by 48 h of acid exposure. In contrast with the recovery in acid–base status, however, during continued acid stress the observed perturbations in all measured haemolymph ion concentrations persisted for much longer periods (Figure 10). $[NO_3^-]_p$ increased to levels significantly (five-fold) above control values within seven days. All other ion contents continued to decrease with $[Na^+]_p$, $[Ca^{2+}]_p$ and $[Cl^-]_p$ reaching contents significantly below control values at seven days. Because of the loss of water from haemolymph the actual losses of ions may be substantially greater than those indicated in Figures 12 and 13. Actual losses of $[Cl^-]_p$ and $[Na^+]_p$ are greatest and may reach 45–50% of the initial haemolymph pool over the initial seven-day period. Losses of $[Ca^{2+}]_p$ are of similar percentage although smaller in absolute magnitude. During the subsequent 53 d of acid exposure individual ions followed separate trends. $[Mg^{2+}]_p$ and $[K^+]_p$ showed continued progressive depletion reaching levels significantly below control at 14 and 21 d acid exposure respectively (Figure 12). All other ion levels showed some signs of recovery to control levels during maintained acid exposure. $[NO_3^-]_p$ decreased 50% but still remained significantly above control levels. $[SO_4^{2-}]_p$ increased slightly but was never significantly elevated over control values. $[Na^+]_p$ and $[Ca^{2+}]_p$ increased progressively and reached levels not significantly below control by 60 d acid exposure. $[Cl^-]_p$ increased more gradually at first but still had risen to levels not significantly below control at 60 d (Figure 8).

Muscle intracellular ion status

From two to seven days' maintained acid exposure all intracellular ions with the exception of $[Cl^-]_i$ had decreased significantly when compared (t-test) with day one values. With the exception of $[Na^+]_i$ and $[Ca^{2+}]_i$ which were elevated initially and decreased to control levels, all other ions were significantly lower than control values at seven days' acid exposure (Figure 14). Intramuscular $[Ca^{2+}]$ increased to control values and $[K^+]_i$ and $[Mg^{2+}]_i$ returned towards control values but remained significantly depressed (Figure 14). $[Na^+]_i$, however, exhibited a further increase above control values. Data for $[X]_p$ for the 21–60 day exposure period (Figure 12) would suggest some further recovery might occur at least for $[Na^+]_i$, $[Cl^-]_i$ and $[K^+]_i$. Combination of the data for both haemolymph and intramuscular ion levels suggests that heavy ion losses occurring initially, especially from the extracellular compartment, are limited after the seven day exposure period and there is evidence in all cases that recovery in both haemolymph and muscle ion status may occur in longer exposures.

Carapace ionic status

The cationic composition of carapace (Figure 15) exhibited extremely high concentrations of Ca^{2+}, moderate amounts of Mg^{2+}, Na^+ and low levels of K^+ as presented for the H_2SO_4 treated group above. Following one day's exposure to reduced pH $[Ca^{2+}]_c$ had decreased by 7% but no further significant change occurred (Figure 15). The losses of carapace calcium, either via the internal environment (i.e. across the gills) or to the medium directly by dissolution of the external carapace were small and temporary. No softening of the carapace was noted even after 60 d exposure in this species.

Role of the renal organs in compensation for acid disturbance

The renal organs (antennary glands) of *P. clarki* are very effective in conservation of Na^+, Cl^- and K^+ and potentially involved in some regulation of Ca^{2+}, Mg^{2+} and in removal of NH_4^+ in water of neutral pH. In acid water there is a slight transient increase in urine production in the first 24 h of acid exposure and a decrease in urine pH varying from 0.1–0.25 units is seen throughout acid exposure. There is no significant change in urinary level of any ion but a slight increase in urinary output of almost all ions [Cl^-, Na^+, Ca^{2+}, K^+, Mg^{2+}] occurred over the first 48 h of acid exposure. Following this period urinary levels of K^+, Ca^{2+}, NO_2^- tended to be lower than those of the control group suggesting some retention of these ions from the urine. The total losses of these ions from the urine of *P. clarki* are only a tiny fraction of those observed for whole animals of *O. propinquus* (Wood & Rogano, 1986) suggesting that urinary conservation by the kidney plays only a very minor role in either the initial ion loss, or in the subsequent recovery period. The urinary output of NH_4^+ increased in the first 24 h of acid exposure was extremely variable for the next 20 d but for the last 30 d of acid exposure NH_4^+ excretion was

Figure 15. Changes in carapace ion content during 21 d exposure to soft water acidified to pH 4.0 with nitric acid and in recovery. Ion contents are expressed per wet wt of carapace and thus show only changes in relative not absolute concentration. *(a)* potassium, *(b)* sodium, *(c)* magnesium, *(d)* calcium content of carapace. Mean data for acid-treated animals are shown as dashed lines. Solid lines denote data from 'control' animals, treated similarly but not acid-exposed. Asterisks show points of significant difference between acid-treated and control values at that interval.

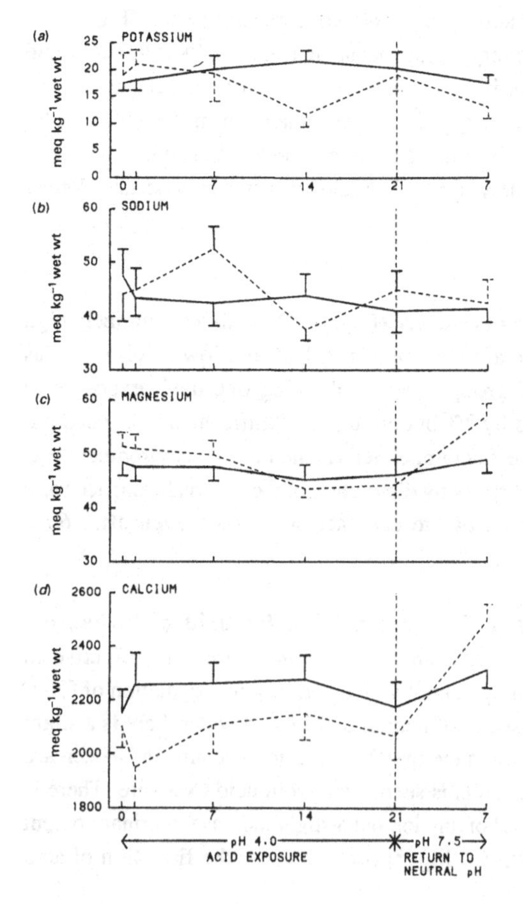

routinely lower in urine of the acid treated group. The overall conclusion here is that the renal structures of crayfish play only a very small role either in removal of acid or in compensation for ion loss, at least in this crayfish.

Haemolymph oxygenation in acute acid exposure

Previous studies on acid exposure in fish have raised the question as to whether a combination of blood osmotic, ionic and acid–base changes resulting from acid exposure may seriously disrupt either oxygen uptake or transport. Measurements of blood oxygen tensions and O_2 capacities from haemolymph of *P. clarki* in acute (one to seven days) HNO_3 exposure show no change in O_2 delivery to tissues (Figure 16) even at day one when acidosis and ion loss were probably most severe. There is almost no change in either Pa_{O_2} and Pv_{O_2} after one day of acid exposure. The only effect noted was a small decrease in content resulting from a small Bohr shift ($d\ \log P_{50}/pH = -0.43$) slightly reducing the venous O_2 reserve (Figure 16). As the acidosis is transient in maintained nitric acid exposure O_2 affinity should return to normal within 14 d. No further change occurs following 60 d acid exposure. Clearly O_2 delivery to tissues is tightly controlled in this species. Similarly, even in *O. propinquus* where substantial mortality accompanied short term exposure, no significant increase in haemolymph lactate concentration occurred (Wood & Rogano, 1986) again suggesting that failure of O_2 supply was not an important factor.

Recovery in neutral water

Both the experiments of Morgan & McMahon (1982) and of McMahon & Stuart (1985); S.A. Stuart & B.R. McMahon (unpublished data) include periods when haemolymph and other samples are taken during a recovery period following acid exposure. Following short term (four days) exposure to H_2SO_4 and subsequent return of *P. clarki* to neutral water both acid–base and ionic status of haemolymph recovered quickly (at 4 d = first measurement point) indicating that no sustained damage to gill or other regulatory tissues resulted from acute acid exposure (Morgan & McMahon, 1982). Similar conclusions can be drawn from the flux data for *O. propinquus* (Figure 4, Wood & Rogano, 1986) except for Ca^{2+} for which net efflux continued even at neutral pH.

The results observed during recovery following chronic acid exposure either to H_2SO_4 or HNO_3 differ in detail (cf. Figures 2, 6, 11). Following long term exposure to nitric acid, a pronounced alkalosis occurs on re-exposure to neutral water (Figure 11). This alkalosis is approximately similar in magnitude to the initial acidosis seen on acid exposure, but of longer duration, with pH remaining elevated for at least seven days. Both the size and the duration of the alkalosis suggest that a new ionic equilibrium has been reached in haemolymph during acid exposure, and that relocation of the animals in neutral water constitutes a new ionic and acid–base disturbance occurring in the opposite direction. Most ion levels in haemolymph return to control values within one day. $[Mg^{2+}]_p$ and $[K^+]_p$, however, require much longer recovery with a time course more similar to that of the pH change. $[Mg^{2+}]$ and $[K^+]$ are also slow to recover in the muscle intracellular pool.

There is no dramatic alkalosis on recovery from long-term sulphuric acid exposure (Figure 7). Increases in pH, PCO_2 occur, but only to return these variables to control values within the first day. A small overshoot as was noticed following HNO_3 exposure may occur but is never significant.

The apparent shift from ECF to ICF is reversed almost exactly within seven days on return to neutral water. The rapid recovery in many ion levels suggests that ion exchange mechanisms are not damaged by even 60 d exposure to pH 4.0 but the

Figure 16. Oxygen transport by haemolymph in *Procambarus clarki* prior to, during and after 60 d exposure to water acidified to pH 4.0 with nitric acid. Oxygen equilibrium curves were constructed *in vitro* at appropriate levels of pH equivalent to those recorded *in vivo* for each condition. Data points are mean levels of haemolymph Pa_{O_2} and Pv_{O_2} and Ca_{O_2} and Cv_{O_2} measured *in vivo* for each condition. No change in O_2 delivery is seen even at one day's acid exposure when haemolymph acid–base disturbance is maximal.

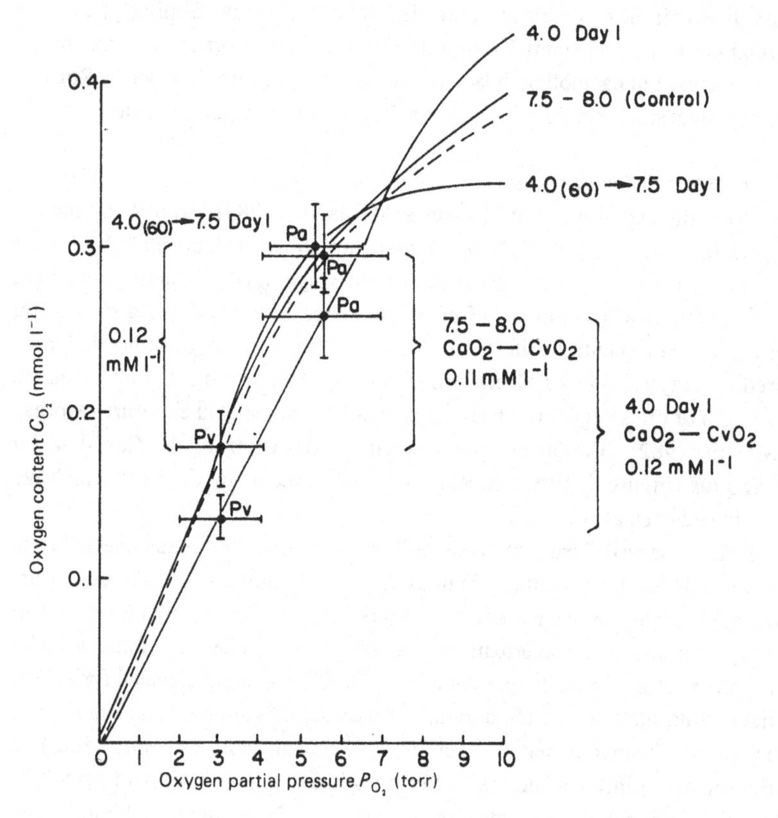

difference in response suggests that NO_3^- and SO_4^{2-} have very different effects either on the ionic balance between extracellular and intracellular compartments, or on ion uptake pumps.

Although adult *Procambarus clarki* show equal tolerance to acid exposure, their responses to sulphuric and nitric acid exposure are surprisingly different. The acidosis seen in soft water H_2SO_4 exposure is lower at 24 h than that seen at the same time in nitric acid. However, in the latter case the acidosis is compensated within 7–14 d while that resulting from H_2SO_4 is more prolonged. The difference is most marked at seven days when the acidosis is no longer significant in HNO_3 but still increasing in H_2SO_4. Additionally there is a distinct respiratory component (i.e. resulting from increase in haemolymph P_{CO_2}) to the acidosis resulting from H_2SO_4 which is totally absent in HNO_3.

Changes in both the haemolymph and intramuscular ion levels accompanying acid exposure also vary dramatically between the two acids. Loss of haemolymph ions and subsequent haemoconcentration is apparently greater in HNO_3^- exposure at least initially (Table 1). However total ion loss is greater at 60 d acid exposure in H_2SO_4 than in HNO_3^-. It is clear that for this species experiments of even greater length are needed to understand clearly this animal's compensatory ability in acid exposure.

It seems evident that the associated anion in acid exposure is extremely important in determining the responses observed. Some differences seen could result from a difference in size of the anion (assuming that SO_4^{2-} is larger than NO_3^-) and resulting differences in gill permeability. Whereas the data for both *O. propinquus* and *P. clarki* above strongly suggest that SO_4^{2-} does permeate the gill it is nonetheless possible that the larger, more highly charged sulphate ion permeates more slowly than does the nitrate ion. Interestingly, Stobbart (1967) found that in larval *Aedes aegyptii* increased NO_3^- (no acid stress) inhibited only Na^+ uptake. If prolonged, this one sided inhibition might explain the more protracted acidosis seen in H_2SO_4 exposure in crayfish.

Several authors have noted increase in haemolymph Ca^{2+} during acidotic conditions in Crustacea and have associated this with buffering of the invading protons using dissolution of the carapace or other internal $CaCO_3$ stores. All species of crayfish tested can boost haemolymph levels of Ca^{2+} or Mg^{2+} the major constituents of the carapace in H_2SO_4 exposure but this either does not (or need not) happen during HNO_3 exposure. A small but significant increase in intracellular [Ca^{2+}] in HNO_3^-, however, suggests that some Ca^{2+} dissolved from the skeleton may be sequestered in the tissues early in acid exposure.

Although long term experiments have been carried out only on the crayfish *Procambarus clarki* it seems apparent that at least intermoult adults have a remarkable ability to withstand acid exposure. In their ability to correct haemolymph acid–base status and apparently to correct the loss of ions from haemolymph, it appears that this species can recover while in maintained severe acid exposure.

The reason for the apparently greater acid tolerance of *P. clarki* is not known. However, this animal is found naturally in somewhat acid water either in marshy areas or when water stagnates in the burrows to which they are commonly forced by drought. Another *Procambarus* species, *P. fallax*, also occurs naturally in acid waters and may also show considerable tolerance. Other species of crayfish are more sensitive and other stages of *P. clarki* are also likely to show increased acid sensitivity. Thus the present experiments show that some crayfish species can withstand considerably greater acid stress than many fish species and that *P. clarki* may be able to survive in waters more acid than other crayfish species. Further work on this and other acid-tolerant species is needed to show the manner and extent of acid tolerance.

References

Abrahamsson, S. (1972). Fecundity and growth of some populations of *Astacus astacus* Linne in Sweden. *Rep. Freshwater Res., Drottingholm*, **52**, 23–37.

Almer, B., Dickson, W., Erkstrom, G., Hornstrom, E. & Miller, U. (1974). Effects of acidification on Swedish lakes. *Ambio*, **3**, 30–6.

Bedford, J.J. & Leader, J.P. (1977). The composition of the haemolymph and muscle tissue of the shore crab *Hemigrapsus edwardsi*, exposed to different salinities. *Comp. Biochem. Physiol.*, **57A**, 341–45.

France, R.L. (1983). Response of the crayfish *Orconectes virilis* to experimental acidification of a lake with special reference to the importance of calcium. In *Freshwater Crayfish V. Papers from the Fifth International Symposium on Freshwater Crayfish Davis California*, ed. C.R. Goldman, 569 pp. Westport, Connecticut: AVI Publishing.

France, R.L. (1985). Preliminary investigation of effects of sublethal acid exposure on maternal behavior in the crayfish *Orconectes virilis*. *Bull Envir. Contam. Toxicol.*, **35**, 641–5.

France, R.L. (1985). Relationship of crayfish (*Orconectes virilis*) growth to population abundance and system productivity in small oligotrophic lakes in the experimental lakes area, Northwestern Ontario. *Can. J. Fish. Aquat. Sci.*, **42**, 1096–102.

France, R.L. & Graham, L. (1985). Increased microsporidian parasitism of the crayfish *Orconectes virilis* in an experimentally acidified lake. *Water Air Soil Pollut.*,, **26**, 129–36.

Graham, M.S. & Wood, C.M. (1981). Toxicity of environmental acid to rainbow trout: Interactions of water hardness, acid type and exercise. *Can. J. Zool.*, **19**, 1518–26.

Haines, T.A. (1981). Acidic precipitation and its consequences for aquatic ecosystems: a review. *Trans. Am. Fish. Soc.*, **110**, 669–707.

Harvey, H.H. (1979). The acid deposition problem and emerging research needs in the toxicology of fish. Proc. Fifth Annual Aquatic Toxicity Workshop Ham. Ont. Nove. 7–9, 1978. *Fish Marine Serv. Tech. Rept.*, no. 862, 115–28.

Harvey, H.H. (1982). Population responses of fish to acidified waters. In *Proceedings of an International Symposium on Acidic Rain and Fisheries Impacts on Northeastern North America*, August 1981, Ed. R.E. Johnson, pp. 227–42.

Harvey, H.H. & Lee, C. (1982). Historical fisheries changes related to surface water pH changes in Canada. In *Acid Raihn/Fisheries*, ed. R.E. Johnson, pp. 45–55. Bethesda: American Fisheries Society.

Hõbe, H., Wood, C.M. & McMahon, B.R. (1984). Mechanisms of acid–base and ionoregulation in white suckers (*Catostomus commersoni*) in natural soft water. I. Acute exposure to low ambient pH. *J. comp. Physiol.*, **104**, 35–46.
Houston, A.H. & Mearow, K.M. (1979). [^{14}C] PEG–4000, chloride/potassium and sodium spaces as indicators of extracellular phase volume in the tissues of the rainbow trout, *Salmo gairdneri* Richardson. *Comp. Biochem. Physiol.*, **62A**, 747–52.
Jarvenpaa, T., Nikinmaa, M., Westman, K. & Soivio, A. (1983). Effects of hypoxia on the haemolymph of the freshwater crayfish, *Astacus astacus* L., in neutral and acid water during the intermolt period. In *Freshwater Crayfish V. Papers from the Fifth International Symposium on Freshwater Crayfish Davis California*, ed. C.R. Goldman, pp. 569. Westport, Connecticut: AVI Publishing.
Leivestad, H. & Muniz, I.P. (1976). Fish kill at low pH in a Norwegian river. *Nature (Lond.)*, **259**, 391–2.
Likens, G.E. & Bormann, F.H. (1974). Acid rain: a serious regional environmental problem. *Science* **184**, 1176–9.
Malley, D.F. (1980). Decreased survival and calcium uptake by the crayfish *Orconectes virilis* in low pH. *Can. J. Fish. Aquat. Sci.*, **37** , 364–72.
McDonald, D.G. (1983). The effects of H^+ upon the gills of freshwater fish. *Can. J. Zool.*, **61**, 691–703.
McDonald, D.G. & Wood, C.M. (1981). Branchial and renal acid and ion fluxes in the rainbow trout (*Salmo gairdneri*), at low environmental pH. *J. exp. Biol.*, **93**, 101–18.
McMahon, B.R. & Morgan, D.O. (1983). Acid toxicity and physiological responses to sublethal acid exposure in crayfish. In: *Freshwater Crayfish 5. Papers from the Fifth International Symposium on Freshwater Crayfish, Davis, California*, ed. C.R. Goldman, pp. 71–85. Westport, Connecticut., AVI Press.
McMahon, B.R. & Stuart, S.A. (1985). Hemolymph and tissue ion and acid–base balance in crayfish during chronic exposure to nitric acid. *Am. Zool.* **25** (4), 48A.
Morgan, D.O. & McMahon, B.R. (1982). Acid tolerance and effects of sublethal acid exposure on iono-regulation and acid–base status in two crayfish *Procambarus clarki* and *Orconectes rusticus*. *J. exp. Biol.*, **97**, 241–52.
Patterson, N.E. (1983). Respiratory and cardiovascular function in Crayfish exposed to low pH. *Abstr. Amer. Zool.*, **23**, 938.
Stobbart, R.H. (1967). The effect of some anions and cations upon the fluxes and net uptake of chloride in the larva of *Aedes aegypti* (L), and the nature of the uptake mechanism for sodium and chloride. *J. exp. Biol.*, **47**, 35–57.
Stuart, S.A. & McMahon, B.R. (1984). Acclimation to Nitric acid exposure in the crayfish. In: *Crayfish 6. Papers from the 6th International Symposium on Astacology*, Lund, Sweden 1984. AVI Press, Westport, Connecticut 1987.
Ultsch, G.R., Ott, M.E. & Heisler, N (1981). Acid–base and electrolyle status in carp (*Cyprinus carpio*) exposed to low environmental pH. *J. exp. Biol.*, **93**, 65–80.
Wood, C.M. & McDonald, D.G. (1982). Physiological mechanisms of acid toxicity to fish. In *Proceedings of an International Symposium on Acidic Rain and Fishery Impacts on Northeastern North America*, August 1981, ed. R.E. Johnson, pp. 197–226.
Wood, C.M. & Rogano, M.S. (1986). Physiological responses to acid stress in crayfish (*Orconectes propinquus*): Haemolymph ions, acid–base status, and exchanges with the environment. *Can. J. Fish. Aquat. Sci.*, **43**, 1017–26.

W.T.W. POTTS AND P.G.MCWILLIAMS

The effects of hydrogen and aluminium ions on fish gills

Introduction

During the last decade a small library of books and papers has accumulated dealing with the effects of acid waters on fish. In this review we shall attempt to discuss and, where possible, interpret a small part of this work in terms of what is known of the physiology of fresh water fishes.

The blood plasmas of freshwater fishes contain around 150 mequiv Na^+ and 130 mequiv $Cl \cdot l^{-1}$ in addition to lower concentrations of other ions, particularly Ca^{2+}, K^+ and HCO_3^-. In contrast, soft fresh waters, in which acidification may be a problem, usually contain less than 0.1 mequiv NaCl l^{-1} and much lower concentrations of other ions. Although fish obtain some salts from their food, most species, including all salmonids, are dependent on the active uptake of salt from the medium to balance diffusion and urinary losses. Freshwater fishes produce a dilute urine and urinary losses are usually less than 10% of the total losses, most of which take place across the body wall, particularly across the gills, where the blood plasma and external medium are separated by only a thin layer of respiratory epithelium. The gills are also the main site of salt uptake, although in some marine teleosts salt transport takes place at other sites, particularly on the inside of the operculum, and the possibility of salt uptake and loss at other sites in freshwater fishes should not be overlooked.

Structure of the gill

Fish gills have a complicated structure. Gill slits divide the side of the throat into gill arches. Gill filaments, or lamellae, project from the gill arches and gill lamellae, or secondary lamellae, project from the filaments, further increasing the surface area (Figure 1). Several different kinds of cell are found on the gill surface. Most of the surface is formed of respiratory epithelial cells which cover the lamellae and most of the filaments. In marine teleosts the mitochondria-rich cells, commonly but inappropriately referred to as "chloride cells", are found mainly on the filaments, between the bases of the lamellae (Fig. 1) but in freshwater fishes they may extend onto the lamellae. Blood from the heart passes into the afferent arterioles and on into the lamellae where it is oxygenated. The blood is then gathered into the efferent arterioles from whence most of it travels on to the body but a proportion drains into the central venous sinus and is returned to the heart (Figure 1).

The lamellar surface, which is covered by the respiratory epithelium, is irrigated by the afferent-efferent arterial system while the filaments are irrigated mainly by the venous drainage (Figure 1). Experiments by Girard & Payan (1980, 1984) showed that in perfused brown trout gills (*Salmo trutta*) in fresh water, sodium and chloride ions abstracted from the medium appeared in the efferent arterial drainage, while calcium ions appeared in the venous drainage. They concluded that the respiratory epithelium was the site uptake of sodium and chloride while calcium ions were taken up by the mitochondria-rich cells. Calcium uptake by mitochondria-rich cells has been confirmed in the rainbow trout, *Salmo gairdneri*, (Perry & Wood, 1985). However, perfused teleost gills are notoriously difficult preparations, the fluxes *in vitro* are much smaller than *in vivo* and decline with time and as mitochondria-rich cells extend onto the lamellae they may well be responsible for sodium uptake. The teleost is exceptional if different epithelia are concerned with salt uptake in fresh water and salt-excretion in sea water. In the anostracans such as *Chirocephalus* and *Artemia*, polyphemids such as *Polyphemus* and *Podon* or water-fleas, such as *Daphnia* and *Bosmina*, the sites of salt uptake and excretion are identical. The curious distinction in the mitochondria-rich cell between sodium uptake and chloride

Figure 1. Diagram of circulation in a gill filament. AFA, Afferent filamental artery; EFA, Efferent filamental artery; CVS, Central venous sinus; Lam., Lamella. Stipple – Epithelium containing mitochondria-rich cells.

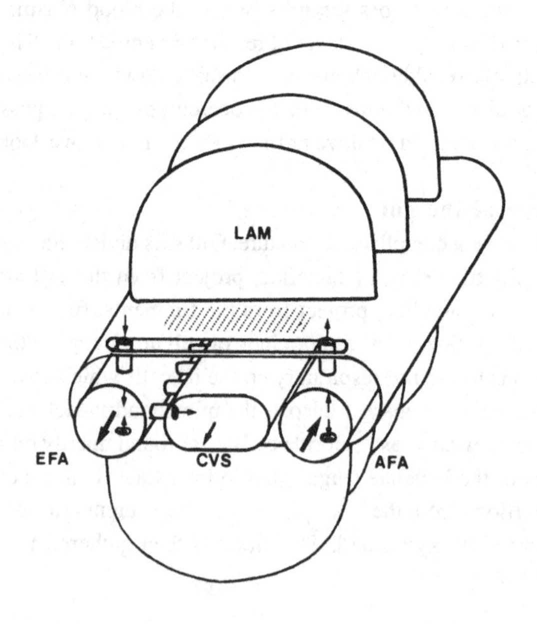

excretion, both driven by ATPase on the baso-lateral membranes, would seem to be a remarkable adaption to allow salt uptake or excretion as circumstances require.

Ionic transport mechanisms in gills

Sodium uptake is driven by Na–K-ATPase from energy supplied by ATP, as in other epithelia which transport sodium ions. The basal cell membrane is the major barrier to sodium movements (Girard & Payan, 1977) and the probable site of the Na–K-ATPase, although little is known about the situation and properties of the Na–K-ATPase in the respiratory cells because it was assumed, until recently, that the mitochondria-rich cells were the main sites of sodium uptake in freshwater fish. Sodium uptake requires either the simultaneous transport of an anion or the export of a cation to balance the electronic charge. As long ago as 1938 Krogh suggested that ammonium ions, of metabolic origin, were exchanged for sodium ions in fish gills. Garcia Romeu & Maetz (1964) confirmed Krogh's suggestion by demonstrating that ammonium ions in the water inhibited sodium uptake in the goldfish *Carassius auratus,* but stimulated uptake when injected into the peritoneum. However, sodium uptake frequently exceeds ammonium excretion and Kerstetter *et al.* (1970) showed that hydrogen ions could also act as counter ions. Sodium influx correlated well with the sum of the efflux of hydrogen and ammonium ions (Maetz, 1973) although there is a tight link between ammonium and sodium ions, as removal of ammonia from the perfusion fluid reduces sodium uptake and removal of sodium reduces ammonium excretion in the rainbow trout (Payan, 1978). Sodium uptake in exchange for NH_4^+ or H^+ takes place in two steps. Movement across the apical membrane takes place through channels which can be blocked by amiloride (Kirschner *et al.*, 1973), transport across the baso-lateral membranes is mediated by Na–K-ATPase and is reduced by ouabain (Payan, 1978). In the rainbow trout (*S. gairdneri*) in fresh water, acidification of the perfusate reduces ammonium efflux suggesting that ammonia crosses the baso-lateral membrane as NH_3 (Payan, 1978); on the other hand, similar experiments with the sculpin and toadfish in sea water show that ammonia crosses the baso-lateral membranes as NH_4^+, (Goldstein *et al.*, 1982).

The details of chloride transport are more obscure. Krogh (1938) proposed that chloride ions were taken up in exchange for bicarbonate ions. Freshwater animals depleted of chloride but replete with sodium can take up chloride ions without sodium ions. The chloride ions absorbed are replaced by bicarbonate ions in the external medium, but this does not exclude the possibility that the primary counter ion is hydroxyl, the bicarbonate arising from metabolic carbon dioxide. Measurement of the transfer coefficients of chloride ions across respiratory cells show that the basal membrane constitutes the major barrier to chloride movements and is therefore likely to be the site of active transport (Girard & Payan, 1977).

The uptake of both sodium and chloride ions follow Michaelis–Menten kinetics where:

$$f = f_{max} \left(\frac{C}{C+K}\right)$$

where f is the flux at an external concentration C, f_{max} is the saturated flux at high external concentrations and K is a constant which is a measure of the affinity of the system for the transported ion. In freshwater fish K is usually around 0.1 or 0.2 mequiv l^{-1}, so that the pump operates at a reasonable level in normal fresh waters, although in the goldfish *Carassius auratus* De Renzis & Maetz (1973) found that K was 40 mequiv Cl l^{-1}.

Salt loss

Sodium and chloride ions may be lost either through cells or between the cells through the intercellular junctions. The multistrand tight junctions between respiratory cells (Sardet *et al.*, 1980) have a lower permeability than the single strand junctions between the mitochondria-rich cells in seawater fish but how the losses in fresh water are partitioned between the cells and the intercellular junctions is not known. Ions may be lost in several ways: by simple diffusion across or between cells or by more complex routes through the cells by carrier-mediated exchange diffusion or by a leaky pump. For example, if an ion pump exchanging sodium for hydrogen ions failed to distinguish exactly between sodium and hydrogen ions it might occasionally excrete a sodium ion. This would appear as a sodium efflux proportional to the rate of sodium uptake. Such a flux, unlike an exchange diffusion, would cease when the uptake stopped.

The rate of loss of ions by diffusion is dependent on the permeability of the epithelium and the transepithelial potential. In fish gills the potential is predominantly of diffusional origin (Potts, 1981). A diffusional potential depends on all permeant ions present and is given by the Goldman equation:

$$E = \frac{RT}{F} \ln \left(\frac{P_{Na}[aNa_m] + P_H[aH_m] + P_a[aCl_p]}{P_{Na}[aNa_p] + P_H[aH_p] + P_a[aCl_m]} \right) .$$

Where E is the potential, P_{Na}, P_H etc. are the permeabilities to sodium and hydrogen ions etc. and $[aNa_m]$ and $[aNa_p]$ are the activities of sodium ions in the medium and plasma, etc. and R, T and F have their usual meaning. In neutral waters only sodium and chloride ions contribute significantly but in acid water hydrogen ions also become significant because of the high permeability of the gill to hydrogen ions (McWilliams & Potts, 1978). The gill therefore behaves like a hydrogen electrode, becoming more positive inside at low pH (Figure 2). These changes in potential are rapid, and reversible, as pH becomes more alkaline again (Figure 3). Such potential shifts in acid conditions would facilitate the efflux of sodium ions and would increase the energy required to maintain the influx. A diffusional flux of a monovalent ion J is dependent on the potential E so that

$$J \propto \frac{\pm EF/RT}{1 - e^{EF/RT}}$$

Figure 2. Response of transepithelial gill potentials in brown trout (*S. trutta*) to changes in pH of external medium. Ca = 0.5 mmol l^{-1}. From McWilliams & Potts, 1978.

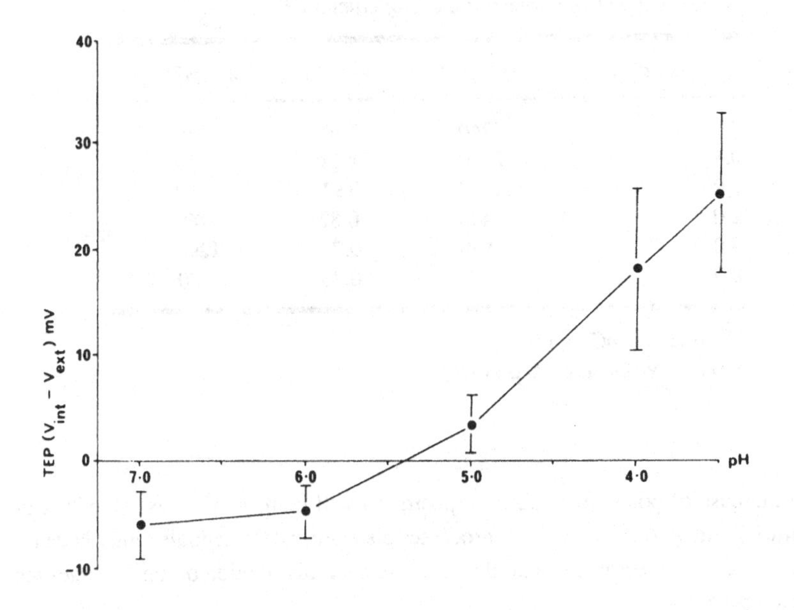

Figure 3. Pen recorder tracing of transepithelial gill potentials in brown trout, *Salmo trutta*, in response to pH changes. Note the rapid, reversible shifts in potential, indicative of the high permeability of the gill to H^+.

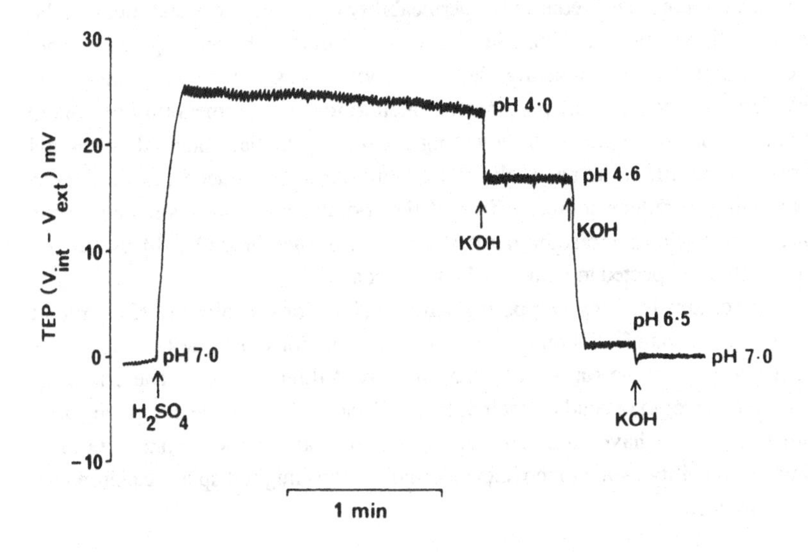

Table 1. *Relative permeabilities (ϕ) of the gills of brown trout (Salmo trutta) to sodium, chloride and hydrogen ions in the presence of various concentrations of calcium*[a]

Calcium (mM)	$\phi H^+/\phi Cl^-$	$\phi Na^+/\phi Cl^-$	$\phi H^+/\phi Na^+$
0	4240	1.67	2540
0.5	2420	1.12	2160
1.0	1780	0.91	1960
2.0	1440	0.82	1760
4.0	980	0.76	1290
8.0	670	0.73	920

[a] Assuming $\phi Cl^- = 1.0$.
From McWilliams & Potts (1978).

An increase of potential of 30 mV (positive), will increase the passive efflux of sodium 1.7 times. Active transport processes also generate potentials either because the pumps are not electrically neutral or because the redistribution of ions changes the diffusion potentials.

The permeability of the gill to the various ions is modulated by calcium ions. The divalent calcium ion can reduce the state of hydration of organic molecules as well as forming stabilising cross links with two molecules at one time, to form more condensed structures. This reduces the permeability of cell surfaces and intercellular junctions to all molecules and ions. In addition calcium ions, by binding on the inner surfaces of pores and increasing their positivity, may selectively reduce the permeability to cations. In the presence of calcium ions the permeability of gills to hydrogen ions is reduced, thereby lowering the transepithelial potential at low pH (Figure 4). In addition, the permeability to sodium ions is reduced (Table 1) both effects tending to reduce sodium efflux. If the sodium pump were situated on the basolateral membranes and sodium crossed the apical membrane by diffusion then calcium might be expected to reduce sodium influx as well.

External sodium ions, by competing with calcium ions at binding sites, might counteract the effects of calcium ions. Certainly sodium ions increase the rate of loss of calcium ions from the surface of trout gills (McWilliams, 1983). The effects of aluminium ions would depend on their valency. Divalent ions, while competing with calcium ions, might have similar effects to calcium at very low concentrations, reducing permeability, while monovalent aluminium ions might displace calcium ions as sodium ions do.

Effects of acid waters

Acid waters, with or without aluminium, may have a variety of adverse physiological effects. Below pH 4.2, acid–base disturbances may become important (see Ultsch *et al.*, 1981, for references), while below pH 3.8, oxygen uptake may be impeded, causing anoxia in the tissues (see Ultsch *et al.*, 1981, for references). These high acidities rarely occur in nature even in areas affected by acid rain. Ionic balance is more susceptible to disturbance, adverse effects becoming apparent at pH 5.5 or even above (Figure 5). The prime cause of fish deaths in acid waters is the loss of ions from the blood. Salt deficient fish can survive with blood concentrations as low as 110 mequiv $Na^+ l^{-1}$ and 90 mequiv $Cl^- l^{-1}$ or even lower (Fugelli & Vislie, 1980). The exact cause of death at low plasma concentrations is uncertain. As the blood concentration falls cell volume will increase and extracellular volume will decline. This can be compensated to some extent by the transfer of intracellular sodium and chloride from the cells to the extracellular compartment but the quantity of intracellular sodium and chloride is limited and any substantial transfer increases the gradients across the cell membrane which must be actively maintained. Swelling of the cells may also be reduced by the loss of organic solutes and in the brown trout intracellular taurine declined from 80 mM kg^{-1} cell water to only 20 mM in fish at pH 4.0–4.6 (Fugelli & Vislie, 1980). Intracellular potassium also fell from around 120 mM kg^{-1} cell water to about 95. Unless the potassium lost from the cells is quickly eliminated from the body it will depolarise nerve, muscle and other cells. Fugelli & Vislie do not provide data on plasma potassium, but in the carp plasma potassium increased from 2.5 to 4 mM after 80 h at pH 4.0. If the intracellular potassium had declined as in the

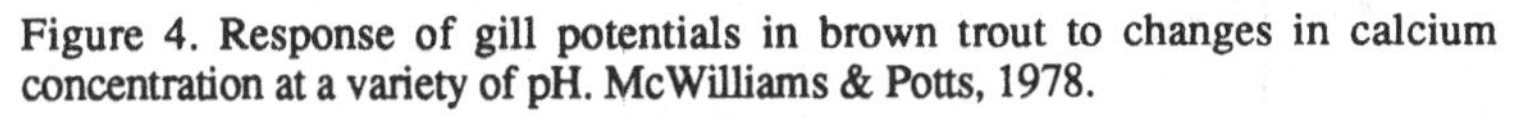
Figure 4. Response of gill potentials in brown trout to changes in calcium concentration at a variety of pH. McWilliams & Potts, 1978.

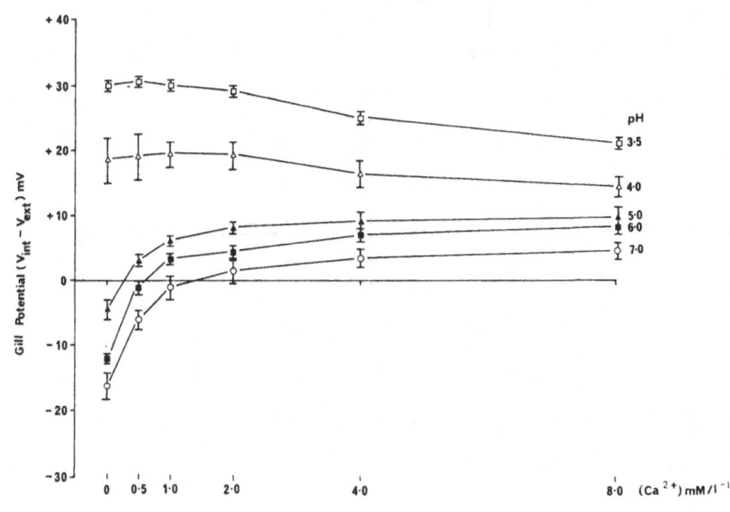

trout the intracellular potentials would have declined from around 94 mV to 77 mV, at which level both nerve and muscle action potentials and enzyme activity might be adversely affected.

Low pH and sodium uptake

Inhibition of sodium uptake at low pH has been demonstrated in many fresh water fishes eg. brook trout, *S. salvelinus,* (Packer & Dunson, 1970); sailfin molly, *Poecilia latipinna,*(Evans, 1975); rainbow trout, *S. gairdneri,* (Kerstetter, Kirschner & Rafuse, 1970); brown trout, *S. trutta,* McWilliams, 1980a), goldfish, *Carassius auratus,*(Maetz, 1973) as well as several invertebrates such as the crayfish *Pacifastacus* (Shaw, 1960). In contrast Dalziel (1985) found no effect of pH on sodium influx in the brown trout down to pH 4.5 except in the presence of aluminium ions. Dalziel attributes the difference to the effects of 'shock' in other experiments but more work is required on this point.

An inhibition of sodium uptake at low pH is to be expected in any system which exchanges sodium ions for hydrogen ions. Several mechanisms may be involved, including competition between Na^+ and H^+ at the binding sites, the increased load on the pump and damage to the ATPase at low pH.

Figure 5. Effect of pH on sodium uptake and loss in the brown trout *S. trutta.* Efflux, solid circles; influx, open circles; Ca = 0.5 mM. From McWilliams & Potts, 1978.

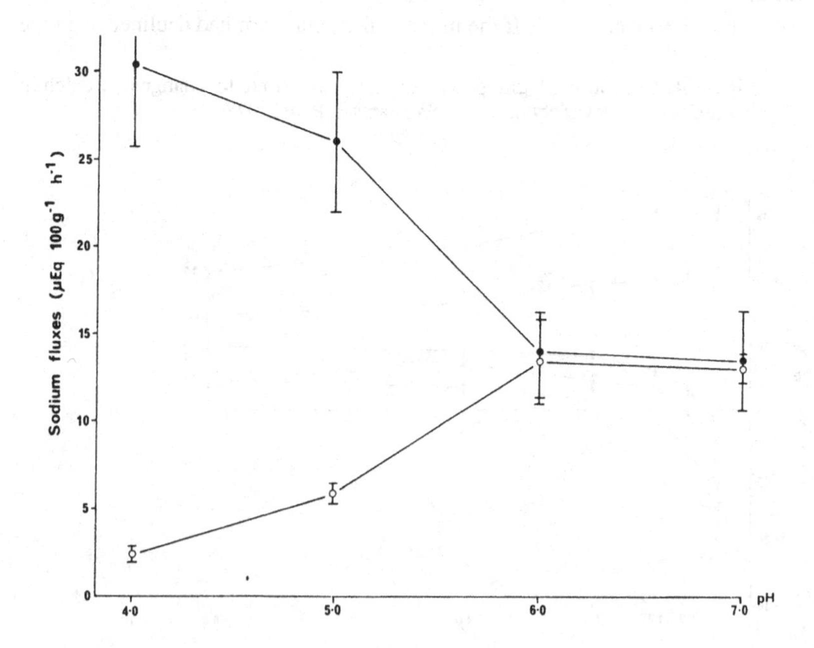

Twitchen (1987) has shown that in stonefly nymphs the reduction of sodium uptake at low pH is due to competitive inhibition. In neutral water sodium uptake in all species examined follows Michaelis–Menten kinetics but when neutral water species, such as *Leuctra moselyi*, are placed in acid water sodium uptake is severely reduced, proportionally more at low sodium concentrations than at high concentrations, when the Na^+/H^+ ratio is higher (Figure 6). In the acid resistant species such as *Amphinemura sulcicollis*, sodium uptake is only slightly reduced at pH 4 (Figure 7). It can be calculated that in *L. moselyi* the carrier has a higher affinity for hydrogen ions than for sodium ions, *Aff* Na^+/*Aff* $H^+ = 0.45$ while *A. sulcicollis* has a higher affinity for sodium ions, *Aff* Na^+/*Aff* $H^+ = 4.5$. In neutral water containing 0.1 mM Na l^{-1} there are 1000 sodium ions for every hydrogen ion so the pump in *L. moselyi* functions adequately. At pH 4, where the proportions of sodium and hydrogen ions are equal, the pump is severely inhibited. The carrier in

Figure 6. The effect of pH on sodium uptake by *Leuctra moselyi* (acid/neutral water). Each point represents the mean ± SE at pH 7, $n = 6$ and pH 4, $n = 6$.

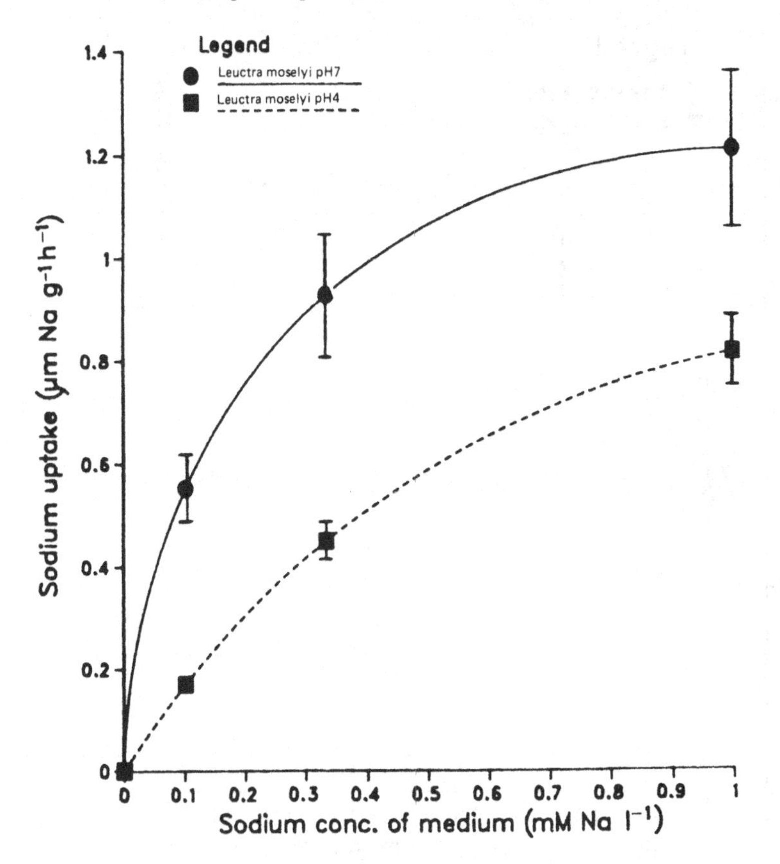

A. sulcicollis also has a greater affinity for sodium ions than that in *L. moselyi*, K_m = 0.09 and 0.15 respectively. The mechanism of sodium uptake in fish also follow Michaelis–Menten kinetics and the different abilities of species to survive in acid waters probably involve similar mechanisms.

There is evidence of the recycling of sodium ions in several species which indicates that the pumps do not always discriminate perfectly between sodium and the counter ion on the serosal side of the epithelium. For example in the crayfish *Pacifastacus* the rate of sodium efflux is higher in water containing sodium than in distilled water and the efflux increases further when the pump is stimulated and decreases if the pump is inhibited. The additional efflux is evidently due to the active excretion of sodium ions by the pump (Bryan, 1960). Approximately one ion of sodium is actively exported for every four ions taken up, thus in deionised water the efflux drops by a quarter while in salt deficient animals, where uptake is enhanced, the efflux is also increased.

Figure 7. The effect of pH on sodium uptake by *Amphinemura sulcicollis* (acid water). Each point represents the mean ± SE at pH 7, $n = 9$ and pH 4, $n = 8$.

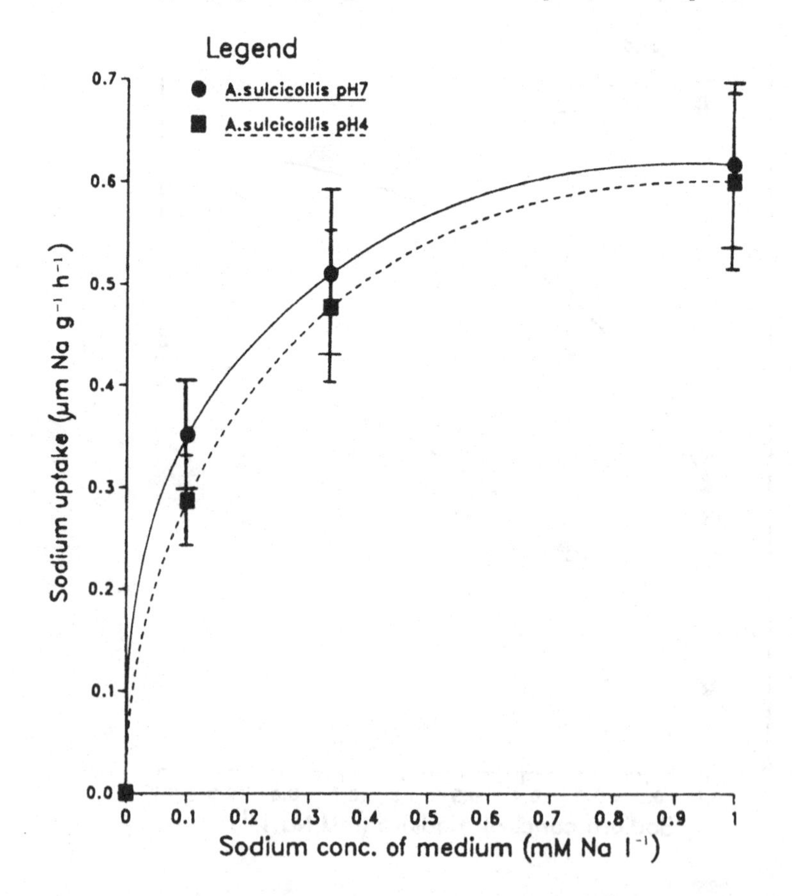

This must be due to imperfect discrimination between sodium ions and the counter ion inside the membrane. Such a failure of discrimination will result in the influx of hydrogen ions rather than sodium ions at low pH thus reducing sodium uptake even if the overall rate of pumping were unaffected.

In addition the work load on the pump will be increased at low pH. In neutral waters the pump will be concentrating sodium ions against a thousand-fold gradient but the concentration of hydrogen ions inside and outside the cell will be similar. At pH 4 the hydrogen ions must also be expelled against a thousand-fold gradient as the cell plasma remains close to neutrality. This might well retard the pump, although the ammonium gradient might still be low. It is possible that at low pH the enzyme Na–K-ATPase is partially inhibited, but the evidence is not strong (Powell & McKeown, 1986).

Although instantaneous transfers to lower pH usually result in a fall in the rate of sodium uptake, as the fish loses sodium the capacity of the transport system may be increased to compensate. In the long term influx may be restored and even increased above normal (McWilliams, 1980b).

Calcium ions have been shown to reduce the rate of influx in several species such as the goldfish, *Carassius auratus* (Eddy, 1975) and the brown trout (Dalziel, 1986). This looks like a direct effect on the permeability of the transporting epithelium.

Effects of low pH on sodium losses

Low pH often has a greater effect on sodium loss than on sodium uptake. Sodium loss may be more than doubled in the brown trout between pH 7 and pH 4 (McWilliams & Potts, 1978; McWilliams, 1980a). Most of the increase is due to the change in transepithelial potential consequent on the high permeability to hydrogen ions but the increase is quantitatively greater than can be accounted for by this effect alone (McWilliams, 1982a) so there must be an associated increase in the permeability of the gill to sodium ions, probably due to the displacement of calcium ions from the gill surface (McWilliams, 1983). Any increase in the external calcium concentration in acid waters is therefore peculiarly beneficial not only because it directly reduces sodium permeability but it also decreases hydrogen ion permeability and hence the transepithelial potential as well, even if the pH is unchanged.

Wild trout taken from acid waters are significantly less permeable to both hydrogen and sodium ions than trout from neutral waters (McWilliams, 1982b) (Table 2). During prolonged adaptation to acid conditions trout from both neutral and acid waters can reduce their permeabilities to sodium ions to a significant extent (McWilliams, 1980b, 1982b) but the same is not true of hydrogen ion permeability (Tables 2, 3). This, perhaps is not an unexpected observation considering the nature of the proton. In order to realise a significant reduction in hydrogen ion permeability of the gill (which may be 2000–3000 times more permeable to hydrogen than to

Table 2. *Branchial permeability characteristics for acid tolerant (Galloway) and non-tolerant (Cumbrian) brown trout, Salmo trutta, and the perch, Perca fluviatilis*[(a)] *(external calcium concentrations (μequiv l^{-1}) in brackets)*

	$\phi Na^+/\phi Cl^-$	$\phi H^+/\phi Na^+$
Cumbrian trout (50)	1.57	2212
[1]Cumbrian trout (250)	1.12	2160
Galloway trout (50)	1.37	654
Galloway trout (250)	0.94	886
Perch (250)	0.84	400
[2]Cumbrian trout (250) (laboratory acclimated to pH 6.0)	0.76	2390

[(a)] Assuming $\phi Cl^- = 1.0$.
[1] From McWilliams & Potts (1978).
[2] From McWilliams (1980b).

Table 3. *Effects of long-term exposure of brown trout to an acid medium of pH 6.0 on the relative permeability of the gill to sodium and chloride ions*[(a)] *(External Ca^{2+} = 0.5 mM)*

Days acid exposure	$\phi Na^+/\phi Cl^-$	$\phi H^+/\phi Na^+$
0	1.12	2160
42	0.76	2390

[(a)] Assuming $\phi Cl^- = 1.0$.
From McWilliams (1980b).

sodium ions; Table 1) a considerable degree of restructuring of the gill membranes would be necessary to reduce permeability to such a small ion. At present there is no evidence that this is taking place during periods of laboratory acclimations to low pH which have so far been investigated. However, it is interesting to note that those teleosts which are generally accepted as being acid tolerant, e.g. perch (*Perca fluviatilis*), do show relatively low hydrogen ion permeabilities, as do acid tolerant strains of the brown trout (*Salmo trutta*) (see Table 2). These would appear to be genetic rather than phenotypic characteristics since long term exposure of these fish to neutral, high conductivity water for up to 12 months does not appear to alter hydrogen ion permeability significantly. It is likely that these characteristics are

Table 4. *Effect of pH on influx and efflux of chloride in S. trutta (% Cl h^{-1}; n = 8)*

	Influx	Efflux
pH 7	0.31± 0.04	0.23 ± 0.01
pH 4	[(a)]0.1± 0.02	[(a)]0.57 ± 0.027

[(a)] Adapted to pH 7.

acquired over many generations in response to naturally declining water pH. It is interesting to speculate whether these characteristics can alter rapidly enough to ensure long term survival in fish populations subject to increasing water acidity as a result of human activity.

Two other effects may help to increase sodium loss at low pH. The ventilation rate in fish increases when the partial pressure of carbon dioxide (P_{CO_2}) is increased (Neville, 1979). The addition of acid to solutions containing bicarbonate ions would increase the P_{CO_2} although the effect should be temporary in well aerated waters. Increased ventilation increases the flux of ions across the gill. pH *per se* does not increase the ventilation rate (Neville, 1979). The proportion of gill lamellae irrigated can also affect the ion fluxes. In well-aerated waters the proportion of lamellae irrigated with blood may be as low as 60% in an inactive fish. Partial anoxia, produced at very low pH, would increase the proportion (Barth, 1979) and therefore lead to an increased loss of sodium, but this effect should only be found in extreme conditions.

The effects of low pH on chloride fluxes

Measurements of chloride fluxes are technically more tedious than measurements of sodium fluxes and have therefore been relatively neglected. A recent examination of both influx and efflux at Lancaster (J.C. Battram, unpublished data) (Table 4) shows that the effects of low pH, calcium ions and aluminium ions on the fluxes are surprisingly similar to their effects on sodium ions. In vertebrate blood plasma the chloride and bicarbonate together are about equivalent to the sum of the cations, of which sodium predominates. If sodium falls then chloride must soon follow. Bicarbonate, the probable counter-ion for chloride uptake, must vary inversely with chloride. If sodium falls and hydrogen ions enter, bicarbonate will be converted to carbon dioxide, the bicarbonate concentration will fall and chloride uptake will be correspondingly reduced.

At low pH chloride uptake is rapidly reduced (Table 4) but calcium ions provide partial protection. Chloride efflux also increases, although the change of potential

Table 5. *Branchial ^{36}Cl effluxes from the brown trout, Salmo trutta, in neutral and acid media (instantaneous transfer from pH 7 to pH 4; efflux rates monitored for three hours only)*

pH	7.0	4.0
Efflux rate μequiv 100 g^{-1} h^{-1}	4.16 ± 0.96	1.58 ± 0.20
n	8	7
p	< 0.01	

From McWilliams (1982a).

should retard chloride loss. McWilliams (1982a) observes that brown trout do in fact show a reduced rate of ^{36}Cl loss immediately following transfer from neutral to acid water (Table 5) but this effect was measured only over a short period (two to three hours following transfer) and is likely to be only a transient effect in response to changes in gill potential, the later rise implying an increase in the absolute permeability to chloride. As would be expected under these circumstances calcium provides some protection against chloride loss.

Effects of calcium ions on ion exchanges

Exchangeable calcium ions, which are rapidly displaceable by lanthanum ions, are bound on the external surfaces of trout gills (McWilliams, 1983). The calcium is bound at two kinds of sites which have different affinities for the ions. Hydrogen ions displace the calcium ions from both sites but sodium ions displace calcium from the higher affinity sites only. Significantly, the gills of trout from naturally acid waters in Galloway have higher affinities for calcium than fish from neutral waters (McWilliams, 1983). Individual fish can increase their affinity to some extent when adapted to acid waters (P.G. McWilliams, unpublished data), but Galloway fish retain a much higher affinity even after six weeks' adaptation and the differences between different stocks may be genetically determined. From the competitive behaviour of hydrogen, sodium and calcium ions at the sites, McWilliams (1983) argues that it is the lower affinity sites which are related to gill permeability. The higher affinity sites may be involved with sodium and/or calcium uptake.

Sulphuric acid is more effective in displacing calcium ions than either nitric or hydrochloric acid of the same pH (McWilliams, 1983). This may be related to the adverse effects of sulphate ions compared with chloride ions reported by various authors. Graham & Wood (1981) found that sulphuric acid was more toxic to rainbow trout than hydrochloric acid, in soft water above pH 3.8. Below pH 3.8 the effects of the two acids were not significantly different but at extremely low pH death is more likely to be due to acid-base imbalance or anoxia, than to osmoregulatory disturbance (Ultsch *et al.*, 1981). On the other hand hydrochloric acid was more toxic than sulphuric acid in very hard water containing 2–4 mequiv Ca l^{-1}. In rather different circumstances De Renzis & Maetz (1973) found that goldfish survived better in deionised water than in 1 mM Na_2SO_4. The greater effectiveness of sulphuric acid in removing calcium ions is probably related to the very low activity coefficients of electrolyte solutions in which both cation and anion are divalent, thus the effective concentration of free calcium ions will be much lower in the presence of sulphate than in the presence of chloride or nitrate ions. The International Critical Tables do not contain activity coefficients of calcium sulphate solutions but the activity coefficients of salts in which both ions are divalent are much lower than those of salts in which one ion is monovalent. A solution containing 0.1 M $MgSO_4$ has an activity coefficient of only 0.15 compared with coefficients of 0.528 and 0.522 for solutions of 0.1 M $MgCl_2$ and 0.1 M $Mg(NO_3)_2$ respectively (Robinson & Stokes, 1959).

The intracellular concentrations of calcium ions are exceedingly low but the concentrations are very important in the regulation of many cellular activities. Most intracellular calcium is tightly bound to the microsomes which regulate the level of free calcium. In view of the very high permeance of hydrogen ions the intracellular pH of the gill epithelium is probably lower in acid waters than in neutral waters. The microsomal fraction of the gills of trout from acid waters bind calcium ions more effectively than those from neutral waters. The accumulation of calcium by microsomes probably involves a Ca-ATPase. Adaptation of the trout to both acid waters and low calcium waters involves an increase in the Ca-ATPase activity of the microsomal fraction. The accumulation displays Michaelis–Menten kinetics and the adaptation involves a fall in K_m, indicating an increased affinity, although J_{max} is unchanged (P.G. McWilliams, unpublished data).

As might be expected from their effect on calcium adsorption, acid waters increase the permeability of gills to water. In fingerling brown trout transfer from pH 6.5 to 5.0 increased the rate of exchange of tritiated water by 50% (Potts, unpublished data). The osmotic permeability of the brown trout, on which urine flow depends, is similar to the diffusional permeability, measured by tritiated water (Oduleye, 1975a) although they differ in some species. Any increase in urine flow would increase the rate of sodium loss further. Both the osmotic permeability and the diffusional permeability are reduced in the presence of calcium in the external medium (Oduleye, 1975a,b).

The effects of aluminium ions on sodium balance

Certain species of aluminium ions act synergistically with hydrogen ions to decrease sodium uptake and to increase sodium loss. Calcium ions help to protect fishes from both effects.

Many experiments have demonstrated that the adverse effects of aluminium are greatest between pH 5.2 and 5.5. Careful experiments on the survival of brown trout fry in various combinations of calcium and aluminium ions showed that aluminium was more damaging at pH 5.4 than at pH 5.1 or 4.5 (Brown, 1983). Muniz & Leivestad (1980b) found that aluminium was more toxic at pH 5.0 than at 4.0 while in another series of experiments it was more toxic at pH 5.1 and 5.5 than at pH 4.3 or 6.0 (Muniz & Leivestad, 1980a). Similar effects were found with the brook trout *Salvelinus fontinalis* where aluminium was most toxic at pH 5.5 and toxicity decreased down to pH 4.5 (Baker & Schofield, 1980). A more detailed analysis of the problem in the brown trout (Dalziel, 1985) showed that aluminium ions stimulated sodium efflux more potently at pH 5.4 than at pH 4.5. The effects of aluminium on influx at pH 5.4 were slight although influx was strongly inhibited at pH 4.5 or 4.0. Even as little as 1 μM Al reduced influx by about 75% at pH 4.5. These results suggest that the deleterious effects of aluminium at pH 5.4 are due to the increased efflux of salt rather than a reduction of influx. However, McWilliams (unpublished data) noted that in the absence of external calcium, aluminium at concentrations above 4 μM (at pH 5.4) depressed Na^+ uptake in brown trout. As external calcium increased aluminium had progressively less effect on Na uptake (Figure 8). Contrary to most previous workers, Dalziel found that in the complete absence of aluminium ions influx was not affected down to pH 4.5 and was only slightly affected at pH 4.0. Dalziel's experiments were carried out in synthetic media. The extreme sensitivity of the influx to aluminium demonstrated by Dalziel raises the question whether some of the previous effects reported of low pH on sodium uptake have in fact been due, at least in part, to the presence of extremely low concentrations of aluminium ions (*c.* 1 μM) in the media used. On the other hand the discrepancy may arise from the different lengths of time of the experiments. Reduction of efflux may be apparent in experiments involving instantaneous transfer but where the fish are adapted to low pH for some time before measurements are made the fish may compensate. Staurnes, Vedagiri & Reite (1984) noted that after two to five days at pH 5 the ATPase activity in the gills of *S. salar* had increased. Once again calcium ions facilitated sodium influx in the presence of aluminium and reduced the efflux, over the range 0.05 to 0.5 mM Ca.

Unfortunately the chemistry of aluminium hydroxides is still a somewhat inexact science but the present opinion is that the proportion of $Al(OH)_2^+$ is maximal around pH 5.4 while below that pH the proportion of $Al(OH)^{2+}$ and Al^{3+} increase (Potts *et al.*, 1985). If this is the case then monovalent $Al(OH)_2^+$ increases efflux while divalent $Al(OH)^{2+}$ decrease it. Conversely Al^{3+} and or $Al(OH)^{2+}$inhibit influx at micromolar concentrations while $Al(OH)_2^+$ has little or no effect on influx.

Aluminium ions at pH 5.0, at a concentration of 200 μg l^{-1} (7 5 μM) produced a 25% inhibition of the gill Na–K-ATPase in salmon and rainbow trout during experiments lasting between two and five days. Similar reductions occurred in the carbonic anhydrase activity. These reductions were associated with marked falls in the sodium and chloride concentrations in the blood plasma (Staurnes, Sigholt & Reite, 1984; Staurnes, Vedagiri & Reite, 1984). In contrast, in the absence of aluminium, the ATPase activity increased at pH 5 over several days, presumably in response to the fall in blood concentration.

Although $Al(OH)^{2+}$ or Al^{3+} are more effective in inhibiting influxes than are $Al(OH)_2^+$, the cytoplasm must be closer to neutrality and the poisonous effects of aluminium ions on ATPase must depend on interactions between $Al(OH)_2^+$ and/or $Al(OH)_3$ and the enzyme. As the enzyme is believed to be situated on the baso-lateral membranes the aluminium ions, whatever the species, must cross the apical membrane, or penetrate the intercellular junctions, more readily at low pH. This picture is consistent with the protective effects of calcium ions.

Aluminium ions at a concentration of 6.5 μM reduce chloride uptake at pH 5.5 but have no significant effect at pH 7.0 and little effect at pH 4.0 (Table 6). On the other hand, they do not increase chloride loss at pH 4.0; if anything they may have a protective effect (J.C. Battram, personal communication.)

At pH 5.4 aluminium ions displace calcium from the microsomes extracted from the gills of both normal and acid-resistant trout. In contrast aluminium had no clear effect on calcium binding on the gill surface (McWilliams, unpublished observations). Studies of the time course of aluminium effects might throw light on

Figure 8. Combined effects of calcium and aluminium on sodium uptake rates at pH 5.4 in the brown trout, *Salmo trutta*.

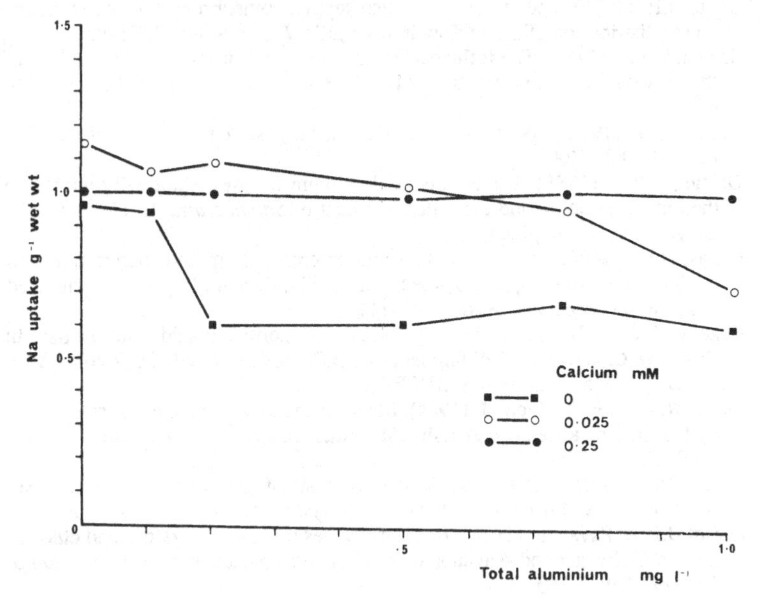

Table 6. *Effect of Al on influx and efflux of chloride in* S. trutta *(% Cl h^{-1} ± SE)*

	0 Al	6.5 μM Al l^{-1}	*n*
Influx			
pH 7.0	0.40 ± 0.016	0.39 ± 0.05	6
pH 5.5	[a]0.44 ± 0.12	[a]0.21 ± 0.07	7
pH 4.0	[b]0.45 ± 0.08	[b]0.45 ± 0.08	7
Efflux			
pH 7.0	0.33 ± 0.08	0.67 ± 0.10	5
pH 5.5	[a]0.76 ± 0.14	[a]1.28 ± 0.42	5
pH 4.0	[b]0.68 ± 0.12	[b]0.74 ± 0.19	6

[a] Adapted to pH 5.5.
[b] Adapted to pH 4.0.

these mechanisms. If aluminium acts intracellularly on the microsomes and ATPase, the effects should develop more slowly than if the aluminium acts on the gill surface.

References

Baker, J.P. & Schofield, C.L. (1980). Aluminium toxicity to fish as related to acid precipitation and Adirondack surface water quality. In *Proc. Int. Conf. Ecological Impact of Acid Precipitation*,ed. D. Drabløs & A. Tollan, pp. 292–3. Oslo–Ås: SNSF Project.

Bouth, J.H. (1979). The effects of oxygen supply, epinephrine, and acetylcholine on the distribution of blood flow in trout gills. *J. exp. Biol.*, **83**, 31–9.

Brown, D.J.A. (1983). The influence of calcium on the survival of eggs and fry of the brown trout (*Salmo trutta*) at pH 4.5. *Bull. Envir. Contam. Toxicol.*, **28**, 664–8.

Bryan, G.W. (1960). Sodium regulation in the crayfish *Astacus fluviatilis*. *J. exp. Biol.*, **37**, 83–128.

Dalziel, T.R.K. (1985). The effects of aluminium in low conductivity waters on the ionic regulation and early development of *Salmo trutta*. L. Ph.D. Thesis, University of Nottingham.

Evans, D.H. (1975). The effects of various external cations and sodium transport inhibitors on sodium uptake by the Sailfin Molly, *Poecilia latipinnia*, acclimated to sea water. *J. comp. Physiol.*, **96**, 111–15.

Fugelli, K. & Vislie, T. (1980). Physiological responses to acid water in fish. In *Proc. Int. Conf. Ecological Impact of Acid Precipitation*, ed. D. Drabløs & A. Tollan, pp. 346–7. Oslo–Ås: SNSF Project.

Garcia Romeu, F. & Maetz, J. (1964). Mechanisms of sodium and chloride uptake by the gills of a freshwater fish, *Carassius auratus* L. *J. gen. Physiol.*, **47**, 1195–207.

Girard, I.P. & Payan, P. (1977). Kinetic analysis of sodium and chloride fluxes across the gills of trout in fresh water. *J. Physiol. (Lond.)*, **273**, 195–209.

Girard, J.P. & Payan, P. (1980). Ionic exchanges through respiratory and chloride cells in freshwater and seawater adapted teleosteams: adrenergic control. *Am. J. Physiol.*, **238**, R260–8.

Goldstein, L., Claiborne, J.B. & Evans, D.H. (1982). Ammonia excretion by the gills of two marine fishes. The importance of NH_4^+ permeance. *J. exp. Zool.*, **219**, 395–7.

Graham, M.S. & Wood, C.M. (1981). Toxicity of environmental acid to the rainbow trout in interactions of water hardness, acid type and exercise. *Can. J. Zool.*, **59**, 1218–26.

Kerstetter, T.H., Kirschner, L.B. & Rafuse, D.D. (1970). On the mechanisms of sodium ion transport by irrigated gills of rainbow trout (*Salmo gairdneri*). *J. gen. Physiol.*, **56**, 342–59.

Kirschner, L.B., Greenwald, L. & Kerstetter, T.H. (1973). Effect of amiloride on sodium transport across body surfaces of freshwater animals. *Am. J. Physiol.*, **224**, 832–7.

Krogh, A. (1938). The active absorption of ions in some freshwater animals. *Z. vergl. Physiol.*, **24**, 656–66.

McWilliams, P.G. (1980a). Effect of pH on sodium uptake in Norwegian brown trout (*Salmo trutta*) from an acid river. *J. exp. Biol.*, **88**, 259–67.

McWilliams, P.G. (1980b). Acclimation to an acid medium in the brown trout, *Salmo trutta*. *J. exp. Biol.*, **88**, 269–80.

McWilliams, P.G. (1982a). The effects of calcium on sodium fluxes in the brown trout, *Salmo trutta*, in neutral and acid water. *J. exp. Biol.*, **96**, 439–42.

McWilliams, P.G. (1982b). A comparison of physiological characteristics in normal and acid exposed populations of the brown trout, *Salmo trutta*. *Comp. Biochem. Physiol.*, **72A**, 515–22.

McWilliams, P.G. (1983). An investigation of the loss of bound calcium from the gills of the brown trout, *Salmo trutta*, in acid media. *Comp. Biochem. Physiol.*, **74A**, 107–16.

McWilliams, P.G. & Potts, W.T.W. (1978). The effects of pH and calcium concentration on gill potentials in the brown trout *Salmo trutta*. *J. comp. Physiol.*, **126**, 277–86.

Maetz, J. (1973). Na^+/NH_4^+,Na^+/H^+exchanges and NH_3 movement across the gills of *Carassius auratus*. *J. exp. Biol.*, **58**, 255–75.

Muniz, I.P. & Leivestad, H. (1980a). Acidification – effects on freshwater fish. In *Proc. Int. Conf. Ecological Impact of Acid Precipitation*, ed. D. Drabløs & A. Tollan, pp. 84–92. Oslo–Ås: SNSF Project

Muniz, I.P. & Leivestad, H. (1980b). Toxic effect of aluminium on the brown trout, *Salmo trutta* L. In *Proc. Int. Conf. Ecological Impact of Acid Precipitation*, ed. D. Drabløs & A. Tollan, pp. 320–1. Oslo–Ås:SNSF Project.

Neville, C.M. (1979). Ventilatory response of rainbow trout (*Salmo gairdneri*) to increased H^+ ion concentration in blood and water. *Comp. Biochem. Physiol.*, **63A**, 373–6.

Oduleye, S.O. (1975a). The effects of calcium on water balance on the brown trout *Salmo trutta*. *J. exp. Biol.*,**63**, 343–56.

Oduleye, S.O. (1975b). The effects of hypophysectomy and prolactin therapy on water balance of the brown trout *Salmo trutta*. *J. exp. Biol.*, **63**, 357–66.

Packer, R.K., Dunson, W.A. (1970). Effects of low environmental pH on blood pH and sodium balance of brook trout. *j. exp. Zool.*, **174**, 65–72.

Payan, P. (1978). A study of the Na^+/NH_4^+ exchange across the gill of the perfused head of the trout (*Salmo gairdneri*). *J. comp. Physiol.*, **124**, 181–8.

Payan, P. & Girard, J.P. (1984). Branchial ion movements in teleosts: the roles of respiratory and chloride cells. In *Fish Physiology*, vol. X, part B, eds. W.S. Hoar & D.J. Randall, pp. 39–63. New York: Academic Press.

Perry, S.F. & Wood, C.M. (1985). Kinetics of branchial calcium uptake in the rainbow trout, effects of acclimation to various external calcium levels. *J. exp. Biol.*,**116**, 411–34.

Potts, W.T.W. (1981). Transepithelial potentials in fish gills. In *Fish Physiology*, vol. X, part B, eds. W.S. Hoar & D.J. Randall, pp. 105–28.New York: Academic Press.

Potts, W.T.W., Hunt, D.T.E., Blake, S. & French, P. (1985). Aluminium in acid waters – chemistry and effects on fish. Report prepared for Surface Water Acidification Programme, SWAP.

Powell, J.F.F. & McKeown, B.A. (1986). The effects of acid exposure on the ion regulation and seawater adaptation of Coho salmon (*Oncorhynchus kisutch*) parr and smolts. *Comp. Biochem. Physiol.*, **83C**, 45–52.

De Renzis, G. & Maetz, J. (1973). Studies on the mechanism of chloride absorption by goldfish gill: relation with acid-base regulation. *J. exp. Biol.*, **59**, 339–58.

Robinson, R.A. & Stokes, R.H. (1959). *Electrolyte Solutions*. London: Butterworth.

Sardet, C., Pisam, M. & Maetz, J. (1980). Structure and function of gill epithelia of euryhaline teleost fish. In *Epithelial Transport in Lower Vertebrates*, ed. B. Lahlou, pp. 56–68. London: Cambridge University Press.

Shaw, J. (1960). The absorption of sodium ions by the crayfish, *Astacus pallipes*, Lereboullet, II. *J. exp. Biol.*, **37**, 548–56.

Staurnes, M., Sigholt, T. & Reite, O.B. (1984). Reduced carbonic anahydrase activity and Na–K-ATPase activity in gills of salmonids exposed to aluminium-containing acid water. *Experientia*, **40**, 226–7.

Staurnes, M., Vedagiri, P. & Reite, O.B. (1984). Evidence that aluminium inhibits Na–K-ATPase and carbonic anhydrase activity in the gills of salmon, *S. salar*. *Acta Physiol. Scand.*, **121**, 8.

Twitchen, I.D. (1987). The physiological basis of acid resistance in aquatic insect larvae. *Surface Water Acidification Programme*, Bergen, Norway, June 1987.

Ultsch, G.R., Ott, M.E. & Heisler, N. (1981). Acid–base and electrolyte status in the carp (*Cyprinus carpio*) exposed to low environmental pH. *J. exp. Biol.*, **93**, 65–80.

D.G. MCDONALD, J.P. READER AND T.R.K. DALZIEL

The combined effects of pH and trace metals on fish ionoregulation

Introduction

In freshwater fish the physiological regulation of the major electrolytes is very sensitive to environmental stressors. Low pH environments in both the laboratory and field cause electrolyte losses in a number of fish species and, indeed, plasma electrolytes have proven to be a fairly reliable indicator of sublethal acid stress (e.g. Leivestad & Muniz, 1976). Similarly, there are now several studies on the toxic trace metals showing that disturbances to ion regulation are either a primary or at least a secondary consequence of exposure to a particular metal. Our objective then is to examine how mixtures of trace metals and H^+ might toxically interact to cause ionic disturbances. We have placed emphasis on sublethal effects upon gill function rather than toxicity *per se*. We first examine the chemical and biological bases for metal and H^+ interactions and then present some examples which illustrate the nature of these interactions. It is not our intention to review exhaustively metal and H^+ toxicity but rather to point out how one might examine or even predict the interactions of untested metal/H^+ mixtures. For a more general and thorough treatment of metal and acid toxicity to aquatic biota the reader is referred to the recent review by Campbell & Stokes (1985).

In terrestrial animals, the toxicity of a particular metal is mainly related to its dose; if a metal is not absorbed then it is not toxic, irrespective of its reactivity in aqueous solution. Aluminium is a particularly good example of this point; although toxic to fish in acid waters it is not normally toxic to man because it is so poorly absorbed across the lungs and gut (Hammond & Beliles, 1980). In aquatic animals the external surfaces are, generally speaking, much more structurally and physiologically delicate than comparable liquid-exposed surfaces in terrestrial animals. Thus a particular metal could be toxic to an aquatic animal because of its surface activity as well as whatever internal effects it might have. Metals which are most likely to be internally toxic only are those that are readily absorbed and have little, if any, surface activity. Most prominent in this category are mercury, lead, arsenic and tin. In natural waters, these are typically found as lipid-soluble, organo-metallic complexes that are readily permeable to biological membranes. Most other metal contaminants of aquatic environments tend to occur as water-soluble cations and so are at least potentially active at the surfaces of fish. Indeed, the surface activity of some metals and probably of H^+ as well (McDonald, 1983), may be all that is needed to explain their respective toxicities.

In order to understand the surface interactions of metals and hydrogen ions it is necessary to consider the following: the structural and physiological targets of surface-active metals, the extrinsic factors governing the concentration and chemical form of metals at the surface, and the differences in chemistry amongst the toxic metals. For reasons outlined shortly the main structural target of surface-active metals is likely to be the gills and the main physiological target is likely to be ion regulation.

The gills of freshwater fish

In pelagic freshwater fish, the gills of adults typically make up more than 50% of the total body surface and are thin and delicate compared to the skin, typically being about 6 μm thick vs 20–40 μm for the skin epidermis. Furthermore, the gill surface probably has a substantial net negative charge because of a variety of anionic ligands and, as a result, is likely to have a relatively high affinity for the cationic forms of metals. The surface anions of fish gills have not been characterized to any extent but, based on studies of membranes and other epithelia (Oschman, 1978), are likely to consist of the following: sialic acid residues of mucus and membrane glycoproteins, carboxyl groups of membrane glycolipids, and various polyelectrolytes of intercellular cements; hyaluronic acid, heparin and chondroitin sulphate with carboxyl, phosphate and sulphate groups. Although there is probably a variety of anions on the surface, their one common feature is the predominance of oxygen as a donor ligand. This point becomes important when considering the binding preferences of metals as discussed below.

The reason that branchial ionoregulation is the most likely physiological target of surface active metals is the variety of potential targets and toxic mechanisms. The gills of freshwater fish contain the machinery for the active transport of at least three (and possibly four) electrolytes; Na^+, Cl^-, Ca^{2+} (and possibly K^+: Eddy, 1985). While many of the details have yet to be worked out, branchial ion regulation in freshwater fish can most likely be characterized as follows: the transport mechanisms are independent of one another and each consists of an ion-specific channel or carrier in the outer facing (apical) membrane and a transport ATPase in the baso-lateral membrane (Kirschner, 1980; De Renzie & Bornancin, 1984; Payan *et al.*, 1984). Likely separate from the transport pathways are pathways for the passive outward leak of electrolytes. These are thought to be less ion-specific than the transport pathways and to be located in the intercellular junctions – the so-called 'paracellular pathways' (Marshall, 1985). One of the major determinants of the transport of Na^+, Cl^-, K^+, and Ca^{2+} is, of course, their respective external concentrations but external

Table 1. *Major species of trace metal present at different water pHs*

	Metal speciation vs water pH		
	at pH 7.5	at pH 5.0	at pH 4.0
Copper	$CuCO_3$, Cu^{2+}	Cu^{2+}	Cu^{2+}
Cadmium	Cd^{2+}, $CdOH^+$	Cd^{2+}	Cd^{2+}
Zinc	$ZnOH^+$, Zn^{2+}	Zn^{2+}	Zn^{2+}
Lead	$PbCO_3$, $PbOH^+$	Pb^{2+}	Pb^{2+}
Aluminium	$Al(OH)_4^-$	$AlOH^{2+}$, $Al(OH)_2^+$	Al^{3+}
Manganese	Mn^{2+}	Mn^{2+}	Mn^{2+}
Nickel	Ni^{2+}	Ni^{2+}	Ni^{2+}
Iron	$Fe(OH)_2^+$, $Fe(OH)_4^-$	$Fe(OH)_2^+$	$FeOH^{2+}$

From Stumm & Morgan, 1981.

Ca^{2+} also has an important surface activity. By forming cross-links with ligands on the surface and in the intercellular cement, calcium acts to increase membrane stability and reduce the permeability of the paracellular pathways (Oschman, 1978). Thus the potential physiological effects of metals could be any or all of the following: disruption of one or more of the transport mechanisms, widening or blockade of the diffusion pathways, or interference with the surface binding of calcium with a concomitant effect on membrane stability.

The specific biochemical action of metals on these sites has also not been fully characterized but is likely to fall into one or more of three categories (after Ochiai, 1977): competitive or non-competitive blockade of functional groups on proteins (e.g. blockade of ion channels or carriers), displacement of essential metal ions from proteins (e.g. displacement of Ca^{2+} from surface ligands), and conformational change in proteins (e.g. oxidation of SH groups to –S–S–).

Water chemistry

The metals we will consider (some in more detail than others) are the following: copper, cadmium, zinc, lead, aluminium, manganese, nickel and iron (Tables 1, 2). These have been chosen because they tend to be elevated in acid, soft water lakes either as a result of atmospheric input from industrial operations or because of leaching from soils and sediments. Most are acutely toxic to fish in low micromolar concentrations and many are known to have at least some degree of surface activity relative to fish gill function.

Since the most environmentally relevant metal/H^+ interactions take place in soft waters it is worthwhile first to consider the origin and character of such waters. By

Table 2. *Trace metals elevated in acid soft waters, ranked according to their acute toxicity to rainbow trout (except where noted) in low hardness, low alkalinity water. Upper limits for natural waters are based on a survey by Reader (1986) of values reported for acid lakes in Scandinavia, UK and North America*

	relative atomic mass (µM)	upper limit (µM)	LC_{50}	Reference
Copper	64	0.3	0.3	1
Cadmium	112	0.01	0.3	2
Zinc	65	1.9	1.4	3
Lead	207	0.3	5.6	4
Aluminium	27	36	7.3	5
Iron	56	40	7.3	8
Manganese	55	10	290	6
Nickel	59	0.9	338	7
Hydrogen ions	1	100	100	9

References
1. Howarth & Sprague, 1978. 96 h LC_{50} at pH 5, hardness = 0.3 mM as $CaCO_3$.
2. Cusimano *et al.*, 1986. 96 h LC_{50} at pH 4.7, hardness = 0.1 mM.
3. Bradley & Sprague, 1985. 96 h LC_{50} at pH 5.6, hardness = 0.3 mM.
4. Davies *et al.*, 1976. 96 h LC_{50} at pH 6.6–7.3, hardness = 0.26 mM.
5. Orr *et al.*, 1986. 96 h LC_{50} at pH 5.1–5.3, hardness = 0.1 mM.
6. England & Cummings, 1971 quoted in Lewis, 1978. 96 h LC_{50}, water composition unspecified.
7. Brown, 1968. 48 h LC_{50} at pH 7.0, hardness = 0.2 mM.
8. Decker & Menedez, 1974. 96 h LC_{50} at pH 5.5 for *brook trout*, water hardness unspecified.
9. Graham & Wood, 1981. 96 h LC_{50}, water hardness = 0.14 mM.

definition, soft waters are low in concentration of the hardness metals (calcium and magnesium). Calcium is the more important of the two; its concentration is typically two to three times that of magnesium (NRCC, 1981). Calcium comes mainly from the dissolution of $CaCO_3$ in calcareous soils and sediments by the action of dissolved CO_2. Consequently, hardness is usually expressed in terms of $CaCO_3$ concentration although, in reality, the soluble product of $CaCO_3$ dissolution is, below pH 8, almost entirely $Ca(HCO_3)_2$ (Stumm & Morgan, 1981). The boundary between hard and soft waters, although somewhat arbitrary and undefined, is generally considered to be a hardness of about 0.5 mM $CaCO_3$ (Marier *et al.*, 1979). Since the major source of calcium is $CaCO_3$, hardness and alkalinity (i.e. the acid-neutralizing capacity due mainly to the bicarbonate concentration) are usually present in a 1:2 molar ratio in those waters not receiving mineral acid or base input (Stumm & Morgan, 1981).

Figure 1 shows the relation between Ca^{2+} and pH for softwater lakes in Canada. Note that soft waters are, in fact, not 'created equal' since the calcium concentration varies over almost two orders of magnitude (<0.01–0.5 mM). Also note that water pH begins to decline below a calcium concentration of about 0.15 mM. This is partially due to titration by atmospheric CO_2 (distilled water in equilibrium with atmospheric CO_2 has a pH of 5.6). Mainly, though, the lowered pH reflects the input of mineral acids (chiefly H_2SO_4 and HNO_3). Many of the lakes in Figure 1 that are above 0.15 mM Ca^{2+} are also receiving acid precipitation (shown as open symbols) but their alkalinity is sufficiently high (and sufficiently rapidly renewed by their watersheds) to buffer the acid input.

Surface activity of metals

In soft waters the surface activity of a particular metal is likely to be highly dependent upon the water chemistry; most importantly, pH, hardness and complexing capacity. In the ensuing discussion we review, in general terms, the nature of the effects of these factors, but one should be wary of making any generalizations applicable to all metals because each metal is probably unique in its water interactions.

Of the three factors, the effects of complexing agents are the simplest to explain. Soft waters may contain a variety of inorganic and organic ligands depending on their watershed chemistry: bicarbonate, fluoride, sulphate, fulvic and humic acids. Many

Figure 1. pH vs calcium concentration for soft water lakes in Canada (redrawn from NRCC, 1981). Lakes in areas receiving atmospheric acid input (lakes mainly in region of Sudbury, Ontario) are shown as open symbols.

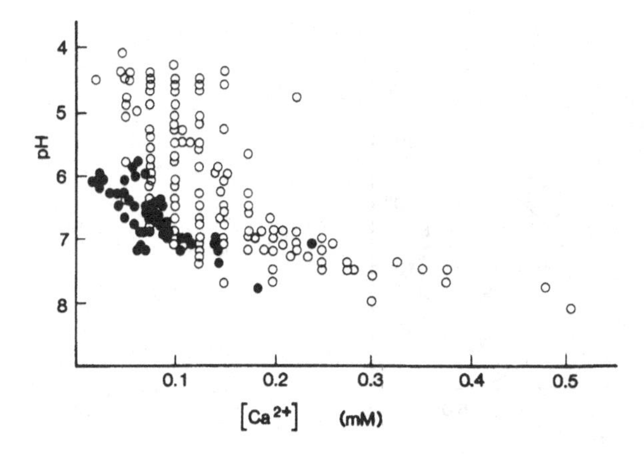

of these ligands will form relatively stable complexes with some metals, complexes which are either neutral or negatively charged and tend to be relatively non-toxic. The stability of the complex varies with the metal and the ligand and the importance of this phenomenon depends, of course, on the concentration of the ligand. For example, of the list of metals in Table 1 or 2, only copper and lead form significant quantities of stable carbonate complexes and these are perhaps only important in waters of relatively high alkalinity. Generally speaking, alkalinity disappears below a pH of about 5.0. Thus in acid waters, the effect of alkalinity on metal availability is likely to be negligible.

The main effect of hardness on metal activity is the reduction of toxicity, an effect that probably can be attributed to competition betwen hardness metals and trace metals for surface ligands (Pagenkopf, 1983). The physiological basis for this phenomenon is not yet well characterized, nevertheless protective effects of the hardness metals, Ca^{2+} and Mg^{2+}, are well established in the metal toxicology literature (e.g. Brown, 1968; Alabaster & Lloyd, 1980). Three main conclusions can be extracted from these studies; calcium is a more effective protecting agent than magnesium (e.g. Potts & Fleming, 1971), the degree of protection varies with the metal (e.g. Cd > Zn > Cu > Ni > Pb, Brown, 1968) and the degree of protection also varies with the log of the hardness concentration (Brown, 1968). The last point is worth emphasizing because

Figure 2. Aluminium chemistry in low ionic strength waters. *(a)* Aluminium speciation diagram. Relative concentrations of aluminium species at 25 °C in freshwater. Calculation based on thermodynamic equilibrium constants for 25 °C and zero ionic strength (Johnson *et al.*, 1981). *(b)* Total soluble aluminium as a function of pH from the dissolution of amorphous $Al(OH)_3$. Calculation based on thermodynamic equilibrium constants for 25 °C and zero ionic strength (Johnson *et al.*, 1981). Superimposed on plot is the experimental matrix referred to in the text (see Figure 5). In natural waters aluminium solubility is usually less than described by this line.

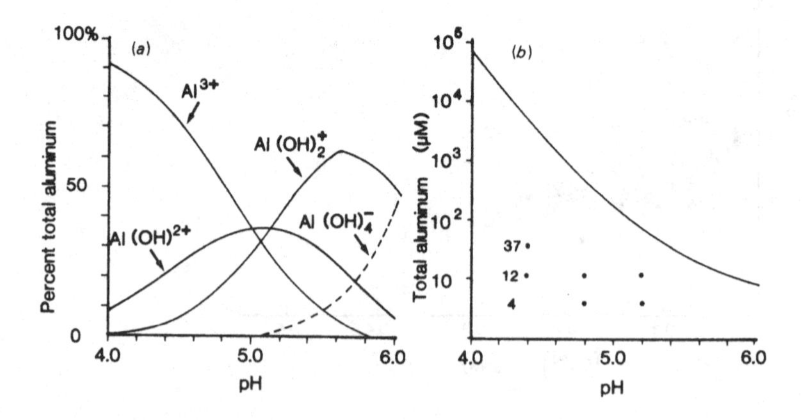

it indicates that small variations in water hardness are likely to have large effects upon metal toxicity, at least for some metals. This has been borne out by field surveys of very soft water lakes in Norway which indicate that variations in calcium concentration were at least as important in determining fishery status as were variations in pH (Wright & Snekvik, 1978).

The effects of water pH on metal activity are the most complex of the three factors to explain since pH can affect either the solubility and/or the speciation of many metals. Table 1 lists the predominant species vs pH for the metals under discussion. Note that the more acidic the pH, the greater the proportion of free cation rather than the first or second hydrolysis species, i.e. Me^{++} vs $MeOH^{+}$ or $Me(OH)_2$. At pH 5.0 and below most of the metals are in the free cation form. The only exceptions are iron and aluminium. By pH 4.0 the latter exists mostly as free cation whereas significant hydrolysis species for iron remain. The effect of pH on aluminium speciation is particularly important because major changes in speciation occur over the typical range in pH for softwater lakes (pH 6.0–4.0, Figure 1, 2). Another important effect of pH is the increasing solubility of metals at lower pH. This is mainly important for aluminium (Figure 2), iron and manganese.

Metal toxicity

Table 2 shows the metals ranked according to their toxicity to rainbow trout (the species for which most data is available). These are acute toxicity values (either 48 or 96 h LC_{50} values) in low alkalinity, low hardness waters at pH ≥ 5.0. According to these data, copper is the most toxic metal to trout and nickel is the least. As far as we are aware, the acute toxicity of iron to rainbow trout has not been established so we have reported a value for brook trout instead. Since brook trout are generally considered to be hardier than rainbow trout (e.g. Grande *et al.*, 1978) this value should be considered an underestimate.

Although our toxic comparison amongst the metals is fairly uniform with respect to choice of test animal and test conditions the ranking of the metals cannot be assumed to hold for lower pH values or more prolonged exposures (i.e. >96 h). Water pH affects the speciation and solubility of some metals more than others and can be expected to modify the accumulation of metals by fish tissues (Campbell & Stokes, 1985). An illustration of this phenomenon is shown in Figure 3 for copper and cadmium uptake by rainbow trout. Note that the uptake of both metals is markedly reduced at pH 5.0 compared to pH ≥ 7.0. In this particular circumstance, where the exposures are relatively short term (24 h or less), the reduction of uptake at lower pH may largely reflect a reduction in surface adsorption of the metal and not true tissue uptake. Nonetheless, over the longer term (up to seven days at least) the toxicity to rainbow trout of these metals, and zinc also, is sharply reduced at acid pH (pH 4.7) compared with more alkaline pH levels (pH 5.7 and 7.0; Cusimano *et al.*, 1986). Since speciation changes are fairly unimportant for these metals over this pH range

(Table 1) the main reason for the reduction in toxicity is likely increased competition from H^+ for binding/uptake sites on the gills. Other metals could, in contrast, become more toxic at lower pH levels because of changes in speciation (e.g. iron and aluminium). Furthermore, if the chronic toxicity of a particular metal depends primarily on true absorption then its toxicity is likely to rise relative to the surface-active metals as the duration of exposure increases. Against this background one must keep in mind that acidification tends to mobilize metals; consequently any ameliorating effect of H^+ on metal toxicity is likely to be counteracted by increasing metal concentrations in the water.

Figure 3. Metal uptake by rainbow trout, *Salmo gairdneri*, in relation to pH, [metal], and [Ca]. *(a)* Total cadmium uptake over 12 h at pH 7.6. Cd exposure was carried out with ^{109}Cd and uptake was determined by whole body counting. Each data point represents the mean of 10 trout (± 1 SEM) of body weight = 1.1 ± 0.2 g. Water t = 14 ± 2 °C, pH 7.6, [NaCl] = 0.20 ± 0.01 mM. Unpublished data of D.G. McDonald & J.L. Ozog-Sloat. *(b)* As in A except carried out at pH 5.0. *(c)* Total copper uptake of trout surviving copper exposure for 24 h. Each point is a mean ± 1 SEM (n = 10–15). , soft water, [Ca^{2+}] = 0.025 mM; , hardwater, [Ca^{2+}] = 1.0 mM; soft water, high alkalinity, [HCO_3^-] = 2 mM. Figure redrawn from Lauren & McDonald, 1986a.

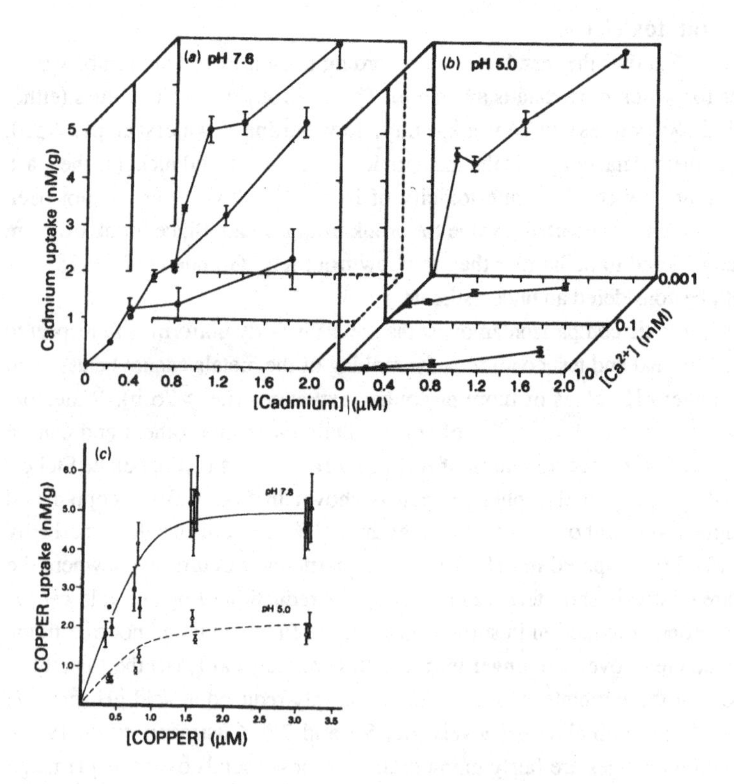

Some reported metal concentrations for acid, soft water lakes are included in Table 2. The values chosen represent the upper levels reported except that unusually high levels have been excluded. Note that the concentrations of copper, zinc, aluminium and iron are close to acutely toxic levels. For the remainder, acute toxicity is not likely to be encountered in natural waters but sublethal effects cannot be ruled out. Note also that on a molar basis, hydrogen ions are less toxic than all of the metals in the table except for manganese and nickel.

Metal chemistry

If one wishes to make predictions about the surface activity and toxicity of a particular metal it is necessary to know something of the metal's chemistry. Two chemical indices particularly useful in this regard are the ionic index and the covalent index (Nieboer & Richardson, 1980). The ionic index (abscissa, Figure 4) is a measure of how strong an ion pair the metal will form with a ligand; the covalent index (ordinate, Figure 4) is a measure of the tendency of a metal to form covalent complexes. It is also of value to know the donor atom preferences of a metal since this will reveal its binding specificity. Class A cations (which include the alkali and alkaline earth metals), have a preference for ligands where oxygen is the donor atom (i.e. ligands that are found on the gill surfaces) whereas the class B cations (mostly

Figure 4. A separation of metal ions into three categories according to binding preferences: class A (oxygen seeking), class B (nitrogen/sulphur seeking) and borderline (intermediate or ambivalent). The class B index, $X_m{}^2r$, is plotted for each ion against the class A index, Z^2/r, where X_m is the metal ion electronegativity, r its ionic radius and Z its formal charge. (Redrawn from Nieboer & Richardson, 1980).

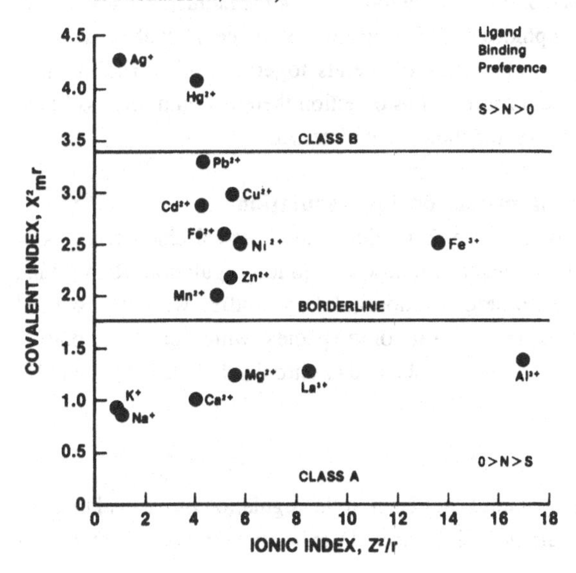

the Period 6 transition metals, Pt to Bi), have a particularly high affinity for biological binding sites containing N/S donor ligands (i.e. surface and subsurface proteins). The borderline cations (most of the remaining transition metals) have less well defined tastes in ligands; generally though, the class B character of these ions increases with covalent index (Nieboer & Richardson, 1980).

Since class A metals have similarly low covalent indices (Figure 4) their interactions with surface ligands are likely to be competitive, forming increasingly stable complexes largely in proportion to their formal charge. In other words, at equimolar concentrations, Al^{3+} would displace Ca^{2+} and Ca^{2+} would displace Na^+, etc. In this context it would be useful to know where H^+ fits in. Unfortunately, H^+ cannot be plotted on Figure 4 because its ionic radius is unknown. Based on biological considerations (cf. McDonald, 1983) it would appear to behave as a class A metal, positioned somewhere between Ca^{2+} and Al^{3+}. However, its chemistry and chemical reactivity suggest that it should be regarded as a borderline metal (Nieboer & Richardson, 1980).

Amongst the borderline metals, note that their toxicity increases approximately with their covalent index (compare Figure 4 with Table 2). With this idea in mind several authors have attempted to correlate metal toxicity with a single chemical characteristic (e.g. Jones, 1939, electrolytic solution pressure; Shaw & Grushkin, 1957, sulphide solubility product; Zitko, 1976, glycine binding constant; Reidel & Christensen, 1979, I_{50} of ATPase inhibition *in vitro*; Turner *et al.*, 1985, Pearson softness parameter). In no case, however, was the correlation exact. This is, perhaps, not surprising since there is likely to be considerable variation amongst the toxic metals as to the chemical characteristic(s) key to their respective toxic mechanisms. The key characteristic could be any one or more of the following: ionic radius, redox potential, tendency to form insoluble sulphides or donor ligand preference. Nonetheless, if one combines a knowledge of the chemistry of metals together with a rudimentary understanding of their biological mechanisms of action then it should be possible to predict how untested metal/H^+ combinations might interact.

Sublethal effects of metals on ion regulation

Having reviewed some of the theoretical and chemical characteristics of metal/H^+ toxicity in natural soft waters we now turn to ion regulation. Rather than discuss all of the now quite numerous ionoregulatory studies we have focused primarily on our own work, largely on salmonid species, which has been mainly concerned with branchial ion fluxes under sublethal or chronically lethal metal and H^+ conditions.

Sodium balance

Although our primary interest has been in the regulation of Na^+ balance we have also to a lesser extent examined Cl^- balance. Despite some minor differences, in general, the two ions tend to be similarly disturbed by toxic exposures so that conclusions drawn concerning one are likely to be generally applicable to the other. To date, the metals which primarily affect Na^+ balance are aluminium and copper, as well as H^+ ions; cadmium, zinc and manganese have either minor or no effects; the remaining metals in Table 2 (Pb, Ni, Fe) have, to our knowledge, not been tested in salmonids for ionoregulatory effects.

Figure 5 shows the effects of pH and aluminium over 48 h on sodium balance in adult brook trout. Increasing acidity in the absence of aluminium increases Na^+ losses in brook trout, an effect that is exacerbated at lower Ca^{2+} (compare Figure 5*(a)* with Figure 5*(b)*). At each test pH, Na^+ losses increase with increasing aluminium concentration and the amount of Na^+ loss is closely correlated with mortality, particularly at low Ca^{2+}, confirming (as with H^+, Wood & McDonald, 1982; Wood, this volume) that with lethal Al exposures at pH levels less than about 5.5, major contributors to death are the physiological consequences of salt imbalance. At each test pH, Al is additive over that of the acid pH alone although on a molar basis, Al has a greater relative potency.

Another manner in which H^+ and Al interact is illustrated by comparing the three different test pH levels at the 12 μM Al concentration (Figure 5*(a)*). Here, the toxicity of aluminium is potentiated by increasing pH. There are two possible explanations for this phenomenon. On one hand, the increasing losses could be explained if the

Figure 5. Effects of aluminium on adult brook trout, *Salvelinus fontinalis* (100–300 g wet weight). *(a)* Cumulative Na^+ losses (1000 μequiv kg^{-1}) over the first 48 h of exposure in low Ca^{2+} water ($[Ca^{2+}]$ = 0.013 mM) at various combinations of pH and Al. Each experiment was conducted for 10 d. Mortalities at end of 10 d are shown on columns expressed as percentages. In each treatment, n = 6 animals. *(b)* Same as A except exposures conducted at higher Ca^{2+} (0.2 mM). Redrawn from Booth, McDonald, Simons & Wood (1988).

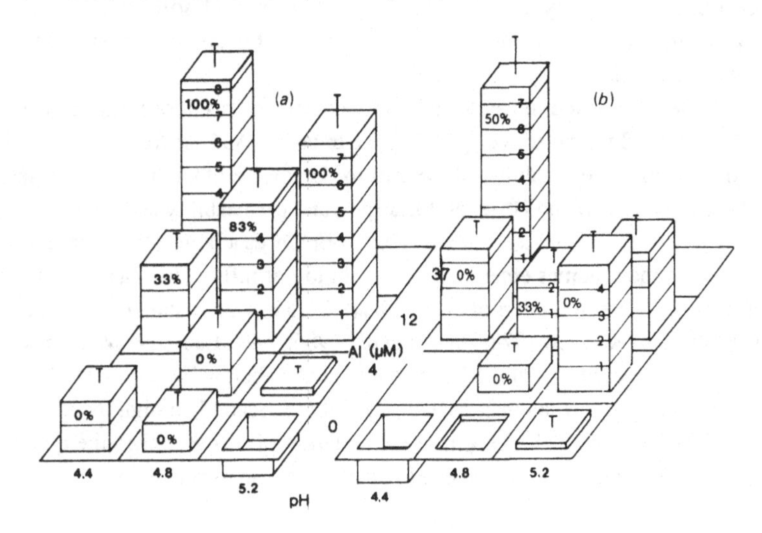

aluminium hydroxides (Table 1) had a more potent effect on the gills than Al^{3+}; the latter is 77% of total aluminium at pH 4.4, but only 22% at pH 5.2 (Figure 2*(a)*). Alternatively, the proximity of Al to its solubility limit may be the key factor in its surface action. Total aluminium solubility decreases sharply over the range of 4.4 to 5.2 (Figure 2*(b)*) and while Al solubility was not exceeded in water, the microenvironment of the gills is likely to be more alkaline (due to buffering by mucus and gill anions and to NH_3 excretion) thus potentially creating precipitating conditions for the metal. It is not possible at this point to conclude which of the two is more likely since little is known of the speciation and solubility of Al in the gill microenvironment. In any case, the two explanations are not mutually exclusive.

These interrelations amongst pH and Al are less clear cut at higher external Ca^{2+} (Figure 5*(b)*). Nonetheless, it is clear that calcium has an overall antagonizing effect on the action of Al, reducing ion loss and percentage mortality.

Copper is similar in its action to H^+ and aluminium in that it produces concentration-dependent sodium losses (Figure 6) but it is different in the nature of its interactions with H^+ and Ca^{2+}. Essentially, neither H^+ or Ca^{2+} has any marked effect on copper's potency as a surface toxicant. Additive effects of H^+ (at 10 μM; pH 5.0) with copper disappear above copper concentrations of 2 μM (Figure 6); copper concentrations above 2 μM produce the same net Na^+ losses over 24 h at pH 7.8 as at pH 5.0. Furthermore, increasing the Ca^{2+} concentration from 0.025 to 1.0 mM (a Ca^{2+} value representative of hard waters) has no protective effect (compare SL to HL in Figure 6). Although hardness has no effect on copper toxicity there is a marked effect of alkalinity. At bicarbonate concentrations of 2 mM (representative of hard waters at circumneutral pH) there is a marked reduction of net Na^+ losses produced by copper. This is likely to be due to the formation of relatively stable copper carbonate complexes having little, if any, surface activity. While the effect of alkalinity is unlikely to be of this magnitude in acid soft waters, it may be of significance at pH $\geq$ 5.0 particularly given the high toxicity of copper relative to other metals (Table 2).

The net loss of sodium at the gills is, of course, the result of an imbalance between influx and efflux, the active uptake by the transport ATPase for Na^+ on the one hand and the passive permeability of the gills to Na^+ on the other. The precise mechanism of action of Cu, Al and H^+ on Na^+ transport and permeability at the gills has yet to be fully characterized and there are no doubt differences amongst the three. Nonetheless, some common themes emerge. First, inhibition of influx (J_{in}) has generally a lower threshold than stimulation of efflux (J_{out}). This has been shown for pH, Al and Cu (McWilliams & Potts, 1978; Dalziel *et al.*, 1986, 1987; Lauren & McDonald, 1986a). For example, copper inhibits J_{in} at concentrations below 0.4 μM and only stimulates J_{out} above 2.0 μM (Lauren & McDonald, 1986a). Second, at high metal or H^+ levels (i.e. those at or near the LC_{50}), the stimulation of efflux makes the quantitatively greater contribution to ion loss. However, if a fish survives the exposure, efflux typically recovers to the same or to lower levels whereas inhibition of influx either persists (e.g. at pH 4.3 and low Ca^{2+}, McDonald *et al.*, 1983) or only slowly recovers (e.g. at 1 μM Cu; Lauren, 1986; Lauren & McDonald, 1986b). Consequently, at sublethal levels of exposure the inhibition of influx will eventually make the quantitativley more important contribution to net ion loss.

Calcium balance

The branchial mechanisms of Na^+ and Ca^{2+} regulation are distinct from one another in a number of ways including the nature of the transport mechanisms, hormonal controls, leak pathways and specific sites in gills for the transport machinery. However, for the purpose of this discussion we will concern ourselves with only two differences. First, branchial influx rates of calcium are usually substantially lower than those of sodium. For example, in adult rainbow trout at external concentrations of Na^+ and Ca^{2+} of 0.6 mM the influx ratio is 20:1 (Reid & McDonald, 1988). Second, recent studies (Spry & Wood, 1985, McDonald & Rogano, 1986) show that salmonids, over a wide range in body weight (< 1 g to

Figure 6. Effects of copper on J_{net} Na^+ at different hardness, alkalinity and pH for juvenile rainbow trout at 15 °C (2–5 g wet weight). Flux rates are in nM g^{-1} h^{-1} measured over 24 h or as long as the fish survived (for fish at 3 and 6 μM the mortality was >50%). [Ca^{2+}] was 0.025 mM for soft water (SL) and 1 mM for hard water (HL and HH). Alkalinity (i.e. bicarbonate concentration) was less than 0.01 mM for low alkalinity tests (SL and HL) and 2 mM for the high alkalinity condition (HH). Figure redrawn from Lauren & McDonald, 1986a.

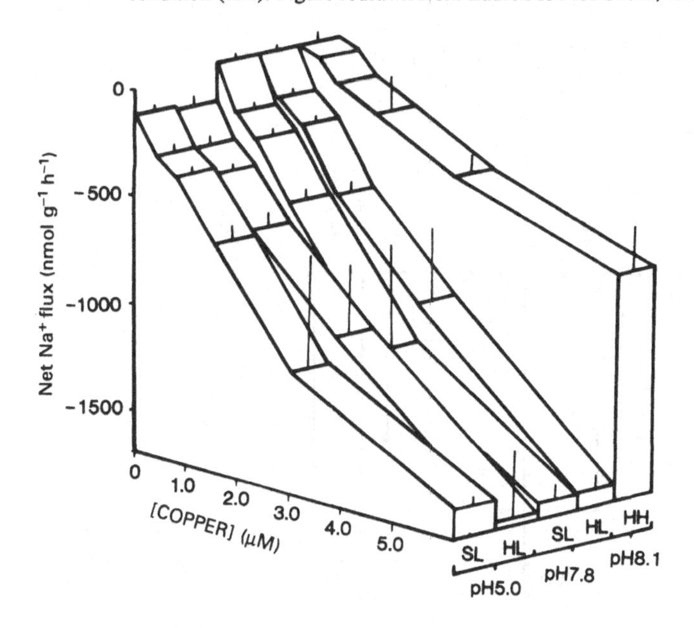

> 300 g) have a persistent net uptake of calcium, probably reflecting growth and calcification of the skeleton. In contrast, normal Na^+ balance is typified by zero net fluxes of sodium (i.e. branchial net uptake = net renal loss).

These differences between Na^+ and Ca^{2+} balance extend to their relative sensitivities to metal poisoning. For example, neither cadmium nor zinc, both of which disturb normal Ca^{2+} homeostasis, has any significant effect upon Na^+ flux rates in rainbow trout over 24 h despite the fact that exposure concentrations employed (6 and 12 μM, respectively) are at or near the 96 h LC_{50} for both metals (Reid & McDonald, 1988; Spry & Wood, 1985).

Based on short term exposures to the metals, the approximate potency order to Ca^{2+} balance is as follows: Cd > Cu > Zn > Mn (Figures 7, 8, 9). Aluminium, tested at 6 μM, has no effect on Ca^{2+} flux rates after five days of exposure (Reader & Morris, 1988) and the other metals in Table 2 (Pb, Ni, Fe) have yet to be tested. All

Figure 7. Acute effects of copper, cadmium and/or H^+ on calcium fluxes in adult rainbow trout (weight = 235 ± 8 g) in soft water ($[Ca^{2+}]$ = 0.04 mM panel A) and hard water ($[Ca^{2+}]$ = 1.0 mM, panel B), n = 6 for each treatment, t = 15 ± 1 °C. [Cd] = 6 μM (707 ppb), [Cu] = 6 μM (400 ppb), $[H^+]$ = 10–12 μM. Animals were exposed for 12 h. Figure redrawn from Reid & McDonald (1988).

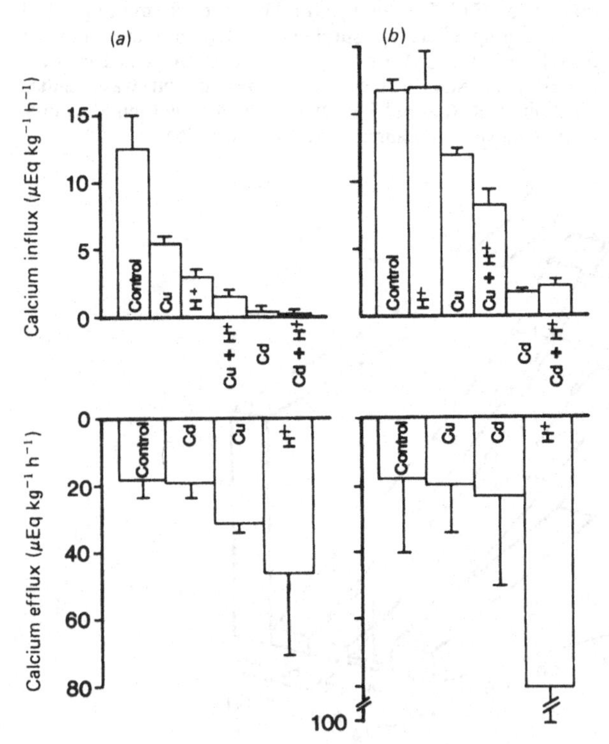

Figure 8. Longer term effects of zinc (12 μM) on net calcium flux μequiv kg^{-1} h^{-1}, means ± 1 SEM) of adult rainbow trout (n = 12, wt = 141–267 g) in soft water (Ca^{2+} = 0.08 mM) at circumneutral pH. Figure redrawn from Spry & Wood, 1985.

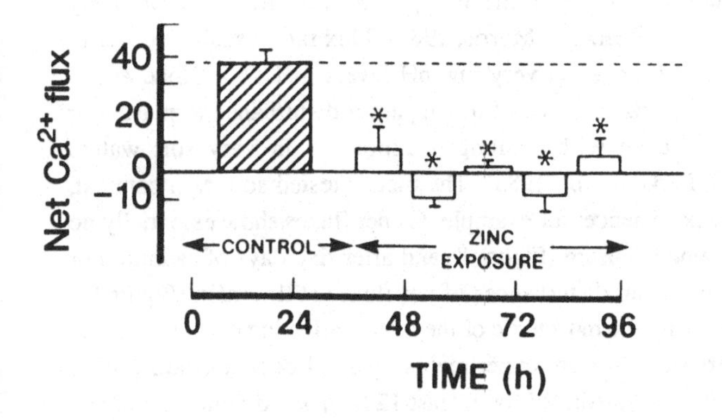

Figure 9. Longer term effects of metals on unidirectional and net calcium fluxes in brown trout, *Salmo trutta* fingerlings (weight *c.* 1g).Shaded area indicates net flux. Fluxes were measured over an 8 h period after animals had been exposed to the metal (or control conditions) for 5 d. The difference between Cd and Mn controls is probably attributable to the use of two genetically different strains. Asterisks indicate means significantly different from controls ($p < 0.001$, t-test). $[Ca^{2+}]$ = 0.02 mM, pH = 6.5. Redrawn from Reader & Morris, 1988.

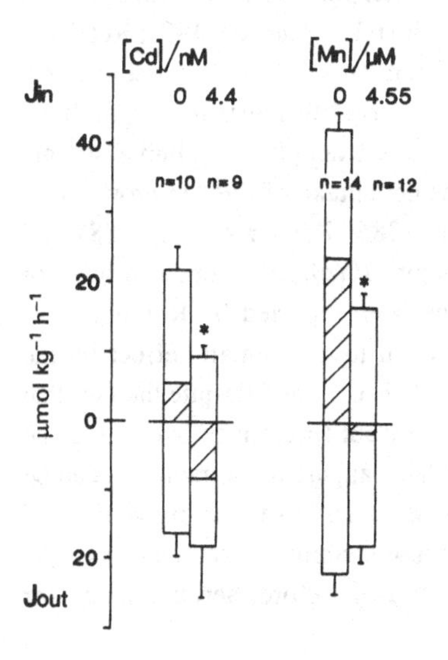

those producing an effect caused net losses of Ca^{2+} or at least a marked reduction in net uptake.

An acute reduction in water pH, to pH 5.0 or less, also disrupts branchial Ca^{2+} fluxes (e.g. Figure 7) but, in contrast to metal exposure, the effects are probably temporary (Hōbe*et al*.., 1984; Reader & Morris, 1988). Flux rates usually recover to normal levels after about 24 h even at very low pH levels (pH ≤4.2; Hōbe *et al*., 1984) and acid exposure, by itself, causes little apparent disturbance to plasma or whole body Ca^{2+} levels even with prolonged exposures in very soft waters (McDonald *et al*., 1980; Parker *et al*., 1985). The metals tested so far, in contrast, show considerably more persistence; for example, Ca net fluxes show essentially no recovery in 2.5 days of zinc exposure (Figure 8) and after five days of cadmium or manganese exposure, significant disturbances of net fluxes still remain (Figure 9). Cadmium is probably not only the most toxic of the metals affecting calcium balance but is also the most persistent. Recent experiments by Reid & McDonald (1988) show that Ca^{2+} fluxes remained disrupted for at least 12 h *after* cadmium is removed from the water. This phenomenon can likely be attributed to the high binding affinity that cadmium has for calcium channels (Giles *et al*., 1983). Generally, the metals affecting Ca^{2+} fluxes are fairly specific in their action in that they primarily inhibit Ca^{2+} uptake rather than stimulating efflux (Figures 7, 9). Acid exposure, on the other hand, leads to both an inhibition of influx and a stimulation of efflux but usually only on a temporary basis (Figure 7). The only metal that also stimulates Ca^{2+} efflux is copper and this effect is also temporary (Reid & McDonald, 1988).

The maintenance of positive Ca^{2+} balance is probably important through most of a fish's life but is critical in early larval development (i.e. post hatch to swim up) when body Ca^{2+} levels increase three- to five-fold (Rombough & Garside, 1984; Reader *et al*., 1988) and all Ca^{2+} intake is by absorption from the water. It is not surprising, therefore, that metals can have dramatic effects in early life stages both upon body calcium content and upon skeletal mineralization. For example, cadmium at 4.4 nM and manganese at 6.5 μM significantly inhibit the net uptake of Ca^{2+} in brown trout alevins over 30 d of metal exposure (Reader, 1986; Reader *et al*., 1988) and markedly reduce calcification of the skeleton (Figure 10*(a), (b)*). Similar results for the effect of cadmium on Atlantic salmon have been reported by Rombough & Garside (1984). Aluminium (2–8 μM) had an even more dramatic effect on the skeletal mineralization in alevins (Figure 10*(c)*). This occurred despite the fact that aluminium has no effect on Ca^{2+} fluxes in brown trout fingerlings over the same range of Al concentrations (Reader *et al*., 1988). This apparent contradiction can be explained if the reduction of Ca^{2+} accumulation is secondary to an inhibition of larval development. The latter comes about possibly because aluminium accumulates on the larval surface and interferes with gas transfer and, therefore, aerobic scope for growth and development.

Summary and conclusions

From the foregoing discussion, it should now be clear that many of the trace metal contaminants of acid soft waters can be expected to have at least some impact upon branchial ion regulation. The mechanisms of action are likely to be different for each metal, with some being primarily disruptive to Na^+ (and Cl^-) balance (Cu, Al, and H^+), others mainly interfering with Ca^{2+} balance (Cd, Zn, Mn), while still others may have little long term impact on gill function and exert their toxic action almost entirely by internal means (e.g. via neural, hepatic or renal disturbances). For metals that are surface-active, the precise effect on the gills depends upon the metal's concentration in the water, upon its ligand preferences and binding characteristics (i.e. the chemistry of the metal) and upon the water chemistry, in particular water pH, hardness and complexing capacity.

Much remains to be learned about the precise role of water chemistry in modulating the surface activity of metals, in particular the precise nature by which H^+ affects speciation and solubility at the gill surfaces and the extent to which H^+ competes for surface ligands. Similar information is required for Ca^{2+}/metal surface interactions. Nonetheless, some general conclusions can be drawn. Based on results presented here, water pH has undoubtedly its greatest impact upon the surface action of

Figure 10. Effects of metals on calcification of the skeleton in brown trout alevins in acid water. Animals were exposed for 30 d starting from 2–4 days post hatch and then preserved. Calcification of the vertebral centra was assessed by alizarin red staining of preserved specimens. Water pH was maintained within ± 0.2 units. t = 10 °C. n = 25–30 in each cell, 15 of which were preserved for analysis. Mortality was zero in most cells and less than 20% in the most affected cells. *(a)* Effect of cadmium at pH 4.8 (redrawn from Reader, 1986). *(b)* Effect of manganese at pH 4.8 (redrawn from Reader *et al.*, 1988). *(c)* Effect of aluminium at pH 4.5 (redrawn from Reader *et al.*, 1988).

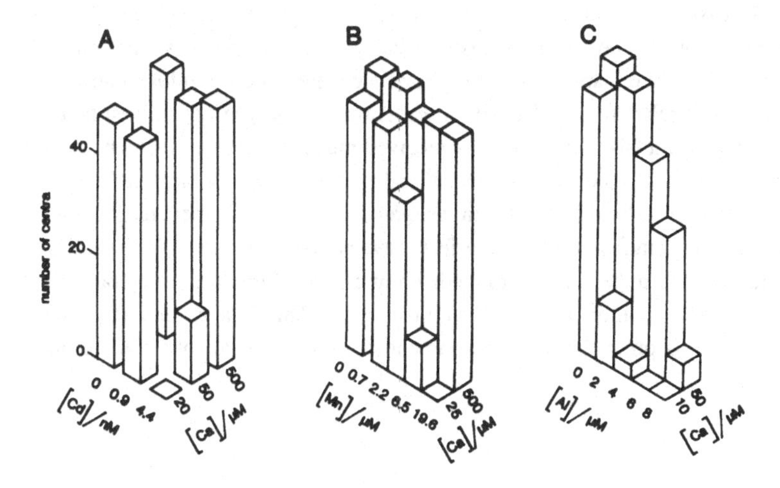

aluminium and its least effect upon the action of copper. For the remaining metals (Cd, Zn, Mn) water pH is likely to have little effect upon surface speciation (Table 1) or surface solubility and little effect upon metal-induced Ca^{2+} flux disturbances (Figure 7) except perhaps for an acute, temporary additive effect. Indeed, over the longer term, lower pHs may reduce metal disturbances to Ca^{2+} balance since, on the one hand, low pH has relatively little effect by itself on Ca^{2+} balance (Figure 10) and on the other may compete with metals for surface ligands. The effect of the latter is seen as reduced metal surface-binding/uptake at lower pH values (Figure 3).

The results presented here also confirm that water calcium concentration is an extremely important variable in metal toxicity. The only metal unaffected by calcium levels typical of acid soft waters is copper and even here Ca^{2+} provides some protection against copper's short term action on calcium balance (Figure 7). Although it is apparent that the protective effect of Ca^{2+} varies with the metal and with the type of disturbance (i.e. Na^+ balance vs Ca^{2+} balance) we have not really attempted to quantify the protective effect of calcium. A useful approximation might be to correlate it with the covalent index of a metal; i.e. the higher the index relative to calcium the less the protective effect. This holds for metals with high indices such as copper (Figures 3, 6) and lead (Brown, 1968) compared to low index metals such as aluminium (Figure 5*(a)* vs 5*(b)*) but calcium protects against the effects of cadmium (Figures 7, 10*(a)*) rather more than one would predict on the basis of covalent index alone. Here, the effect can likely be explained by the chemical similarities between cadmium and calcium (e.g. nearly identical ionic radii). This type of phenomenon would clearly have to be considered for other metals if one wished to accurately quantify the protective effect of calcium.

Finally the present survey, we think, emphasizes the need for continued research on physiological and biochemical aspects of metal toxicity to fish. In these studies there should be an emphasis of exploring mechanisms of toxic action at sublethal or chronically toxic metal concentrations. Much of our current understanding of the interactions of metals and water chemistry is based on acutely lethal studies and it is always risky and often inappropriate to extrapolate these findings to lower and more environmentally relevant metal levels. Furthermore, this type of study should be extended to those metals which have not received much attention and which are also elevated in acid waters, namely lead, nickel and iron. Extremely toxic metals such as silver and mercury should also be investigated even though they might not be elevated, because significant sublethal effects have not been ruled out. With studies of this nature, we should then be able to predict how low level mixtures of metals might interact with one another in low pH environments. This is a particularly useful exercise since it is precisely this environment many fish species now face in their native habitats.

Acknowledgements

The senior author thanks D.J. Lauren, S.D. Reid, D.J. Spry and C.M. Wood for their helpful comments and useful discussions made either in our various research collaborations or in the preparation of this chapter. The original research quoted here has been funded in Canada by the NSERC strategic grants programme in environmental toxicology and the Electrical Power Research Institute (D.G.M.) and in the U.K. by The Science and Engineering Research Council (SERC) (J.P.R.) and by SERC and the Central Electricity Generating Board (T.R.K.D.).

References

Alabaster, J.S. & Lloyd, R. (1980). *Water Quality Criteria for Freshwater Fish.* London: Butterworths, 361 pp.

Booth, C.E., McDonald, D.G., Simons, B.P. & Wood, C.M. (1988). The effects of aluminium and low pH on net ion fluxes and ion balance of the brook trout, *Salvelinus fontinalis. Can. J. Fish. Aquat Sci.*, **45**; in press.

Bradley, R.W. & Sprague, J.B. (1985). The influence of pH, water hardness and alkalinity on the acute lethality of zinc to rainbow trout (*Salmo gairdneri*). *Can. J. Fish. Aquat. Sci.*, **42**, 731–6.

Brown, V.M. (1968). The calculation of the acute toxicity of mixtures of poisons to the rainbow trout. *Water Res.*, **2**, 723–33.

Cusimano, R.F., Brakke, D.F. & Chapman, G.A. (1986). Effects of pH on the toxicities of cadmium, copper and zince to steelhead trout (*Salmo gairdneri*). *Can. J. Fish. Aquat. Sci.*, **43**, 1497–503.

Campbell, P.G.C. & Stokes, P.M. (1985). Acidification and toxicity of metals to aquatic biota. *Can. J. Fish. Aquat. Sci.*, **42**, 2034–49.

Dalziel, T.R.K., Morris, R. & Brown, D.J.A. (1986). The effects of low pH, low calcium concentrations and elevated aluminium concentrations on sodium fluxes in brown trout, *Salmo trutta* L. *Water Air Soil Pollut.*, **30**, 569–77.

Dalziel, T.R.K., Morris, R. & Brown, D.J.A. (1987). Sodium uptake inhibition in brown trout, *Salmo trutta* exposed to elevated aluminium concentrations at low pH. *Annls. Soc. R. Zool. Belg.*, **117**, Supplement 1, 421–34.

Davies, P.H., Goettl, J.P., Sinley, J.R. & Smith, N. (1976). Acute and chronic toxicity of lead to rainbow trout, *Salmo gairdneri*, in hard and soft water. *Water Res.*, **10**, 199–206.

Decker, C. & Menendez, R. (1974). Acute toxicity of iron and aluminum to brook trout. *Proc. West Virginia Acad. Sci.*, F–18–R, **46**(2), 159–67.

De Renzis, G. & Bornancin, M. (1984). Ion transport and gill ATPases. In *Fish Physiology, vol. X, Part B*, eds. W.S. Hoar & D.J. Randall. pp. 65–104. New York: Academic Press.

Eddy, F.B. (1985). Uptake and loss of potassium by rainbow trout. (*Salmo gairdneri*) in fresh water and dilute sea water. *J. exp. Biol.*, **118**, 277–86.

Giles, W., Hume, J.R. & Shibata, E.F. (1983). Presynaptic and postsynaptic actions of cadmium in cardiac muscle. *Fed. Proc.*, **42**, 1994–7.

Graham, M.S. & Wood, C.M. (1981). Toxicity of environmental acid to the rainbow trout: interactions of water hardness, acid type, and exercise. *Can. J. Zool.*, **59**, 1518–26.

Grande, M., Muniz, I.P. & Anderson, S. (1978). The relative tolerance of some salmonids to acid waters. *Verh. Inter. Ver. Limnol.*, **20**, 2076–84.

Hammond, P.B. & Beliles, R.P. (1980). Metals. In *Toxicology: The Basic Science of Poisons, 2nd edn*, ed. J. Doull, C.D. Klaassen & M.O. Amdur, pp. 409–67. New York: MacMillan.

Hōbe, H., Laurent, P. & McMahon, B.R. (1984). Whole body calcium flux rates in freshwater teleosts as a function of ambient calcium and pH levels: a comparison between the euryhaline trout, *Salmo gairdneri*, and the stenohaline bullhead, *Ictalurus nebulosus*. *J. exp. Biol.*, **113**, 237–52.

Howarth, R.S. & Sprague, J.B. (1978). Copper lethality to rainbow trout in waters of various hardness and pH. *Water Res.*, **12**, 455–62.

Johnson, N.M., Driscoll, C.T., Eaton, J.S., Likens, G.E. & McDowell, W.H. (1981). 'Acid Rain', dissolved aluminum and chemical weathering at the Hubbard Brook Experimental Forest, New Hampshire. *Geochim. Cosmochim. Acta*, **45**, 1421–38.

Jones, J.R.E. (1939). The relation between the electrolytic solution pressures of the metals and their toxicity to the stickleback (*Gasterosteus aculeatus* L.). *J. exp. Biol.*, **16**, 425–37.

Kirschner, L.B. (1980). Control mechanisms in crustaceans and fishes. In *Mechanisms of Osmoregulation in Animals*, ed. R. Gilles, pp. 157–219. Chichester: John Wiley & Sons Inc.

Lauren, D.J. (1986). Mechanisms of copper toxicity and acclimation to copper in rainbow trout, *Salmo gairdneri* R. Ph.D. thesis, McMaster University.

Lauren, D.J. & McDonald, D.G. (1986a). Interactions of water hardness, pH, and alkalinity with the mechanisms of copper toxicity in juvenile rainbow trout, *Salmo gairdneri*. *Can. J. Fish. Aquat. Sci.*, **43**, 1488–96.

Lauren, D.J. & McDonald, D.G. (1986b). Acclimation to copper by rainbow trout, *Salmo gairdneri*: I. Physiology. *Can. J. Fish. Aquat. Sci.*, **44**, 99–104.

Leivestad, H. & Muniz, I.P. (1976). Fish kill at low pH in a Norwegian river. *Nature (Lond.)*, **259**, 391–2.

Lewis, M. (1978). Acute toxicity of copper, zinc and manganese in single and mixed salt solutions to juvenile longfin dace, *Agosia chrysogaster*. *J. Fish. Biol.*, **13**, 695–700.

Marier, J.R., Neri, L.C. & Anderson, T.W. (1979). Water hardness, human health, and the importance of magnesium. *National Research Council of Canada, Assoc. Comm. on Sci. Criteria for Env. Quality, NRCC Report no. 17581*, 119 pp.

Marshall, W.S. (1985). Paracellular ion transport in trout opercular epithelium models osmoregulatory effects of acid precipitation. *Can. J. Zool.*, **63**, 1816–22.

McDonald, D.G. (1983). The effects of H^+ upon the gills of freshwater fish. *Can. J Zool.*, **63**, 691–703.

McDonald, D.G., Hōbe, H. & Wood, C.M. (1980). The influence of calcium on the physiological responses of the rainbow trout, *Salmo gairdneri*, to low environmental pH. *J. exp. Biol.*, **88**, 109–31.

McDonald, D.G. & Rogano, M.S. (1986). Ion regulation by the rainbow trout, *Salmo gairdneri*, in ion-poor water. *Physiol. Zool.*, **59**, 318–31.

McDonald, D.G., Walker, R.L. & Wilkes, P.R.H. (1983). The interaction of environmental calcium and low pH on the physiology of the rainbow trout, *Salmo gairdneri*. II. Branchial ionoregulatory mechanisms. *J. exp. Biol.*, **102**, 141–55.

McWilliams, P.G. & Potts, W.T.W. (1978). The effects of pH and calcium concentration on gill potentials in the brown trout. *J. comp. Physiol.*, **126**, 277–86.

Nieboer, E. & Richardson, D.H.S. (1980). The replacement of the nondescript term 'heavy metals' by a biologically and chemically significant classification of metal ions. *Envir. Pollut. B*, **1**, 3–26.

NRCC (1981). Acidification in the Canadian Environment. *National Research Council of Canada, Assoc. Comm. on Sci. Criteria for Env. Quality, NRCC Report no.18475*, 369 pp.

Ochiai, E. (1977). *Bioinorganic Chemistry – An introduction*. Boston: Allyn & Bacon, Inc., 513 pp.

Orr, P.L., Bradley, R.W., Sprague, J.B. & Hutchinson, N.J. (1986). Acclimation-induced change in toxicity of aluminum to rainbow trout. (*Salmo gairdneri*). *Can. J. Fish. Aquat. Sci.*, **43**, 243–6.

Oschman, J.L. (1978). Morphological correlates of transport. In *Membrane Transport in Biology. vol. III. Transport across Multi-membrane Systems*, ed. G. Giesbisch, D.C. Tosteson & H.H. Ussing, pp. 55–84. New York: Springer Verlag.

Parker, D.B., McKeown, B.A. & MacDonald, J.S. (1985). The effects of pH and/or calcium-enriched freshwater on gill Ca^{2+}-ATPase activity and osmotic water inflow in rainbow trout (*Salmo gairdneri*). *Comp. Biochem. Physiol.*, **81A**, 149–56.

Pagenkopf, G.K. (1983). Gill surface interaction model for trace-metal toxicity to fishes: role of complexation, pH and water hardness. *Envir. Sci. Technol.*, **179**, 342–7.

Payan, P., Girard, J.P. & Mayer-Gostan, N. (1984). Branchial ion movements in teleosts: the roles of respiratory and chloride cells. In *Fish Physiology, vol. X, Part B*, ed. W.S. Hoar & D.J. Randall, pp. 39–60. New York: Academic Press.

Potts, W.T.W. & Fleming, W.R. (1971). The effects of prolactin and divalent ions on the permeability to water of *Fundulus kansae*. *J. exp. Biol.*, **53**, 317–27.

Reader, J.P. (1986). Effects of cadmium, manganese and aluminium in soft acid water on ion regulation in *Salmo trutta* L. Ph.D. thesis, University of Nottingham.

Reader, J.P., Dalziel, T.R.K. & Morris, R. (1988). Growth, mineral uptake and skeletal calcium deposition in brown trout (*Salmo trutta* L.) yolk sac fry exposed to aluminium and manganese in soft acid water. *J. Fish Biol.* (In press).

Reader, J.P. & Morris, R. (1988). Effects of aluminium and pH on calcium fluxes, and effects of cadmium and manganese on calcium and sodium fluxes in brown trout (*Salmo trutta* L.). *Comp. Biochem. Physiol.* (In press).

Reid, S.D. & McDonald, D.G. (1988). The effects of cadmium, copper and low pH on calcium fluxes in the rainbow trout, *Salmo gairdneri*. *Can. J. Fish. Aquat. Sci.*, **45**, (In press).

Riedel, B. & Christensen, G. (1979). Effect of selected water toxicants and other chemicals upon adenosine triphosphatase activity *in vitro*. *Bull. Envir. Contam. Toxicol.*, **23**, 365–8.

Rombough, P.J. & Garside, E.T. (1984). Disturbed ion balance in alevins of Atlantic salmon *Salmo salar* chronically exposed to sublethal concentrations of cadmium. *Can. J. Zool.*, **62**, 1443–50.

Shaw, W.H.R. & B. Grushkin (1957). The toxicity of metal ions to aquatic organisms. *Arch. Biochem. Biophys.*, **67**, 447–52.

Spry, D.J. & Wood, C.M. (1985). Ion flux rates, acid-base status, and blood gases in rainbow trout, *Salmo gairdneri*, exposed to toxic zinc in natural soft water. *Can. J Fish. Aquat. Sci.*, **42**, 1332–4 1.

Stumm, W. & Morgan, J.J. (1981). *Aquatic Chemistry – An introduction emphasizing chemical equilibria in natural waters*. New York: John Wiley and Sons. 780 pp.

Turner, J.E., Williams, M.W., Jacobson, K.B. & Hingerty, B.E. (1985). Correlations of acute toxicity of metal ions and the covalent/ionic character of their bonds. In *Quantitative Structure Activity Relationships in Toxicology and Xenobiochemistry*, ed. M. Tichy, pp. 1–8. New York: Elsevier.

Wood, C.M. & McDonald, D.G. (1982). Physiological mechanisms of acid toxicity to fish. In *Acid Rain/Fisheries*, ed. R.E. Johnson, pp. 197–226. Bethesda: American Fisheries Society.
Wright, R.F. & Snekvik, E. (1978). Chemistry and fish populations in 700 lakes in southernmost Norway. *Verh. Inter. Ver. Limnol.* **20**, 765–75.
Zitko, V. (1976). Structure-activity relations and the toxicity of trace elements to aquatic biota. In *Toxicity to Biota of Metal Forms in Natural Water*, ed. R.W. Andrew, P.V. Hodson & D.E. Konasewich, pp.9–32. Windsor, Ontario: International Joint Commission, Great Lakes Advisory Board.

S.E. WENDELAAR BONGA AND P.H.M. BALM

Endocrine responses to acid stress in fish

Introduction

There is some consensus now that the death of many fish species, exposed to acid water, is caused by a chain of events starting with the loss of body electrolytes and eventually leading to osmoregulatory and cardiovascular failure (Muniz & Leivestad, 1980; McDonald & Wood, 1981; McDonald, 1983). Sublethal exposure to acidified water often leads to transient or chronic hypo-osmolarity of the blood plasma, mainly caused by reduced Na^+ and Cl^- levels (McWilliams, 1980; McDonald, 1983; Wendelaar Bonga, Van der Meij & Flik, 1984a). The severity and duration of these effects are determined by both external and internal factors. External factors, such as the calcium and aluminium concentration of the water and the presence of heavy metals, are dealt with by Wood, Potts & McWilliams, and McDonald *et al.* (this volume). The rate and degree of change of the environmental pH is also important. Internal factors, in particular hormones, are the subject of this chapter.

The endocrine system is of pre-eminent importance for the control of physiological processes that enable animals to adjust to changes in their environment. Since acidification of the water deeply affects many aspects of fish physiology, pronounced and multiple responses of the endocrine system may be envisaged. When it is taken into account that a predominant deleterious effect of acid water on fish is disturbed water and ion balance, it is not surprising that the hormones with osmoregulatory actions in fish, in particular cortisol, ACTH and prolactin, are given greatest consideration in this chapter. Studies on the effects of acid on fish endocrines are still scarce, and limited to a few species. One of these, the African cichlid fish *Oreochromis (Sarotherodon) mossambicus* (tilapia), will receive especial attention, since our studies on tilapia, although incomplete, are probably the most extensive that are available for a single species. Substantial information is also available on salmonid species, in particular the brook trout.

The study of McWilliams (1980) on brown trout was the first to show restoration of initially depressed plasma Na^+ concentrations under low pH conditions. We have reported the same phenomenon after transfer of tilapia to pH 4 (Wendelaar Bonga

Table 1. *Total volume per fish of the interrenal cells, determined in serial light microscope sections of 7 μm thickness and stained with haematoxylin and eosin (means ± SEM; controls: n = 6; low pH, n = 7) of tilapia exposed for five days to pH 3.5*

	controls pH 7.4	acid treated pH 3.5	P(a)
Total volume of interrenal cells per fish (mm^3 10^{-4})	91 ± 14	696 ± 98[a]	<0.001

(a) Student's t–test

et al., 1984a). These observations illustrate the existence of physiological adaptation to low pH in both species. In tilapia, however, the adaptive capacities, in particular with respect to osmoregulation, are very impressive when compared to other species. We therefore selected this species for our studies on the role of the endocrine system during adaption to acid water.

Endocrine tissues

Interrenal cells

The main steroid produced by the interrenal cells, which are homologous with the cells of the adrenal cortex of tetrapods, is cortisol. The interrenal cells are concentrated around the large blood vessels in the head kidneys. Plasma cortisol levels are generally used as an indicator of the involvement of the interrenal tissue in teleosts during physiological adaptation and many different kinds of stress (Donaldson, 1981). Of the many actions of this versatile hormone its stimulating effect on gluconeogenesis, on protein catabolism, and on active ion transport mechanisms in the gills can be considered of primary importance. Traditionally, cortisol has been implicated in teleost osmoregulation in seawater, but there is ample evidence that the hormone is of equal importance for osmoregulation in freshwater fish (Hirano & Mayer Gostan, 1978; Dharmamba, 1979; Balm, 1986).

Mudge *et al.* (1977) and Ashcom (1979), working on brook trout, were the first to demonstrate that acid exposure leads to a pronounced but transient rise in plasma cortisol levels. This has been confirmed by others, e.g. for rainbow trout (Lee, Gerking & Jezierska, 1983) and tilapia (Balm, 1986). In the latter species, plasma cortisol concentration increased from 7.6 ± 1.5 μg% (*n* = 15) to 61.2 ± 4.4% (*n* = 5) within five hours after reduction of the water pH from 7.4 to 3.5. The cortisol concentration returned to control levels within two days, although the low pH continued (Figure 1). This phenomenon has also been reported for trout (Mudge *et al.*, 1977; Ashcom, 1979). Brown *et al.* (1984) noticed elevated cortisol levels in rainbow trout exposed to low pH for three weeks, but this may be related to the high mortality in their experiments. Moribund salmonids generally display high plasma cortisol levels (Donaldson & Fagerlund, 1968).

Although, in tilapia, plasma cortisol was back to control values 48 h after acidification of the water, the number of interrenal cells increased rapidly and continuously for at least the first week (Table 1), and remained elevated for at least the first two months. The activity of the interrenal cells also remained higher than in the controls, as was established during superfusion *in vitro* of freshly dissected head kidneys. The initial release of cortisol by head kidneys of fish from acid water was significantly higher than that of controls (Table 2). We interpret the initial *in vitro* release rate as a reflection of the *in vivo* release. During the first hours *in vitro* the

Figure 1. Plasma cortisol concentration of tilapia after exposure to pH 3.5 (H_2SO_4 was used throughout our studies). Exposure started at $t = 0$ and the final pH was reached at $t = 1$ h. Means ± SEM; $n = 5-15$.

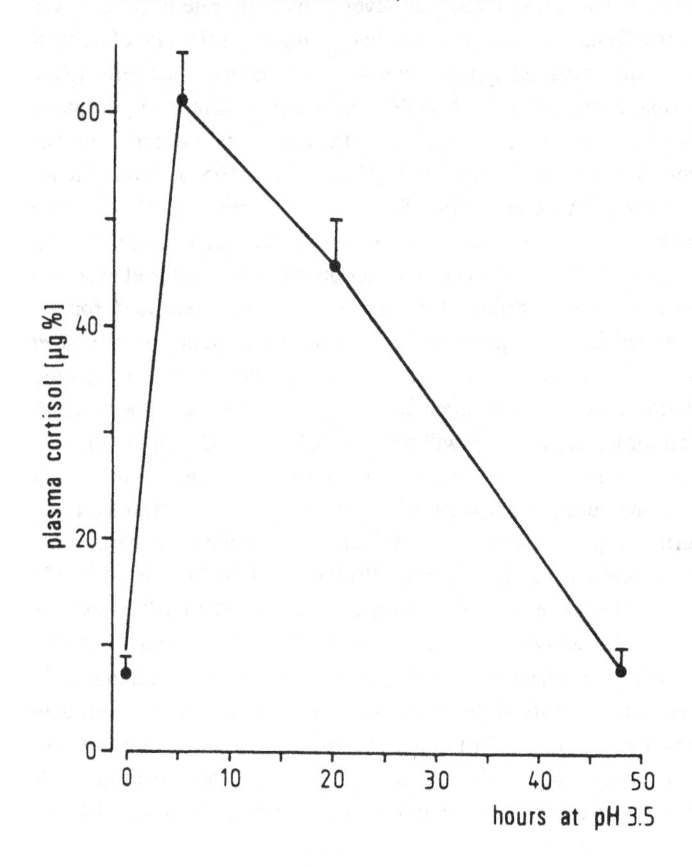

Table 2. *In vitro cortisol release by superfused head kidneys of tilapia exposed for five days to pH 3.5. For technical details, see Balm (1986). Initial release rates (determined immediately after the start of superfusion of freshly dissected head kidneys) and basal release rates (determined two hours after the start of the superfusion) are presented. Means ± SEM of five superfusions. Four head kidneys (from two fish) were used per superfusion*

	controls pH 7.4	acid treated pH 3.5	$P^{(a)}$
Initial cortisol release (pg min^{-1} g^{-1} body weight)	3.8 ± 0.6	8.8 ± 1.4[a]	< 0.01
Basal cortisol release (pg min^{-1} g^{-1} body weight)	1.5 ± 0.3	2.0 ± 0.3	n.s.

[a] Mann-Whitney U-test.

cortisol secretion rate decreased to a constant level. This basal rate of release was similar in both groups (Table 2). This indicates that the higher initial rate of cortisol release in acid treated fish is caused by tonic stimulation of the interrenal cells, likely by adrenocorticotropic hormone (ACTH) and/or melanocyte-stimulating hormone (MSH) (see below). Activation of interrenal cells after acidification of the water has also been reported for trout (Ashcom, 1979; Lee *et al.*, 1983), although no hyperplasia was observed (Ashcom, 1979). Since plasma cortisol levels return to normal after a few days in acid water, whereas the activation of these cells proceeds, the peripheral cortisol clearance rates must be higher during prolonged acid exposure than they are under control condition. This suggests an increased need for the hormone in acid-stressed fish. In contrast to the reports mentioned above, Mudge *et al.* (1977) concluded that in brook trout exposure to acid water leads to impaired steroid synthesis in the interrenal cells, after the initial surge of secretory activity. A possible explanation for this discrepancy will be given below (see Conclusions).

The rapid increase of plasma cortisol levels and the apparently increased peripheral turnover of the hormone during the first days in acid water can be interpreted as an appropriate endocrine response. Acid exposure leads to a dramatic increase in the branchial effluxes of ions, such as Na^+, Cl^- (McWilliams, 1980; McDonald, 1983) and Ca^{2+} (Flik *et al.*, 1985), and to reduction of ion-dependent ATPases: the branchial mechanisms for active ion uptake from the water. Branchial specific Na^+/K^+ ATPase activity of salmonids were increased four to seven days (albeit only in the absence of aluminium in the water, Staurnes *et al.*, 1984), and 21 d after acid exposure (McKeown *et al.*, 1985). In tilapia, specific and total Na^+/K^+ ATPase activities were significantly reduced after five days, but slightly above control levels after 28 d at low pH. An enzyme activity with the characteristics of Na^+/H^+ ATPase

Table 3. *Total volume of ACTH cells per fish, as determined in immunocytochemically stained light microscopical serial sections (anti-ACTH antiserum) of pituitary glands of Tilapia exposed for five days to pH 3.5. Means ± SEM*

	controls pH 7.4	pH 3.5	$P^{(a)}$
Volume of ACTH cells (mm^3 10^{-4}; $n = 5$)	31 ± 2	74 ± 9	< 0.01

(a) Student's *t*-test.

was already increased after five days. The stimulation of these branchial Na^+-dependent ATPase activities is most likely regulated by cortisol (Balm, 1986). This hormone further promotes tissue catabolism, which leads to hyperglycemia and increased nitrogen excretion, phenomena that have been reported for other species, including rainbow trout (Brown *et al.*, 1984; McDonald *et al.*, 1983) and tilapia (Balm, 1986) after acidification of the water. These actions of cortisol are obviously important elements of the adaptive response of fish to acid stress. The hyperglycemia not only reflects the mobilization of energy sources, but also contributes to the maintenance of plasma osmolarity. Nitrogen excretion is for the larger part branchial efflux of NH_3. Under acid conditions, NH_3 will be protonated to NH_4^+ immediately upon entry of the water, and this may change the pH, directly at the gill surface, to a higher level than the ambient pH (Neville, 1986). This side effect of cortisol may be beneficial for tilapia in particular, since NH_4^+ efflux rates at low pH in this species are up to four times as high as in trout under similar conditions (Balm, 1986).

The pituitary gland

The pituitary gland of teleosts contains seven types of endocrine cells, located in the rostral pars distalis (the adrenocorticotropic or ACTH cells, and the prolactin cells), in the proximal pars distalis (the growth hormone cells, the gonadotropic cells, and the thyrotropic cells) and in the pars intermedia (the melanotropic or MSH cells, and the Periodic-Acid-Schiff or PAS-positive cells). For most of these cell types responses to water acidification are known.

ACTH cells The ACTH cells usually form a layer of one or two cells that separates the prolactin cells from the tissue of the neural part of the pituitary gland. In tilapia exposed for five days to acid water, the region consisting of ACTH cells was significantly enlarged compared to control fish (Table 3), mainly due to an increase in the number of ACTH cells. Ultrastructural signs of increased cellular activity, such as extension of the granular endoplasmic reticulum and Golgi areas, were also noticeable, but the changes were less conspicuous than in, for example, prolactin cells (see below). The marked and rapid hyperplasia of the ACTH cells suggests that the rate of secretion of ACTH is enhanced at low pH.

Although intrinsic osmoregulatory actions have been attributed to ACTH (Langdon, Thorpe & Roberts, 1984), it is clear that the primary action of this hormone is indirect and concerns the stimulation of the cortisol production of the interrenal cells. These cells become activated at low pH, as was shown in the

Figure 2. Effect of ACTH on cortisol release during *in vitro* superfusion of head kidneys of tilapia (15–20 g body weight) exposed for five days to pH 3.5 (solid lines; controls at pH 7.4: broken lines). Two pairs of head kidneys (from two fish) were used for each superfusion. $ACTH_{1-39}$ was administered as a five minute pulse (hatched bar). Basal cortisol release rates at $t = 200$ min represent 100% (controls: 2.6 ± 0.6, experimentals: 3.7 ± 0.3 pg min^{-1} g^{-1} body weight; n.s.). The black dots represent means (± SEM) of three superfusions. Areas under the curves (controls: 501 ± 65; experimentals: 895 ± 101; $p < 0.05$) show that ACTH-induced stimulation is significantly higher for the head kidneys of acid-treated fish.

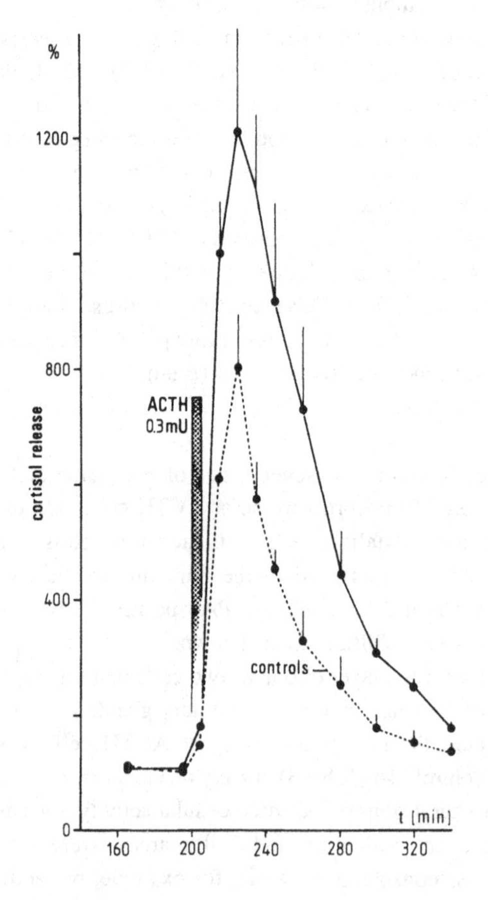

previous section. To investigate whether or not this activation could be mediated by ACTH, the effects of ACTH on the release of cortisol *in vitro* were compared in head kidneys from acid treated fish and controls (Figure 2). The results indicate that tilapia head kidneys become more sensitive to ACTH-stimulation after exposure of the fish to low pH. It remains to be established whether the increased release of cortisol is caused by the higher number of interrenal cells present in the kidneys of acid-treated fish, the higher sensitivity to ACTH of the individual interrenal cells, or both. The data clearly show, however, that the pituitary interrenal axis in tilapia is activated during low pH acclimation.

Prolactin cells Prolactin is an important osmoregulatory hormone in freshwater fish. Its primary action probably concerns the control of the permeability to ions and water of the epidermal layers of the skin and, in particular, the gills (Bern, 1983; Hirano & Mayer Gostan, 1978; Wendelaar Bonga *et al.*, 1984a; Flik *et al.*, 1986). Acidification of the water increases the permeability of the gills, which leads to massive efflux of ions (McWilliams, 1980; McDonald, 1983; Flik *et al.*, 1985) and to increased osmotic influx of water (Wendelaar Bonga *et al.*, 1984a). Thus, a response of the prolactin cells may be anticipated when fish are exposed to acid water.

Notter *et al.* (1976) demonstrated for brook trout that acid exposure for three to five days increased the metabolism of the prolactin cells, as reflected by the rate of RNA synthesis. The prolactin cells of teleost fish are concentrated in the rostral pars distalis. They are only separated by non-secretory stellate cells. During adaptation of tilapia to acid water, the size of the rostral pars distalis increases significantly, which is mainly caused by increase in volume and number of the prolactin cells (Wendelaar Bonga *et al.*, 1984a). The ultrastructure of the prolactin cells also reflects their activation. Increased release of secretory granules is noticeable a few hours after the start of water acidification: the cells become almost completely degranulated within four to ten hours. Later on, granularity is partially restored. Granular endoplasmic reticulum and Golgi areas increase in extent during the first 10 d (Figure 3*(a)*,*(b)*) and the rate of prolactin synthesis and release, as reflected during *in vitro* incubation of freshly dissected pituitary glands, is significantly higher than in controls (Wendelaar Bonga *et al.*, 1984a,b).

The extent of the response of the prolactin cells is not only determined by the pH of the water but also by the experimental protocol used. We observed a 40% increase in volume of the rostral pars distalis in ten days when the fish were transferred directly to water of pH 3.5. When the pH was reduced gradually, and the final pH reached in an hour instead of a few minutes, the increase of the rostral pars distalis volume was more pronounced and occurred more rapidly: a 50% increase was noticed already after five days (Balm, 1986). Ultrastructural examination of the skin showed that

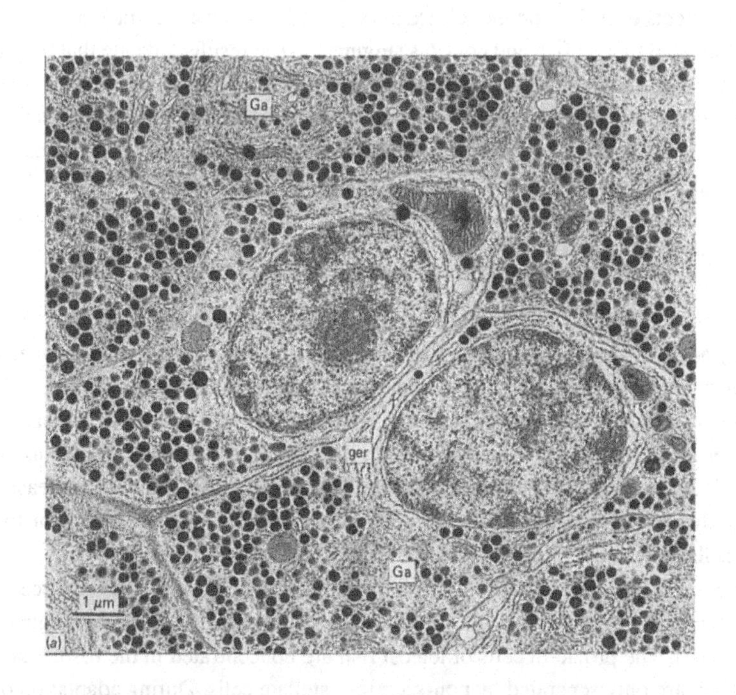
Ga
ger
Ga
1 μm
(a)

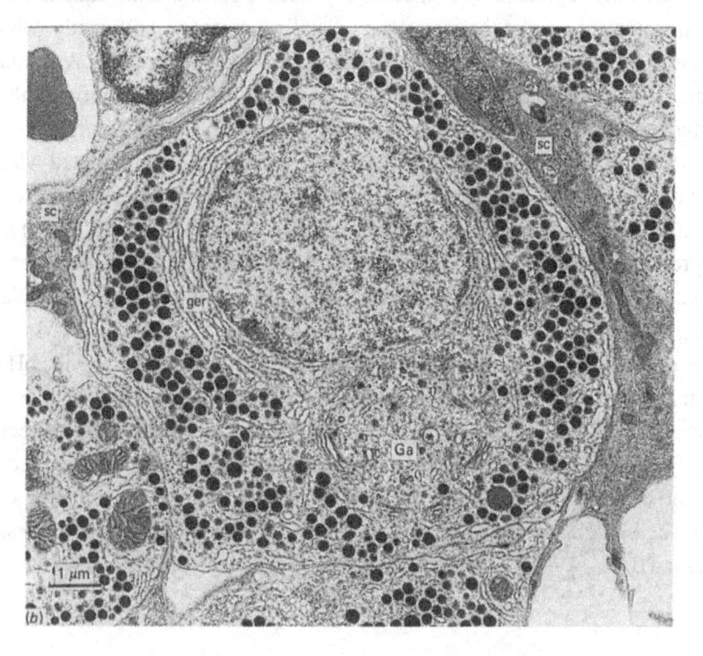
sc
sc
ger
Ga
1 μm
(b)

after a sudden drop in water pH the superficial layer of the epidermis, including many chloride cells in the gill epithelium, were severely damaged. This may contribute to the severe drop in plasma osmolarity (from 320 ± 12 to 265 ±18 mOsmol l^{-1}) noticed after 24 h, and the 20% mortality in the first week. When the reduction of water pH took an hour or longer, the drop in plasma osmolarity was insignificant and mortality did not occur. The absence of a severe decrease in osmolarity, even though Na^+ efflux was significantly increased under these conditions (our unpublished observations), may be connected with our finding that there was substantially less damage to the upper epidermal cell layers in the first 24 h after acidification of the water. We suggest that the osmoregulatory stress imposed by a sudden drop in water pH affects the condition of the fish so dramatically that the response of the prolactin cells, and possibly other endocrine responses, is impaired.

Whereas the prolactin cells in tilapia become highly active after acidification of the water, and remain activated for at least two months (unpublished data), no structural signs of activation of these cells could be observed in goldfish (Wendelaar Bonga *et al.*, 1984b). This phenomenon might be connected with the finding that in goldfish, in contrast to tilapia, there was no restoration of the severe drop in plasma osmolarity that immediately followed acidification of the water (Figures 4, 5; Wendelaar Bonga *et al.*, 1984b). On the other hand, clear ultrastructural signs of cellular activation were observed in a study on the prolactin cells of eels (*Anguilla anguilla*) exposed for 10 d to pH 3.5 (our unpublished data).

Growth hormone Sublethal acid stress requires greater energy expenditure for maintaining elementary physiological functions such as osmoregulation, which may be to the detriment of growth and reproduction. In some experimental studies, environmental pH had little direct effect on growth (Jacobsen, 1977; Sadler & Lynam, 1985) while in others reduced growth rates have been reported (Menendez, 1976; Muniz & Leivestad, 1979). In our experiments with tilapia, in which experimental and control fish were fed the same fixed food rations, growth stopped immediately after acidification and body weight decreased during the first week at pH 4 or lower. Within a few weeks, however, growth was resumed, although at a reduced rate: at pH 4 the growth rate was less than 50% of that at pH 7.

Figure 3.(*a*) Prolactin cells of control tilapia (water pH: 7.4); ger, granular endoplasmic reticulum: Ga, Golgi area (from Wendelaar Bonga *et al.*, 1984b). (*b*) Prolactin cell of tilapia exposed for 14 d to acid water (pH 4.0). Compared with controls (*a*), the cells are enlarged and the granular endoplasmic reticulum (ger) and the Golgi areas (Ga) are extended significantly, which reflects increased prolactin secretion; sc, stellate cells.

Figure 4. Plasma osmolarity and total plasma calcium in tilapia before ($t = 0$; control value at water pH 7.4) and after acute exposure to acid water (pH 3.5) for 1, 7 or 14 d; means ± SD; $n = 6$ (from Wendelaar Bonga *et al.*, 1984b).

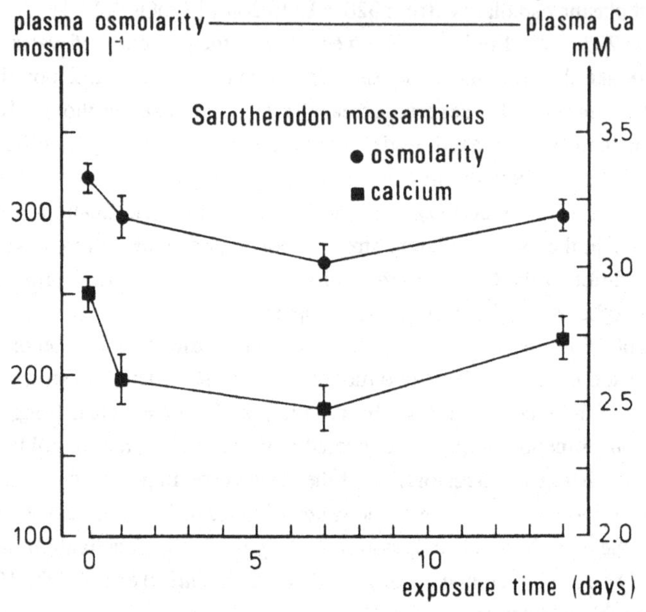

Figure 5. Plasma osmolarity and total plasma calcium in goldfish before ($t = 0$; control value at water pH 7.4) and after acute exposure to acid water (pH 3.5) for 1, 7 or 14 d; means ± SD; $n = 6$ (from Wendelaar Bonga *et al.*, 1984b).

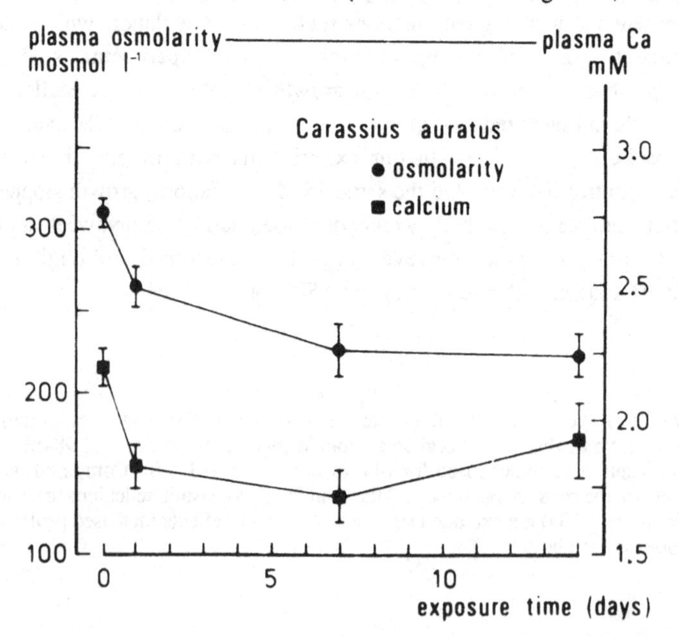

The inhibition of growth immediately after acidification of the water was reflected by the growth hormone cells in the proximal pars distalis of the pituitary gland. After two days at pH 4, the number of secretory granules was increased, whereas the activity of the Golgi areas, as indicated by the absence of presecretory granules, was reduced. Thus, synthesis and release of growth hormone seems decreased in acid water. This conclusion was supported by our observation that the rate of synthesis of tilapia growth hormone *in vitro*, by freshly dissected pituitary glands incubated in the presence of labelled amino acids, was significantly below the control level when pituitary glands were obtained from tilapia exposed to water of pH 3.5 for two days. After two weeks in water of this pH, the ultrastructure of the growth hormone cells also reflected reduced secretory activity: the cells were smaller, and the extent of the granular endoplasmic reticulum and of the Golgi areas was decreased (unpublished observations).

Gonadotropic and thyrotropic cells To our knowledge these cell types have not been studied in acid-stressed fish.

The pars intermedia MSH cells The MSH cells of teleosts are comparable to the MSH cells of other vertebrate groups and produce a variety of melanotropic and non-melanotropic peptides such as MSH, ACTH, endorphins and γ-lipotropic hormone (γ–LPH) (Van Eys, Löwik & Wendelaar Bonga, 1983; Balm, 1986). The function of the MSH cells in fish is, as in mammals, unclear. The MSH cells of eels, trout or tilapia are activated when the fish are kept on a black background (Van Eys, 1980; Ball & Batten, 1981) and, in the latter species, infusion of α–MSH stimulates melanin synthesis and melanophore proliferation of the skin (Van Eys & Peters, 1981).

Baker & Rance (1981) suggested a role for α–MSH in the control of cortisol secretion, since trout and eels showed higher plasma cortisol levels when kept on a black background. Initially, this could not be confirmed since the steroidogenic potency of α–MSH, when tested on interrenal tissue of trout *in vitro,* proved to be very low (Rance & Baker, 1981). However, the desacetylated form of α–MSH (desac–α–MSH) was 100 times more potent than the acetylated form (α–MSH). Desac–α–MSH is considered the principal storage form of α–MSH in higher vertebrates and Balm (unpublished data) has recently shown that MSH–cells of tilapia, when superfused *in vitro,* secrete substantially more desac–α–MSH than α–MSH into the medium. The high potency of this form to stimulate cortisol secretion was confirmed for this species. Therefore, the desacetylated form of the peptide might be a physiological stimulator of cortisol production. MSH cells have recently been implicated in the control of aldosterone and corticosterone production of the mammalian adrenal gland (Szalay & Stark, 1982; Shenker *et al.*, 1985). In line with this, Sumpter *et al.* (1985, 1986) have recently reported a rise of plasma cortisol and MSH levels (MSH includes des–, mono–, and probably diacetyl–α–MSH; Follenius,

1986) after stressful treatments of fish. Thus, there is growing evidence that the MSH cells are implicated in the control of the interrenal cells in fish.

Acidification of the water leads to activation of the MSH cells of tilapia. This conclusion is based on biochemical and ultrastructural evidence. The rate of secretory activity of the MSH cells was estimated by HPLC-analysis of the products that were synthesized and released by freshly dissected pars intermedia tissue during incubation *in vitro* (for technical details see Van Eys, Löwik & Wendelaar Bonga, 1983). It appeared that the production of desac–α–MSH and endorphins in pituitary glands of acid-treated fish was increased when compared to the glands of control fish kept at pH 7.4 (Figure 6). The total volume of the MSH cells per fish, as determined

Figure 6. High performance liquid chromatograms of pars intermedia homogenates of control tilapia (*a*) and tilapia exposed to pH 3.5 for two days (*b*). Pars intermedia lobes were dissected and incubated in the presence of ^{3}H–lysine (75 μCi in 75 μl teleost Ringer for three lobes incubated simultaneously for 90 min). Afterwards the lobes were homogenized in ice-cold 0.1 N HCl, and the entire volume of supernatant was subjected to reversed phase HPLC (Spectra Physics 8000), with application of an elution gradient consisting of 0.5 M formic acid and 0.14 M pyridine (pH 3.0) as the primary solvent, and 1–propanol as the secondary solvent. During the 90 min incubation only small amounts of newly synthesized products were released into the incubation medium. Body weight of controls: 26.8 ± 3.6 g; of acid-treated fish: 25.3 ± 3.3 g (means ± SEM). The three products eluting between 10 and 20 min possess α–MSH immunoreactivity, and represent the desacetylated (des), monoacetylated (mono), and diacetylated (di) forms of the peptide, respectively. The peaks leaving the column at 38 min and 49 min stem from proopiomelanocortin (POMC) produced by the MSH cells, and from the pars intermedia PAS-positive (PIPAS) cells, respectively.

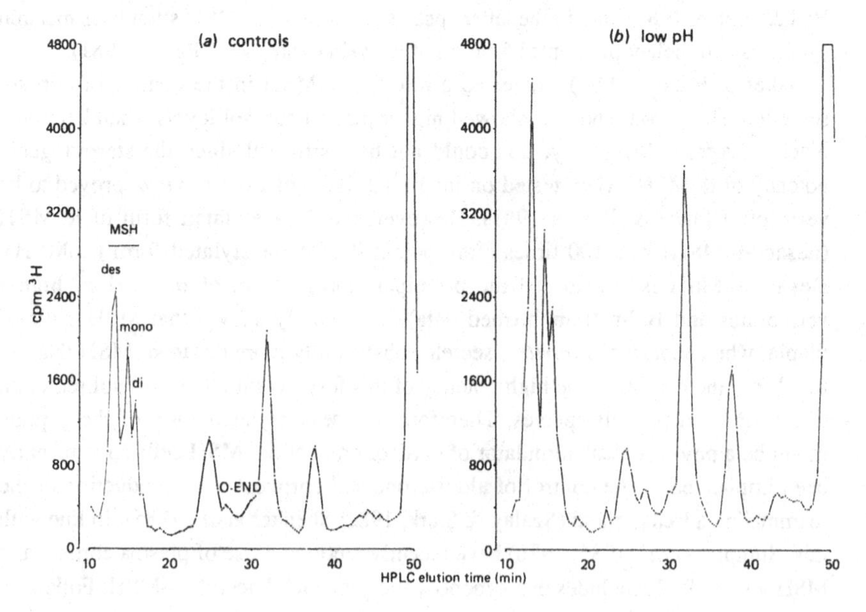

morphometrically in light microscope sections, was almost twice that of the controls. Exposure of the fish for three weeks to pH 3.5 induced visible signs of activation in the ultrastructure of the MSH cells, such as increased extent of the membranes of the granular endoplasmic reticulum and of the Golgi lamellae (Figure 7*(a),(b)*). In view of the above mentioned evidence in favour of a role of MSH in the control of the interrenal tissue, we propose that the MSH cells are implicated in the endocrine control of acclimation to acid stress, probably by stimulating cortisol secretion.

The pars intermedia PAS-positive cells (PIPAS cells) These cells, which occur specifically in fish, derive their name from their affinity to the Periodic-Acid-Schiff stain (Ball & Batten, 1981). Their function is in dispute. In tilapia, these cells appeared to produce proteins with apparent molecular weights of 22 and 26 kdaltons (Van Eys *et al.*, 1983). In several species, including tilapia, the PIPAS cells are activated when fish are kept on a black background. In other species these cells do not respond to background reflectivity to light, but to osmoregulatory challenges (Ball & Batten, 1981). For instance, in eels and goldfish the PIPAS cells increase in size and number when the fish are exposed to calcium-deficient water (Olivereau, Aimar & Olivereau, 1980a,b; Olivereau, Olivereau & Aimar, 1981).

We have studied the effects of water acidification on the PIPAS cells in tilapia and goldfish. In tilapia, there was no noticeable effect on size, number or ultrastructure of the cells after acute exposure of the fish to pH 3.5 for periods up to two weeks (Wendelaar Bonga *et al.*, 1984b). In contrast, similar treatment of goldfish had a very pronounced effect on the PIPAS cells. They increased in number and showed a five-fold increase in individual cell volume (Figures 8*(a),(b)*). The acid stress induced reduction of plasma osmolarity and of plasma Na^+, Cl^- and Ca^{2+} levels (Figure 5). Since the PIPAS cells become activated when goldfish are transferred from normal freshwater to demineralized water, and this activation can be prevented by the addition of calcium ions to the water, Olivereau *et al.* (1980b, 1981) have concluded that these cells produce a hormone with hypercalcaemic properties. However, we found that the PIPAS cells of goldfish were stimulated to the same extent in acidified water when the hypocalcaemia of the fish was prevented by the addition of calcium ions to the water. Acidification of water containing Na^+ and Cl^- to a concentration that made the water iso-osmotic with the blood, activated the PIPAS cells to the same extent as acidification of normal freshwater, even though plasma osmolarity, Na^+ and Cl^- were increased instead of reduced (Wendelaar Bonga, Van der Meij & Flik, 1986). Thus, there is no evidence that the PIPAS cells in goldfish produce a hypercalcaemic hormone, or a hormone implicated, directly or indirectly, in the maintenance of plasma osmolarity, Na^+ or Cl^-. The role of these cells in acclimation of the fish to acid stress is therefore unclear. It is possible that they are involved in acid-base regulation (Wendelaar Bonga *et al.*, 1986).

Figure 7.(*a*) MSH cell of control tilapia (water pH 7.4). (*b*) Part of the cytoplasm of MSH cell of tilapia exposed for three weeks to acid water (pH 3.5). The Golgi areas (Ga), which are small and insignificant in control fish (in many cross sections of MSH cells they are absent, see (*a*)), are prominent in MSH cells of fish from acid water. The accumulation of electron dense presecretory material within the Golgi lamellae (arrows) indicate a high rate of granule synthesis.

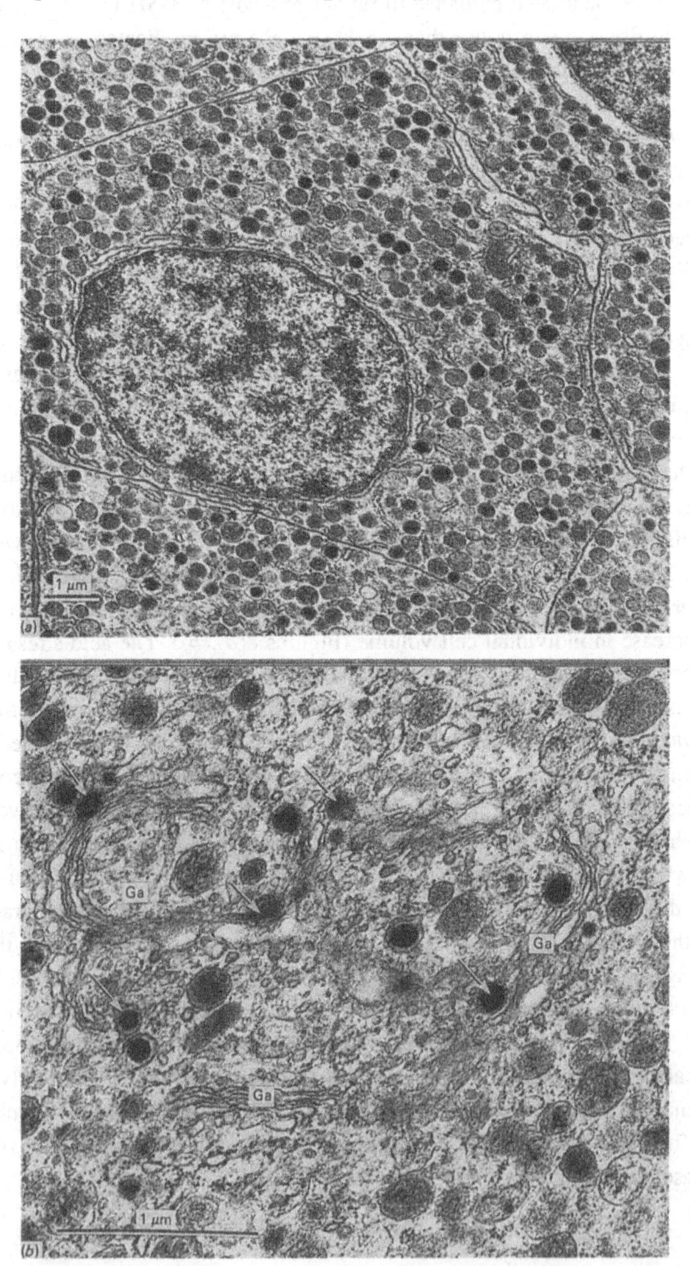

Figure 8.(*a*) PIPAS cells of control goldfish (PIPAS), surrounded by MSH cells (MSH). The cells are small and contain some granular endoplasmic reticulum (ger), small Golgi areas (Ga), and electron dense secretory granules (from Wendelaar Bonga *et al.*, 1984b). (*b*) PIPAS cells of tilapia exposed for 14 d to acid water (pH 3.5). Compared to controls (*a*) the cells are enlarged significantly and the granular endoplasmic reticulum (ger) and Golgi areas (Ga) are extended, which reflects increased hormone secretion (from Wendelaar Bonga *et al.*, 1984).

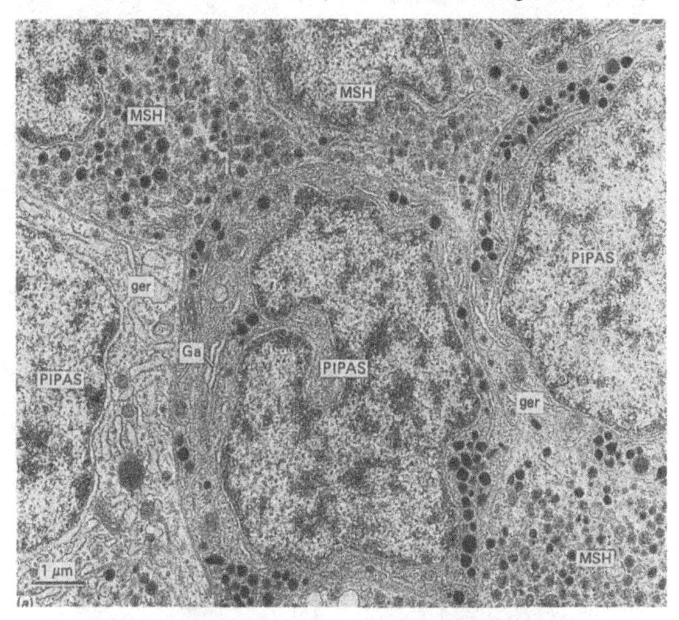

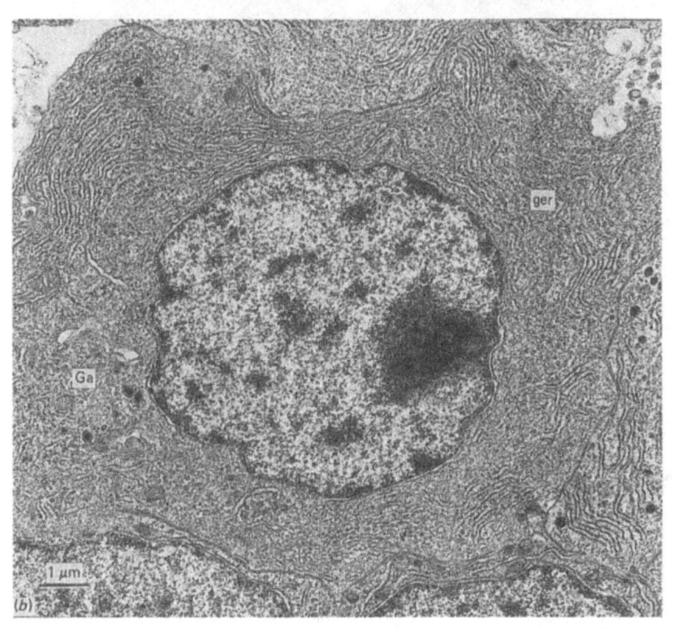

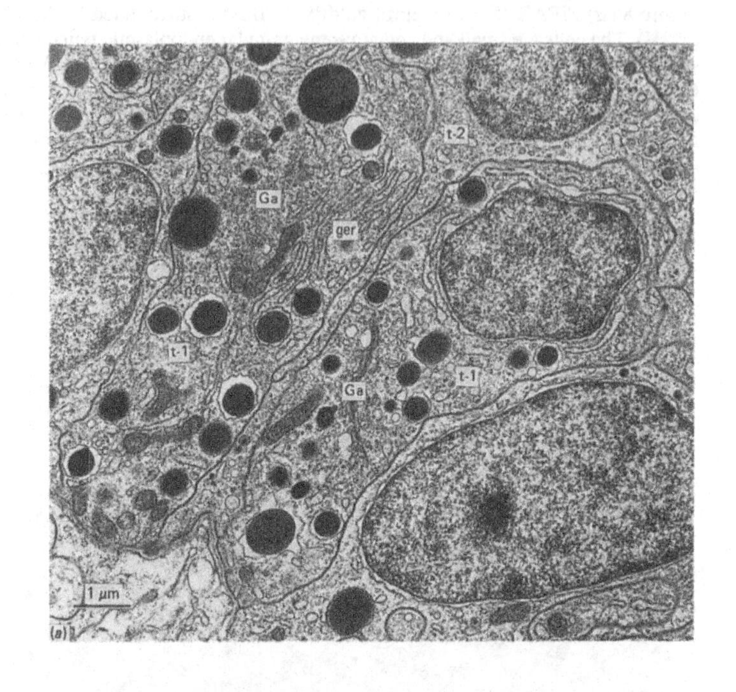
t-2
Ga
ger
t-1
t-1
Ga
1 μm
(a)

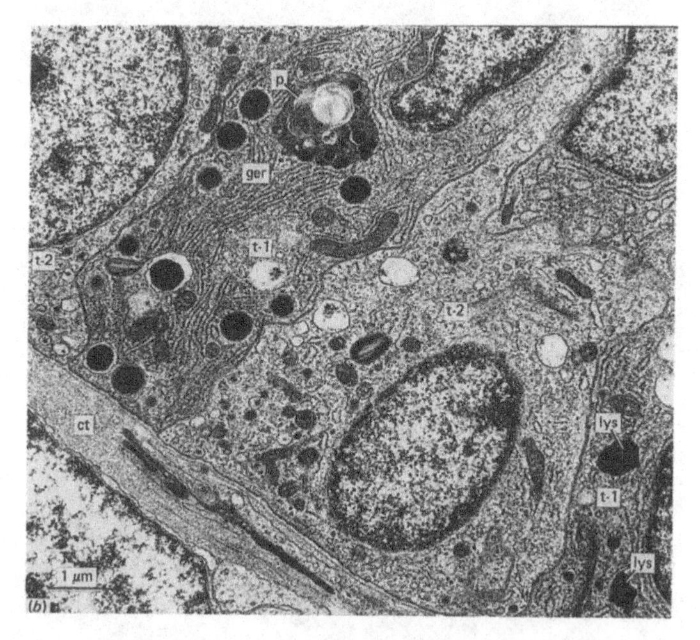
p
ger
t-1
t-2
t-2
ct
lys
t-1
lys
1 μm
(b)

Stannius corpuscles

The Stannius corpuscles (SC) located in the kidneys of teleostean and holostean fishes are endocrine glands implicated in the control of calcium metabolism. The major endocrine product of the SC is a protein called hypocalcin by Pang *et al.* (1974), that lowers plasma calcium levels. Since acid water affects calcium metabolism in fish, the calcium-regulating endocrine glands may contribute to the physiological acclimation to low pH.

The study on brook trout by Notter *et al.* (1976) is the only report known to us on the effects of acid water on the SC. These authors showed increased cellular RNA and protein contents in the SC cells of trout exposed for two or three days to acid water, and concluded that synthesis and release of the hormonal factors of the SC was stimulated. This response is difficult to understand, as in trout most studies have shown a decrease in the plasma concentrations of most ions – with the exception of H^+ – during the first days of exposure to acid water (Giles *et al.*, 1984). Although some authors did not observe such an effect (McKeown *et al.*, 1985), a reduction of the activity of the SC or no change at all might be expected.

In tilapia and goldfish plasma calcium levels are reduced immediately after reduction of water pH (Figures 4, 5). For tilapia we have shown that this is accompanied by an increase in Ca^{2+} efflux, probably across the gills (Flik *et al.*, 1985). In line with the observed hypocalcaemia, we found that the predominant cell type of the SC, the type-1 cells, showed ultrastructural signs of decreased secretory activity when tilapia were kept for five days at pH 3.5 (Figure 9*(a)*,*(b)*). The type-1 cells are the presumptive hypocalcin-producing cells (Urasa & Wendelaar Bonga, 1985). After a week at pH 3.5, plasma calcium levels in tilapia gradually rise to control levels (Figure 4). The apparent inhibition of the type-1 cells of the SC may contribute to this restoration.

Conclusions

The present survey of the endocrine responses of fish to acid water illustrates that our knowledge is incomplete and limited to a few species. The function of the interrenal cells and of the pituitary gland has been studied rather extensively, but mainly with respect to osmotic and ionic regulation. It is surprising that hormones

Figure 9.*(a)* Type-1 cells (t–1), characterized by large secretory granules, and type-2 cells (t–2), with small secretory granules, of Stannius corpuscles of control tilapia; ger, granular endoplasmic reticulum; Ga, Golgi area. The type-1 cells predominate. *(b)* Type-1 (t–1) and type-2 (t–2) cells of Stannius corpuscles of tilapia exposed for five days to pH 3.5. Type-1 cells are less prominent than in the controls; type-2 cells (of unknown function) seem enlarged. The type-1 cells contain lysosomes (lys) and a large phagosome (p), in which secretory granules can be recognized, that are indicative for cellular involution; ger, granular endoplasmic reticulum; ct, connective tissue.

involved in reproduction have not received much attention. It is well known that the reduction of fish populations in acid water is partly caused by impaired ovarian maturation and reduced ability of female fish to release their eggs.

Since the number of fish species studied so far is low, the question arises as to whether the available data are representative for fish in general. With respect to tilapia it is clear that the intensity and speed of the endocrine responses shown by this strong and highly adaptive species is not typical for fish in general. The high survival rate and prompt restoration of the drop in osmolarity after sudden and severe water acidification seems to be an exception rather than the rule (Figure 4). The type of responses observed, however, in particular the activation of cells producing osmoregulatory hormones and the reduction of growth hormone cells, will undoubtedly occur in many other species. All available evidence indicates that at least the involvement of the pituitary–interrenal axis in the physiological adaptation of fish to acid water is a general phenomenon.

On the other hand, our observations on prolactin cells and PIPAS cells of tilapia and goldfish illustrate that species-specific differences in the endocrine responses do occur. The difference reported for these cells however, are consistent with well known differences in the control mechanisms of the osmoregulatory system between both species (Wendelaar Bonga *et al.*, 1984b).

Some of the discrepancies reported in this chapter may be connected with differences in the experimental protocols. This point is illustrated by our observations on the prolactin cells of tilapia. Our finding that the response of these cells to acid water was more intense when the reduction of the water pH was gradual instead of acute indicates that severe physiological and structural damage may slow down the rate of physiological adaptation. It is possible that the differences reported above for the long term effect of water acidification on the activity of the interrenal cells reflect a similar phenomenon. Notwithstanding these differences and inconsistencies in the present data, it is clear that water acidification deeply affects the endocrine system.

References

Ashcom, T.L. (1979). Serum cortisol and electrolyte responses in acid-stressed brook trout (*Salvelinus fontinalis*). Ph.D. thesis, Pennsylvania State University.

Baker, B.I. & Rance, T.A. (1981). Differences in concentrations of plasma cortisol in the trout and the eel following adaptation to black or white backgrounds. *J. Endocrinol.*, **89**, 135–40.

Ball, J.N. & Batten, T.F.C. (1981). Pituitary and melanophore responses to background in *Poecelia latipinna* (Teleostei): role of the pars intermedia PAS cell. *Gen. Comp. Endocrinol.*, **44**, 233–48.

Balm, P. (1986). Osmoregulation in teleosts by cortisol and prolactin-adaptation to low pH environments. Ph.D. thesis, University of Nijmegen, The Netherlands.

Bern, H.A. (1983). Functional evolution of prolactin and growth hormone in lower vertebrates. *Am. Zool.*, **23**, 663–71.

Brown, S.B., Eales, J.G., Evans, R.E. & Hara, T.J. (1984). Interrenal, thyroidal and carbohydrate responses of rainbow trout (*Salmo gairdneri*) to environmental acidification. *Can. J. Fish. Aquat. Sci.*, **41**, 36–45.

Dharmamba, M. (1979). Corticosteroids and osmoregulation in fishes. *Proc. Indian Natn. Sci. Acad.*, B, **45**, 515–25.

Donaldson, E.M. (1981). The pituitary–interrenal axis as an indicator of stress in fish. In *Stress and Fish*, ed. A.D. Pickering, pp. 11–47. London: Academic Press.

Donaldson, E.M. & Fagerlund, U.H.M. (1968). Changes in the cortisol dynamics of sockeye salmon (*Oncorhynchus nerka*) resulting from sexual maturation. *Gen. Comp. Endocrinol.*, **11**, 552–61.

Flik, G., Wendelaar Bonga, S.E., Kolar, Z., Mayer-Gostan, N. & Fenwick, J.C. (1985). Environmental effects on Ca^{2+}-uptake in the cichlid teleost *Oreochromis mossambicus*. In *Fish Culture, Proc. Seventh Conf. Comp. Physiol. Biochem.*, pp. B6.4–9. Barcelona: Promociones Publicaciones Universitarias.

Flik, G., Fenwick, J.C., Kolar, Z., Mayer-Gostan, N. & Wendelaar Bonga, S.E. (1986). Effects of ovine prolactin on calcium uptake and distribution in the cichlid teleost *Oreochromis mossambicus*. *Am. J. Physiol.*, **250**, R161–70.

Follenius, E., Van Dorsselaer, & Meunier, A. (1986). Circulating forms of α–MSH in the carp and trout blood: an HPLC and RIA study. *Gen. Comp. Endocrinol.*, **62**, 185–92.

Giles, M.A., Majewski, H.S. & Hobden, B. (1984). Osmoregulatory and hematological responses of rainbow trout (*Salmo gairdneri*) to extended environmental acidification. *Can. J. Fish. Aquat. Sci.*, **41**, 1686–94.

Hirano, T. & Mayer–Gostan, N. (1978). Endocrine control of osmoregulation in fish. In *Comparative Endocrinology*, eds. P.J. Gaillard & H.H. Boer, pp. 209–12. Amsterdam: Elsevier–North Holland Biomedical Press.

Jacobsen, O.J. (1977). Brown trout (*Salmo trutta* L.) growth at reduced pH. *Aquaculture*, **11**, 81–4.

Langdon, J.S., Thorpe, J.E. & Roberts, R.J. (1984). Effects of cortisol and ACTH on gill Na^+/K^+/ATPase, SDH and chloride cells in juvenile Atlantic salmon *Salmo salar* L. *Comp. Biochem. Physiol.*, **77A**, 9–12.

Lee, R.M., Gerking, S.D. & Jezierska, B. (1983). Electrolyte balance and energy mobilization in acid-stressed rainbow trout, *Salmo gairdneri*, and their relation to reproductive success. *Envir. Biol. Fish.*, **8**, 115–23.

McDonald, D.G. (1983). The effects of H^+ upon the gills of freshwater fish. *Can. J. Zool.*, **61**, 691–703.

McDonald, D.G. & Wood, C.M. (1981). Branchial and renal acid and ion fluxes in the rainbow trout, *Salmo gairdneri*, at low environmental pH. *J. exp. Biol.*, **93**, 101–18.

McDonald, D.G., Walker, R.L. & Wilkes, P.R.H. (1983). The interaction of environmental calcium and low pH on the physiology of the rainbow trout, *Salmo gairdneri*. II. Branchial ionoregulatory mechanisms. *J. exp. Biol.*, **102**, 141–55.

McKeown, B.A., Geen, G.H., Watson, T.A., Powell, J.F. & Parker, D.B. (1985). The effect of pH on plasma electrolytes, carbonic anhydrase and ATPase activities in rainbow trout (*Salmo gairdnerii*) and largescale suckers (*Catostomus macrocheilus*). *Comp. Biochem. Physiol.*, **80A**, 507–14.

McWilliams, P.G. (1980). Effects of pH on sodium uptake in Norwegian brown trout (*Salmo trutta*) from an acid river. *J. exp. Biol.*, **88**, 259–67.

Menendez, R. (1976). Chronic effects of reduced pH on brook trout (*Salvelinus fontinalis*). *J. Fish. Res. Bd Can.*, **33**, 118–23.

Mudge, J.E., Dively, J.L., Neff, W.H. & Anthony, A. (1977). Interrenal histochemistry of acid-exposed brook trout, *Salvelinus fontinalis* (Mitchell). *Gen. Comp. Endocrinol.*, **31**, 208–15.

Muniz, J.P. & Leivestad, H. (1979). Long term exposure of brook trout to acidic water. SNSF Project, IR 44/79, 32 pp. Oslo–Ås.

Muniz, J.P. & Leivestad, H. (1980). Acidification effects on freshwater fish. In *Proc. Int. Conf. Ecological Impact of Acid Precipitation*, ed. D. Drabløs & A. Tollan, pp. 84–92. Oslo–Ås: SNSF Project.

Neville, C.M. (1986). Studies on the mechanism of toxicity of acid and aluminium to juvenile trout in a low ion environment. Abstr. Water Air Soil Pollut. (In press.)

Notter, M.F.D., Mudge, J.E., Neff, W.H. & Anthony, A. (1976). Cytophotometric analysis of RNA changes in prolactin and Stannius corpuscle cells of acid-stressed brook trout. *Gen. Comp. Endocrinol.*, **30**, 273–84.

Olivereau, M., Aimar, C. & Olivereau, J.M. (1980a). Response of the teleost pituitary (goldfish, eel) to deionized water. *Cell Tiss. Res.*, **208**, 389–404.

Olivereau, M., Aimar, C. & Olivereau, J. (1980b). PAS-positive cells of the pars intermedia are calcium-sensitive in the goldfish maintained in an hyposmotic milieu. *Cell Tiss. Res.*, **212**, 29–38.

Olivereau, M., Olivereau, J. & Aimar, C. (1981). Specific effect of calcium ions on the calcium-sensitive cells of the pars intermedia in the goldfish. *Cell Tiss. Res.*, **214**, 23–31.

Pang, P.K.T., Pang, R.K. & Sawyer, W.H. (1974). Environmental calcium and the sensitivity of killifish (*Fundulus heteroclitus*) in bioassays for the hypocalcemic response to Stannius Corpuscles from killifish and cod (*Gadus morhua*). *Endocrinology*, **94**, 548–58.

Rance, T.A. & Baker, B.I. (1981). The in vitro response of the trout interrenal to various fragments of ACTH. *Gen. Comp. Endocrinol.*, **45**, 497–503.

Sadler, K. & Lynan, S. (1985). Some effects of low pH and calcium on the growth and tissue mineral content of yearling brown trout (*Salmo trutta*). CERL Report TPRD/L/2789/N84. Leatherhead, Surrey, UK: Central Electricity Research Laboratories.

Shenker, Y., Villarea, J.Z., Sider, R.S. & Greking, R.J. (1985). α Melanocyte-stimulating hormone stimulation of aldosterone secretion in hypophysectomized rats. *Endocrinology*, **116**, 138–41.

Staurnes, M., Sigholt, T., Vedagiri, P. & Reite, O.B. (1984). Inhibition of Na–K-ATPase and carbonic anhydrase in gills of the salmon, *S. salar*, exposed to Al-containing acid water. Abstr. First Congr. Comp. Biochem. Physiol., Liège, p. B130.

Sumpter, J.P., Pickering, A.D. & Pottinger, T.G. (1985). Stress-induced elevation of plasma α-MSH and endorphin in brown trout, *Salmo trutta* L. *Gen. Comp. Endocrinol.*, **59**, 257–65.

Sumpter, J.P., Dye, H.M. & Benfey, T.J. (1986). The effects of stress on plasma ACTH, α-MSH and cortisol levels in salmonid fishes. *Gen. Comp. Endocrinol.* (In press.)

Szalay, K.Sz. & Stark, E. (1982). Effect of alpha-MSH on the corticosteroid production of isolated zona glomerulosa and zona fasciculata cells. *Life Sci.*, **30**, 2101–8.

Urasa, F.M. & Wendelaar Bonga, S.E. (1985). Stannius corpuscles and plasma calcium levels during the reproductive cycle in the teleost fish *Oreochromis mossambicus*. *Cell Tiss. Res.*, **241**, 219–27.

Van Eys, G.J.J.M. (1980). Structural changes in the pars intermedia of the cichlid teleost *Sarotherodon mossambicus* as a result of background adaptation and illumination. II. The PAS positive cells. *Cell Tiss. Res.*, **210**, 171–9.

Van Eys, G.J.J.M. & Peters, P.T.W. (1981). Evidence for a direct role of α-MSH in morphological background adaptation of the skin in *Sarotherodon mossambicus*. *Cell Tiss. Res.*, **217**, 361–72.

Van Eys, G.J.J.M., Löwik, C.W.G.M. & Wendelaar Bonga, S.E. (1983). Isolation of the biosynthetic products of the PAS-positive pars intermedia cells in the cichlid teleost *Sarotherodon mossambicus*. *Gen. Comp. Endocrinol.*, **49**, 277–85.

Wendelaar Bonga, S.E., Meij, J.C.A. van der & Flik, G. (1984a). Prolactin and acid stress in the teleost *Sarotherodon mossambicus*. *Gen. Comp. Endocrinol.*, **55**, 323–32.

Wendelaar Bonga, S.E., Van der Meij, J.C.A., Van der Krabben, W.A.W.A. & Flik, G. (1984b). The effect of water acidification on prolactin cells and pars intermedia PAS positive cells in the teleost fish *Oreochromis* (formerly *Sarotherodon*) *mossambicus* and *Carassius auratus*. *Cell Tiss. Res.*, **238**, 601–9.

Wendelaar Bonga, S.E., Meij, J.C.A. van der & Flik, G. (1986). Response of PAS-positive cells of the pituitary pars intermedia in the teleost *Carassius auratus* to acid water. *Cell Tiss. Res.*, **243**, 609–17.

GWYNETH HOWELLS AND R. MORRIS

Conclusions

Introduction

'Acid Rain' as a scientific and political issue has advanced since early claims of ecological damage were first formulated in the 1950s and 1960s from the initial, rather simplistic, assumptions and interpretations that could be formulated as a sequence of events initiated by the emission of SO_2 from fossil fuel combustion, the formation of acidic radicles in the atmosphere which were then removed by rain and deposited; the rain falling in unbuffered waters rendering them acid by progressively consuming their alkalinity.

Following this early conceptualisation, the contributions of N gases and ammonia were later accepted, as well as the complexity of chemical reactions in the atmosphere and the alternate mechanisms of deposition. Hydrological and geological characteristics of catchments were also recognised as major influences determining the quality of surface waters which were thus distinguished from atmospheric deposition. The 'susceptibility' of a lake or stream was seen to be dependent not only on the rate of acidic deposition but also on the supply and/or depletion of alkaline materials leached from soil and weathered from rock. Where loss is not matched by supply from leaching and weathering, these materials become depleted in surface waters. It is recognised that this may eventually become intolerable for some aquatic species although others may be able to exploit the resultant loss of competition. Initial concern was expressed for fish, reported as lost from the affected lakes and rivers: other biological components of the aquatic ecosystem are reported to be similarly restricted.

Further understanding of the processes of acid transfer and the mechanisms of acid toxicity has been achieved in the past decade. In particular, recent studies have helped us to understand the relationship between water quality and the observed 'status' of the fisheries of acid waters (Brown & Sadler, Turnpenny, this volume) allowing a better assessment of minimal water quality needed for a fishery. But while fish dislike acid waters, the threshold of their tolerance is still debated and clearly depends on a variety of other biological and chemical conditions. It is still not clear what is the mechanism of H^+ toxicity in many animals and there are many puzzling features about the development and progressive impoverishment of acid waters which hamper consideration of remedial measures, particularly in the quantitative relationships between emissions and deposition, and between deposition and water quality, as well as in the timescale of progressive acidification or recovery. Moreover, the degree to which acidity alone can be blamed for fisheries decline is not clear.

The studies reported in this volume focus on the physiological or biochemical responses of fish and some other aquatic animals to acid stress, as evidenced by both field and laboratory studies. In this area there is reasonable consensus of view; this represents an important step forward, since water quality conditions necessary for species survival can be reasonably specified as a guide to the necessary remedial measures for restoration of acidified aquatic systems.

Acid deposition and transfer

Earlier models of dispersion and deposition (OECD, 1979; Granat, 1978) were based on relatively simple mathematical concepts using estimates of national sulphur emissions, assumed rates of deposition of SO_2 and SO_4^{2-}, and averaged wind directions. This provided an emittor/receptor relationship between western European nations, internationally accepted even though the estimate carried a large error. Basic shortcomings have been recognised, especially where wet deposited material predominates, as in the western margins of the modelled area, for example in western Norway or the western seaboard of Britain. In these areas, model predictions fail to match observations and a substantial 'background' (in some cases exceeding 50%) has to be added to the predicted values. This 'background' component is now thought to represent a contribution from global transport above the boundary layer (EMEP, 1984) as well as a significant contribution from natural biogenic processes in the ocean and coastal zones (Charlton *et al.*, 1987). It is also still debated whether deposition, especially wet deposition, is directly and linearly related to emissions upwind (Cocks *et al.*, 1983; Calvert *et al.*, 1985). These uncertainties are of special significance in susceptible areas which are often in western zones with high rainfall, and where emission control can at best only improve that fraction (50%) of deposition attributable to identified sources. If control of this component cannot reduce deposition to an acceptable level, considered to be less than 0.5 g sulphur m^{-2} yr^{-1} (Swedish Ministry of Agriculture, 1982), then additional measures will be needed if aquatic ecosystems are to be restored.

Soil reactions

The recognition that the characteristics of soil and bedrock geology are crucial determinants of runoff water quality has also changed our view of how surface waters might respond to changes in deposition. These chemical and hydrological influences are evident in the composition and variety of surface waters (Johnson & Reuss, 1984; Bache, 1984). Moreover, although soils can leach to provide a source of neutralising bases and bedrock weathering mobilises minerals, conversely, the soils themselves can retain base metals in exchange for hydrogen ions (Rosenquist, 1978; Braekke, 1981a). The degree to which these interacting processes influence drainage waters is dependent on the routes of water flow and rate of hydrological processes. In times of heavy rain, or when deeper soils are frozen or saturated, water flow is predominantly superficial (Braekke, 1981b; Bergstrom, 1986), picking up additional acidity, and having less opportunity for neutralisation. These temporal variations in surface water quality are crucial for the survival of fish resident in streams in susceptible areas. Lakes are less subject to these sudden hydrological changes and also may offer refuges of preferred water quality; in lakes the spatial variations in pH may be as much as two pH units (Howells, 1983).

Modelling soil processes and responses is difficult (Cosby, 1986), since both concentration dependent and rate dependent processes are involved, yet it is a necessary requirement for retrospective or prospective estimates of change. Information on the hydrological and geological components of an affected catchment is needed, but is often inadequate. It is clear that base supply is primarily an effect of CO_2 weathering of the basic minerals but this is not necessarily constant nor is it geographically uniform, thus the concept of 'at risk' areas being those of low potential yield of alkalinity is too broad scale. Thus adjacent lakes draining similar geologies and receiving the same deposition may have quite different water qualities. Acid deposition together with the acidity generated within soils leads first to the leaching of bases and then to the mobilisation of aluminium, but ion exchange (Rosenquist, 1976; Braekke, 1981a; Bache, 1984) and biological uptake and harvesting also retain or sequester bases over long periods. Further, soils may retain or release sulphate (a major anion carrier of aluminium) depending on soil conditions (including seasonal variations in temperature, redox potential, saturation levels) (Dovland & Semb, 1978; Braekke, 1981a; Johnson *et al.*, 1981; Singh, 1984). Thus, although influx of sulphur in a catchment may be matched by efflux over the short term (e.g. a year or two; Wright & Johanneson, 1980), the measured parameters are small compared with the soil sulphur reservoir which may exceed the accumulated anthropogenic deposition over the last 50 years. These temporal inequalities make it difficult to anticipate changes that might follow changed rates of deposition. Recent field experiments (Wright, 1986) indicate how difficult it is to predict what response will be seen in drainage water as a consequence of reduced deposition. Since the degree to which aluminium is dissolved or mobilised from soils is thought to be largely determined by the solubility of basic aluminium sulphates (Nilsson, 1985) this is a crucial factor. But other anions can also play a part – seasonal peaks in nitrate are associated with aluminium peaks in surface waters, and it is becoming clear that silicates also have a role to play (Farmer, 1986; Birchall & Espie, 1986). Soil composition as well as soil conditions must thus imply that current run-off quality is determined only in part by current deposition of sulphur.

These changes in thinking have important consequences for our understanding of how water quality may affect fisheries; further, many early analyses of water quality were imprecise and thus misleading. It is certainly now well established that an occasional water sample cannot represent the environment to which fish are exposed over time and that early field measurements, especially of pH, were often imprecise. It is only recently, however, that acidic surface waters have been sampled with sufficient frequency and accuracy to characterise their short term fluctuations. 'Acid episodes' have still not been adequately recorded (Reader & Dempsey, this volume), nor have those of different cause (snowmelt, spates, sea salt 'rains') been adequately distinguished in terms of their chemistry and fluxes. Acid pulses have been attributed by some to deposition from polluted air masses originating in high source areas (Rosseland *et al.*, 1986), although others have argued that acidity in a single rain event is of no special direct significance for the acidity of surface water (Rosenquist & Seip, 1986). It is now recognised that such acid episodes are also evident in northern Scandinavia (Jacks *et al.*, 1986) where deposition is less than 0.5 g sulphur m^{-2} yr^{-1} and where drainage waters are usually adequately buffered. In Wales and in southwest Scotland they may follow high pH, high conductivity rains, the entrained sea salts releasing protons stored or generated within the typically peaty soils of these western areas (Hornung, 1985; Solway River Purification Board, 1986).

Fisheries and other aquatic life

Early studies of acidic lakes showed that as pH was lower some species of fish were lacking, and that the number of species was inversely related to the degree of acidity (Harvey, 1975). At the same time it became clear that acidity alone could not always explain the limits of distribution of particular species and field tolerances may even exceed those derived from laboratory exposures (Howells, 1983). There are also some acid environments with thriving fish communities, notably the waters of the Amazon basin, where 20 fish species and a full range of invertebrates are maintained in water of pH 3.5 or less, about half attributed to organic acids (Henderson & Walker, 1986). It must also be acknowledged that many other factors besides water quality can determine the success of a fishery – extreme physical and climatic conditions can wipe out resident populations and temporarily demolish migratory ones (e.g. Cowx *et al.*, 1984; Howells *et al.*, 1983; Elliot, 1985). Since many acid waters are also low-order streams draining uplands these adverse conditions are often associated with the acidic water quality that is characteristic of run-off where rainfall is heavy, soils thin and poor, and geology unreactive. There is evidence, too that coniferous afforestation, developed in the past four or five decades on a regional scale, especially in acid-sensitive western Britain, has had adverse effects on both water quality and fisheries (Milner *et al.*, 1981; Ormerod & Edwards, 1985; Hornung, 1986; Egglishaw *et al.*, 1986). These effects include both physical (changed temperature and light regimes, greater sediment loading) and chemical (greater variability of composition, decreased productivity) effects.

Field observations as well as better specified and controlled 'through-flow' exposures to acid waters with a range of 'hardness' or calcium concentrations show that at least many common salmonid species can live at environmentally relevant acid pH levels of 4.5 and above (Brown, this volume). The physiological explanation for this seems to be that calcium reduces the permeability of the gills to other ions and hence helps to combat the increased efflux of ions brought about by hydrogen ion exposure (Potts & McWilliams, this volume). This enables the influx rates, which may also be reduced by acid exposure, to restore ion balance at rates which are within the fish's capacity. It also follows that, since fish show little or no change in acid-base balance under these circumstances, i.e. where there is a steep hydrogen ion gradient, there must be a mechanism for excreting the incoming hydrogen ions, perhaps by combining them with ammonium ions which can be excreted by both gills and kidneys (Wood, this volume). There are marked differences in response between fish which are gradually acclimatised (Stuart & Morris, 1985; Potts & McWilliams, this volume) and these may result from the time taken for adaptive processes (reduced gill permeability, increased ion carrier activity in the gills, increased hydrogen ion excretion) to be initiated, perhaps by hormonal mechanisms (Wendelaar Bonga & Balm, this volume). This might explain the discrepancy between experimental findings and field observations. At pH levels below 4.5, or as a result of low calcium concentrations or abrupt exposure, ion loss and hydrogen ion incursion affect the processes regulating the normal ionic and osmotic balance between extracellular and intracellular fluids, causing loss of potassium from the cells, reduced extracellular fluid and blood volume, increased haematocrit, and eventual death due perhaps to cardiac failure (Stuart & Morris, 1985, Wood, this volume). It is at this stage that acid-base balance may start to break down (Heisler, Wood, this volume).

It is recognised that some potentially toxic trace metals are made more readily available by acid soil conditions but their toxicity is not always greater (e.g. copper, cadmium and zinc; McDonald *et al.*, this volume); in particular, aluminium released from acid soils in which the base reserve has been exhausted can be seriously toxic to fish at pH 5.2–5.5; the upper end of the pH range characteristic of many low conductivity, poorly buffered natural waters (Schofield & Trojnar, 1980; Brown & Lynam, 1982; Brown, Potts & McWilliams, this volume). Aluminium can also be toxic in extremely low concentration (2 μM) in more acid waters (pH 4.5) where the effect is to disturb sodium balance by inhibiting active uptake (McDonald *et al.*, this volume; Dalziel, 1987). Over a range of pH levels at higher aluminium concentration (3 μM or more), respiration is affected due to mucus clogging of the gills (Muniz & Leivestad, 1980; Schofield & Trojnar, 1980; Neville, 1985) but anoxia does not seem to be the direct cause of death at pH levels above 6 where exercise or metabolism are more potent agents. It seems more likely that death may arise as a result of ion imbalance leading to the sequence of physiological effects described above. There is still debate, however, as to which aluminium species are present and toxic. The simplest view would implicate the less soluble polymeric species, especially when supersaturated at high pH levels, as being responsible for the respiratory problems, and the soluble monomeric species, which increase in concentration as pH falls, as causing the ion regulatory problems. Other chemical components in natural waters may moderate the toxicity of aluminium, such as fluorides, silicates and humic and fulvic acids (Schofield & Trojnar, 1980; Neville, 1985; Birchall & Espie, 1986) emphasising the need for more information on aluminium chemistry and for more sophisticated analyses in the field. There is growing evidence that the ratio of aluminium and silicon is important, and even that both elements might be essential at low concentrations (Carlisle, 1986).

Although aluminium is most commonly invoked to explain fish absence or fish kills at otherwise tolerable pH, recent field observations have suggested that other transition metals such as iron and manganese, and copper, cadmium and zinc, may also be important (Andersen & Nyberg, 1985; McDonald *et al.*, this volume). In the case of copper, cadmium and zinc there is no doubt that they affect ion regulation, if present in sufficient concentration. Cadmium and manganese are also known to affect calcium metabolism, development and skeletal calcification, as well as long term effects associated with spawning and recruitment. In some special circumstances, e.g. lakes close to metalliferous smelters as at Sudbury, Ontario, high concentrations of toxic trace metals deposited locally have certainly played a role in the loss of fish (Beamish *et al.*, 1975; Almer *et al.*, 1978).

Thus the picture that emerges is a complex one where fish kills and fishless lakes could be the result of a number of interacting conditions which subsequently affect ion regulation, respiration and development and which may be exacerbated if low pH stress is also involved. The key role of calcium in controlling gill permeability is well recognised (McWilliams & Potts, 1978; Brown & Lynam, 1982; Potts & McWilliams, this volume) and it is also acknowledged that this may be the critical factor which limits many fisheries (Chester, 1984).

The study of invertebrate species is much less well advanced with pH thresholds established experimentally for relatively few species (Sutcliffe & Hildrew, Vangenechten *et al.*, this volume) although field surveys suggest that lake and stream faunas are 'impoverished' in acid conditions. It is, however, uncertain what community might be judged 'normal' for these upland streams, since they will not provide the diversity of habitat and stability of physical and chemical conditions associated with greatest species diversity (Vanotte *et al.*, 1980). Some groups, such as crustaceans and molluscs, are clearly restricted to waters with sufficient calcium or alkalinity but there is also evidence that some invertebrates are limited by potassium or sodium, rather than calcium (Minshall & Minshall, 1978; Sutcliffe & Hildrew, this volume). They are apparently tolerant of aluminium at much higher concentrations than fish (McMahon & Stuart, Vangenechten *et al.*, this volume). In some cases, however, there is parallel evidence that ion exchange is impaired in acid conditions, perhaps surprising in those species lacking a gill membrane analagous to that of fish.

With the exception of freshwater crayfish, rather little is known about how low calcium affects ion exchange in invertebrates.

Although observations have been made that the invertebrate fauna of acid waters is impoverished (Almer *et al.*, 1978; Sutcliffe & Hildrew, this volume), there has been little evidence to suggest that this effect is due to lack of appropriate diet (for fish or other organisms) in acid waters – indeed a characteristic of such fish is that they have full stomachs and good condition. In at least one riparian species, however, food availability may be a limiting factor. The abundance of dippers seems related to the invertebrate species needed for nutrition of the nestlings and along the banks of acidic streams, dipper territory is extended over that seen in comparable non-acidic streams (Ormerod *et al.*, 1986).

In recent years more evidence has been brought forward to show that different species of fish show more or less sensitivity to acid conditions, although response also varies with season, age and size, as well as with water quality variables and previous exposure. More physiological work is now reported, but there is still a lamentable lack of information for many fish species which are reported from field surveys to have been lost with progressive acidification over the past two decades. There are puzzling features, too – although some species, e.g. perch, are not classed as sensitive on the basis of laboratory studies, their disappearance from Swedish west coast lakes during the 1950s and 60s is well documented and associated with increasing acidity (Almer *et al.*, 1978; Hultberg, 1985). In contrast, 96% of the population of Lake Windermere perch were lost in 1976 as a result of disease (Craig, 1982).

Although physiological and biochemical mechanisms have now been well established, at least for a number of salmonid species, these anomalies await better explanation. It is not even certain that acidity alone is toxic at the levels encountered in the field; a field experiment (Ormerod *et al.*, 1987) demonstrated that reducing the pH from 7 to 4.28 caused little mortality of resident brown trout, but when aluminium was added, downstream mortality was severe (50–87%). It is evident that sudden acid exposure alone was not sufficient for high mortality until toxic aluminium species were added. It is also clear that a threshold for acid toxicity cannot be defined without specifying calcium concentrations at the least. The early survey of more than 700 brown trout lakes in southern Norway (Wright & Snekvik, 1979) demonstrated that lakes with good fish populations have a Ca/H ratio greater than 3 whereas those without fish have a lower Ca/H ratio (Mason, this volume), Figure 4); the role of aluminium was not clearly identified at the time. Similar findings are indicated since by field surveys elsewhere. The reasons for this are now evident from physiological studies in which the role of calcium in controlling the permeability of the fish gill is evident (Potts & McWilliams, this volume). Work with other species has established that while thresholds of acid sensitivity may vary with species, the principle is of general application. There is also growing evidence that some capacity for recovery from acid shock is present, and that previous exposure to acid conditions may moderate response. Whether this is achieved via a hormonal response (Wendelaar Bonga *et al.*, this volume), or is genetically determined (Potts & McWilliams, this volume) is still debatable.

What is still puzzling is the timing of the changed water quality and of fish disappearance – southern Sweden seems to have been affected several decades ago, southern Norway after 1950, and North America only since the 1970s (Howells, 1983). Since soils were laid down at the retreat of the glaciers, there must have been a continuous and progressive leaching of soil and rock-weathered bases. Given that increased acid deposition over the past 150 years will have increased the rate of base depletion and that sulphur stores in soils will have built up more quickly, the expected sequence and timing of fishery loss is inconsistent, with similarly sensitive waters becoming fishless over several decades, but not in any order that could be attributed solely to changes in deposition, or to hydrological and geological factors. Could selection of tolerant populations and/or genetic mechanisms be sufficient to explain this discrepancy?

Perspectives

There is general consensus that past acid deposition has significantly influenced present water quality in sensitive areas and that fish communities, and other aquatic fauna, are in part determined by this water quality. The importance of hydrological and geological influences on drainage water composition is now evident, including the long term leaching of bases and the accumulation of sulphate in the soils. While the washout of stored sulphate *per se* is of little biological significance, the yield of bases, as well as calcium and aluminium mobilised by anions from soils are now regarded as critical. What factors other than increased acid deposition could have influenced the calcium budget so significantly? Land use changes such as afforestation have been demonstrably damaging on a regional scale in Britain but in some affected areas, notably southern Norway, fish have been lost from lakes above the level of afforestation and in others there has been no widespread change of use. Changes in the level of the water table have been reported in Sweden as a result of recent climatic variability, especially a drier than normal period in the 1960s which preceded the reported acidification of the late 1960s and early 1970s (SNV, 1985). Both a reduced groundwater supply of bases, and the probability that soil sulphur was mobilised, have to be taken into account.

All of these developments have important consequences for an effective remedial strategy. In areas where the 'background' component of deposition is in excess of 50%, control of emissions upwind will have, inevitably, a less than 50% benefit on deposition. If the unattributed deposition exceeds 0.5 g sulphur m^{-2} yr^{-1}, sensitive areas will still be subject to acid events. If calcium and aluminium are the prime determinants of fish survival, will they be, respectively, increased or decreased by a reduction in deposition? For calcium, the evidence suggests that concentrations might in fact fall; for aluminium, present theory suggests that stored sulphates in soil would have to be significantly reduced, inevitably over a long time scale, before aluminium release would drop to acceptable levels. These conclusions point to the need to improve water quality by other means, notably by restoring the base reserve in catchment soils so that the yield of calcium is improved, and aluminium retained. Land use/management changes might be sufficiently effective in some marginal areas to bring about the necessary improvements, but probably not in others. The reversibility of biological effects of acidification in surface waters is also debatable – natural recolonisation can be effective in some situations but is likely to be very slow in low-order streams or upland lakes. Should a strategy of restocking be considered? If so we need to know much more about the effects of water quality on aquatic plants and invertebrates as well as more information on the effects of pH and trace metals on fish species other than salmonids which might be used for recolonisation.

The Conference on Acid Toxicity and Aquatic Animals was called to bring together specialists with new results to report on some of these challenging questions; information gained from field studies can be as revealing as that from laboratory experimentation, the two approaches complementing and stimulating each other. The degree to which this strategy has succeeded can be judged by our readers, but all must be encouraged by the progress made and stimulated by the paradoxes still evident. The questions raised seem likely to instigate further scientific investigation and to provide the basis for remedial action and for lively discussion for a good many years ahead.

References

Almer, B., Dickson, W., Ekstrom, C. & Hornstrom, E. (1978). Sulphur pollution and the aquatic ecosystem. In *Sulphur in the Environment, part II*, ed. J.R. Nriagu, pp. 271–311. New York: Wiley.

Andersson, P. & Nyberg, P. (1984). Experiments with brown trout (*Salmo trutta* L.) during spring in mountain streams at low pH and elevated levels of iron, manganese and aluminium. *Rep. Inst. Freshwater Res. Drottningholm*, **61**, 34–47.

Bache, B.W. (1984). Soil-water interactions. *Phil. Trans. R. Soc. Lond. B*, **305**, 393–407.

Beamish, R.J., Lockhart, W.L., Van Loon, J.C. & Harvey, H.H. (1975). Long term acidification of a lake and resulting effects on fisheries. *Ambio*, **4**, 98–102.

Bergstrom, S. (1986). Simulation of pH, alkalinity, and residence time in natural river systems. In *Report of EEC (COST 612) Workshop on Reversibility of Acidification*, Grimstad, Norway, 9–11 June 1986.

Birchall, D. & Espie, A.W. (1986). Biological implications of the interaction (via silanol groups) of silicon with metal ions. *CIBA Foundation Report no. 121, Silicon Biochemistry*, pp.140–53. Chichester: Wiley & Sons.

Braekke, F.H. (1981a). Hydrochemistry in low pH soils of south Norway; 2. Seasonal variation in some peatland sites. *Rept. Norweg. Forest Research Inst., As 36(12)*, 1–22.

Braekke, F.H. (1981b). Hydrochemistry of high altitude catchments in south Norway; 3. Dynamics in water flow and in release-fixation of sulphate, nitrate and hydronium. *Rept. Norweg. Forest Research Inst., As 36 (10)*, 1–21.

Brown, D.J.A. & Lynam, S. (1982). The effect of calcium and aluminium concentrations on the survival of brown trout (*Salmo trutta*) at low pH. *Bull Envir. Contam. Toxicol.*, **30**, 582–7.

Calvert, J.G., Lazrus, A., Kok, G.L., Heikes, B.G., Walega, J.G., Lind, J. & Cantrell, C.A. (1985). Chemical mechanisms of acid generation in the troposphere. *Nature (Lond.)*, **317**, 27–35.

Carlisle, E.M. (1986). Silicon as an essential trace element in animal nutrition. *CIBA Foundation Symposium no. 121, Silicon Biochemistry*, pp.123–36. Chichester: Wiley & Sons.

Chester, P.F. (1984). General Discussion. *Phil. Trans. R. Soc. Lond. B*, **305**, 564–5.

Cocks, A.T., Kallend, A.S. & Marsh, A.R.W. (1983). Dispersion limitations of oxidation in power plant plumes during long-range transport. *Nature (Lond.)*, **305**, 122–3.

Cosby, B.J. (1986). Modelling reversibility of acidification with mathematical models. In *Report of EEC (COST 612) Workshop on Reversibility of Acidification*, Grimstad, Norway, 9–11 June 1986.

Cowx, I.G., Young, W.O. & Hellawell, J.M. (1984). The influence of drought on the fish and invertebrate populations of an upland stream in Wales. *Freshwater Biol.*, **14**, 165–77.

Craig, J.F. (1982). Population dynamics of Windermere perch. *Freshwater Biol. Assoc. Ann. Rep.*, **50**, 49–59.

Dovland, H. & Semb, A. (1978). Deposition and run-off of sulphate in the Tovdal river: a study of mass balance for September 1974–August 1976. *SNSF Rept. IR 38/78*, 22pp.

Egglishaw, H.E., Gardiner, R. & Foster, J. (1986). Salmon catch decline and forestry in Scotland. *Scott. Geogr. Mag.*, **102**(1), 57–61.

Elliot, J.M. (1985). Population dynamics of migratory trout, *Salmo trutta*, in a Lake District stream, 1966–83, and their implications for fisheries management. *J. Fish. Biol.*, **27** *(Suppl. A)*, 35–43.

EMEP (1984). Summary report for the chemical coordinating centre for the second phase of EMEP. *EMEP/CCC 2/84 Norwegian Inst. for Air Research (NILU).*

Farmer, V.C. (1986). Sources and speciation of aluminium and silicon in natural waters. *CIBA Foundation Symposium no. 121, Silicon Biochemistry*, pp.4–19. Chichester: Wiley & Sons.

Granat, L. (1978). Sulphate in precipitation as observed by the European Atmospheric Chemistry Network. *Atmos. Environ.*, **12**, 413–24.

Harvey, H.H. (1975). Fish populations in a large group of acid stressed lakes. *Verh. Int. Ver. Limnol.*, **19**, 2401–17.

Henderson, P.A. & Walker, I. (1986). The leaf litter community of the Amazonian blackwater stream Tarumazinho. *J. Trop. Ecol.*, **2**, 1–17.

Hornung, M. (1985). Acidification of soils by trees and forests. *Soil Use Mgmt.*, **1**, 24–8.

Howells, E.J., Howells, M.E. & Alabaster, J.S. (1983). A field investigation of water quality, fish and invertebrates in the Mawddach river system, Wales. *J. Fish. Biol.*, **22**, 447–69.

Howells, G. (1983). Acid waters – the effect of low pH and acid associated factors on fisheries. *Adv. Appl. Biol.*, **9**, 143–255.

Hultberg, H. (1985). Changes in fish populations and water chemistry in Lake Gardsjon and neighbouring lakes during the last century. *Ecol. Bull.*, **37**, 64–72.

Jacks, G., Olofsson, E. & Werme, G. (1986). An acid surge in a well-buffered stream. *Ambio*, **15**, 282–5.
Johnson, D.W., Henderson, G.S. & Todd, D.E. (1981). Evidence of modern accumulations of adsorbed sulfate in an East Tennessee forested ultisol. *Soil Science*, **132**, 422–6.
Johnson, D.W. & Reuss, J.O. (1984). Soil mediated effects of atmospherically deposited sulphur and nitrogen. *Phil. Trans. R. Soc. Lond. B*, **305**, 383–92.
Nilsson, S.I. (1985). Budgets of aluminium species, iron and manganese in the Lake Gardsjon catchment in SW Sweden. *Ecol. Bull.*, **37**, 120–32.
Milner, N.J., Scullion, J., Carling, P.A. & Crisp, D.T. (1981). The effects of discharge on sediment dynamics and consequent effects on invertebrates and salmonids in upland rivers. *Adv. Appl. Biol.*, **6**, 153–220.
Minshall, G.W. & Minshall, J.N. (1980). Further evidence of the role of chemical factors in determining the distribution of benthic invertebrates in the River Duddon. *Arch. Hydrobiol.*, **83**, 324–55.
Muniz, I.P. & Leivestad, H. (1980). Acidification – effects on freshwater fish. In *Proc. Int. Conf. Ecological Impact of Acid Precipitation*, ed. D. Drabløs & A. Tollan, pp. 84–92. Oslo–Ås: SNSF Report.
Neville, C.M. (1985). Physiological response of juvenile rainbow trout, *Salmo gairdneri*, to acid and aluminum – prediction of field responses from laboratory data. *Can. J. Fish. Aquat. Sci.*, **42**, 2004–19.
OECD (Organisation for Economic Cooperation and Development) (1979). *The OECD Programme on Long Range Transport of Air Pollutants: Measurements and Findings*. Paris: OECD.
Ormerod, S.J., Boole, P., McCahon, P., Weatherley, N.S., Pascoe, D. & Edwards, R.W. (1986). Short term experimental acidification of a Welsh stream: comparing the biological effects of hydrogen ions and aluminium. *Freshwater Biol.* **17**, 341–56.
Ormerod, S.J. & Edwards, R.W. (1985). Stream acidity in some areas of Wales in relation to historical trends in afforestation and the usage of agricultural limestone. *J. Env. Mgmt.*, **20**, 189–97.
Ormerod, S.J., Tyler, S. & Lewis, J.M.S. (1985). Is the breeding distribution of Dippers influenced by stream acidity? *Bird Study*, **32**, 33–40.
Rosenquist, I.Th. (1978). Alternative sources for river acidification in Norway. *Sci. Total Envir.*, **10**, 39–49.
Rosenquist, I.Th. & Seip, H–M. (1986). Acidification of water courses – how great is the disagreement? *CEGB Trans. T.16668(R)*.
Rosseland, B.O., Skogheim, O.K. & Sevaldrud, I.H. (1986). Acid deposition and effects in Nordic Europe. Damage to fish population in Scandinavia continues apace. *Water Air Soil Pollut.*, **36**, 65–74.
Schofield, C. & Trojnar, J.R. (1980). Aluminium toxicity to the brook trout (*Salvelinus fontinalis*) in acidified waters. In *Polluted Rain*, ed. T.Y. Toribara, M.W. Miller & P.E. Morrow. pp. 341–362. New York: Plenum Press.
Singh, B.R. (1984). Sulfate sorption by forest soils. *Science*, **138**, 189–97.
Solway River Purification Board (1986). Report of the Loch Dee project (F.M. Lees, mimeo and personal communication).
Stuart, S. & Morris, R. (1985). The effects of season and exposure to reduced pH (abrupt and gradual) on some physiological parameters in brown trout (*Salmo trutta*). *Can. J. Zool.*, **63**, 1078–83.
Swedish Environmental Protection Board (SNV). Monitor 1985:90–1.
Swedish Ministry of Agriculture (1982). Proceedings of the 1982 Stockholm Conference on Acidification of the Environment 21–30 June 1982.

United Kingdom Acid Water Review Group (UKAWRG) (1986). *Acidity in United Kingdom Fresh Waters: Interim Report* London: Dept. of the Environment.

Vanotte, R.L., Minshall, G.W., Cummins, K.W., Sedell, J.R. & Cushing, C.E. (1980). The river continuum concept. *Can. J. Aquat. Fish. Sci.*, **37**, 130–7.

Wright, R.F. (1986). RAIN project – results after 2 years of treatment. In *Report of EEC (COST 612) Workshop on Reversibility of Acidification*. Grimstad, Norway, 9–11 June 1986.

Wright, R.F. & Johanneson, M. (1980). Input-output budgets of major ions at gauged catchments in Norway. In *Proc. Int. Conf. Ecological Impact of Acid Precipitation*, ed. D. Drabløs & A. Tollan, pp. 250–1. Oslo–Ås: SNSF Report.

Wright, R.F. & Snekvik, E. (1979). Acid precipitation: chemistry and fish populations in 700 lakes in southernmost Norway. *Verh. Int. Ver. Limnol.*, **20**, 765–75.

Index

Figures in bold type indicate page numbers beginning the relevant chapters.
f indicates following pages also.